Lecture Notes in Mathematics 1783

Editors:
J.-M. Morel, Cachan
F. Takens, Groningen
B. Teissier, Paris

T0236690

Springer
Berlin
Heidelberg
New York
Barcelona
Hong Kong
London
Milan
Paris
Tokyo

Lars Grüne

Asymptotic Behavior of Dynamical and Control Systems under Perturbation and Discretization

Springer

Author

Lars Grüne

J. W. Goethe University
Department of Mathematics
Postfach 111932
D-60054 Frankfurt, Germany

e-mail: gruene@math.uni-frankfurt.de
http://www.math.uni-frankfurt.de/~gruene

Cataloging-in-Publication Data applied for

Die Deutsche Bibliothek - CIP-Einheitsaufnmhme

Grüne, Lars:
Asymptotic behavior of dynamical and control systems under perturbation and
discretization / Lars Grüne. - Berlin ; Heidelberg ; New York ; Barcelona ;
Hong Kong ; London ; Milan ; Paris ; Tokyo : Springer, 2002
  (Lecture notes in mathematics ; 1783)
  ISBN 3-540-43391-0

Mathematics Subject Classification (2000):
37B25, 93D30, 65P40, 93B05, 93B35, 49L25

ISSN 0075-8434
ISBN 3-540-43391-0 Springer-Verlag Berlin Heidelberg New York

Springer-Verlag Berlin Heidelberg New York a member of BertelsmannSpringer
Science + Business Media GmbH

http://www.springer.de

© Springer-Verlag Berlin Heidelberg 2002
Printed in Germany

Typesetting: Camera-ready TEX output by the author

SPIN: 10874150   41/3142/DU - 543210 - Printed on acid-free paper

Preface

During the years in which the research results compiled in this monograph were obtained, many people contributed directly or indirectly to these results. There is, first of all, the head of our research group "Numerics, Dynamics and Optimization", Peter E. Kloeden. I want to thank him not only for his support and his scientific interest in this work, but also for the pleasant working atmosphere and for leaving me so much time for the realization of my own research. Many thanks also go to all the other members of the group and of the Fachbereich Mathematik of the J.W. Goethe–Universität, especially to Helga Ambach for her help with all the common problems in daily university life and to Peter Bauer for keeping my computer running.

A big "grazie" goes to all the members of the Dipartimento di Matematica "Guido Castelnuovo" of the Università di Roma "La Sapienza", who made my one–year visit not only scientifically successful. In particular, I am grateful to Maurizio Falcone for his great hospitality and all the things he did for us during this year. I would also like to thank Martino Bardi for the possibility to enjoy the stimulating atmosphere of the Dipartimento di Matematica Pura ed Applicata of the Università di Padova during my one–month visit.

Special thanks go to Fritz Colonius, Eduardo Sontag and Fabian Wirth for taking great interest in my research and providing me with lots of suggestions, remarks and comments which considerably helped to improve this book. I am also grateful to Albert Marquardt and Christine Schweinem, who proofread parts of this manuscript. Finally, I would like to thank all the other people who in numerous ways helped me to understand one or the other aspect of the behavior of perturbed and discretized systems and thus contributed to the results that can now be found in this monograph, as there are Fabio Camilli, Roberto Ferretti, Gerhard Häckl, Oliver Junge, Christopher Kellett, Viktor Kozyakin, Laurent Praly, Ludovic Rifford, Udo Schmidt, Pierpaolo Soravia, Dietmar Szolnoki and Andrew Teel. I apologize to all those people who are missing in this list although their names should have been included.

This book would have been impossible without the results of several fruitful collaborations. In particular, this concerns the construction of high–order numerical schemes for systems with affine input in Section 5.2, which were developed in collaboration with Peter Kloeden (see also [55]), and the generalization of Zubov's method to systems with input in Section 7.2, which was investigated together with Fabio Camilli and Fabian Wirth, cf. also [15, 16, 17, 58].

These collaborations and the invaluable exchanges with other people was only made possible by the constant funding of several organizations, research programs and networks. First of all I would like to thank the Deutsche Forschungsgemeinschaft (DFG), which not only supported several trips to international conferences but in particular funded the one-year visit at the Università di Roma "La Sapienza". The participation at a number of conferences would not have been possible without the support of the Hermann Willkomm–Stiftung of the J.W. Goethe–Universität. Last but not least, I would like to express my special thanks to the European Union's TMR network "Nonlinear Control Network", to the groups in Rome and Padua of the TMR network "Viscosity Solutions and their Applications" and to the DFG priority research program "Ergodentheorie, Analysis und effiziente Simulation dynamischer Systeme (DANSE)", as well as to their respective coordinators Françoise Lamnabhi–Lagarrigue, Italo Capuzzo Dolcetta, Martino Bardi and Bernold Fiedler. The numerous workshops and conferences within these programs as well as the generous funding of visits and guests have considerably contributed to the research which is documented in this monograph.

Finally, and most importantly, I want to thank Brigitte Grüne for her constant support and understanding, which helped me in many ways. I dedicate this work to her.

Frankfurt am Main, October 2001   LARS GRÜNE

Table of Contents

1 Introduction: Dynamics, Perturbation and Discretization

If anything can go wrong, it will go wrong.

Murphy's Law

Many people—and among them many scientists and mathematicians—will certainly agree to this well known saying. In fact, a constant scepticism towards the things that one expects to be true is probably one of the important driving forces in any kind of scientific development. In contrast to this slightly pessimistic attitude, when turning on a computer many people—and again among them many scientists and mathematicians—are willing to believe in whatever the machine tells them to be true.

It was only several years after the first observations of complicated dynamical behavior by means of numerical methods (like, e.g., the famous discovery of the Lorenz attractor [90]) that mathematicians started to ask whether the basic qualitative features of dynamical systems are correctly represented by numerical approximations. Fortunately, during the last two decades this question has been recognized as an important problem and many contributions have been made during this time. Dynamical objects for which the discretization and approximation behavior have been investigated are, for instance, invariant manifolds, (Beyn [10], Beyn and Lorenz [12], Lorenz [91], Zou and Beyn [128]), homoclinic orbits (Beyn [9], Fiedler and Scheurle [37]), attracting sets and attractors (Kloeden and Lorenz [77, 78], Lorenz [91], Garay and Kloeden [42]) and Morse–Smale systems (Garay [38, 39, 40, 41]). In addition, several survey articles (e.g., by Beyn [11] or by Stuart [111]) and monographs (like the one by Stuart and Humphries [113]) have been published and a number of specialized algorithms has been designed like, for instance, subdivision techniques for the computation of attractors, unstable manifolds and invariant measures by Dellnitz and Hohmann [29], Dellnitz and Junge [30] and Junge [68, 69] or methods for the computation of reachable sets and domains of attraction, see, e.g., Häckl [59, 60], Abu Hassan and Storey [1] or Genesio, Tartaglia and Vicino [43]. Of course, this list of references and topics is far from complete and can only give a short impression about which dynamical features have been addressed.

In this monograph we want to investigate several aspects of long time or asymptotic behavior under numerical discretization. More precisely, we want to consider asymptotically stable attracting sets, attractors and their respective domains of attraction. We will do this not only for classical dynamical systems (as induced, e.g., by the solutions of an autonomous ordinary differential equation), but also for systems with inputs, i.e., control systems or systems subject to some perturbation, for which these "asymptotic objects" can be generalized in a natural way. We will investigate several techniques which on the one hand allow us to conclude convergence (and related convergence rates) of the numerical approximations of these sets and on the other hand help identifying the cases in which convergence does not hold. Thus, in the context of Murphy's Law, the main intention of this book is to give a number of reasons why things do *not* go wrong, even if they could, and try to explain how we can tell the situations where things go wrong from those where things go well.

1.1 Starting Point

The result which can be considered as the starting point of our investigations was published in 1986 by Kloeden and Lorenz [77]. It states that if an ordinary differential equation has a compact attracting set A then any reasonable numerical one–step approximation (or, more precisely, the discrete time dynamical system induced by this discretization) with sufficiently small time step $h > 0$ has a nearby attracting set A_h which converges to A in the Hausdorff metric as the time step h tends to 0. One of the key contributions of this result is that it provides the right setting for obtaining such a general convergence statement. The crucial observation is that one has to formulate this result for the right definition of *attracting sets*, which here are chosen to be compact forward invariant sets A which uniformly attract a neighborhood $B \supset A$ under the respective dynamical system.

The following simple example (which is a slight modification of Example (0.12) in Garay and Kloeden [42]) illustrates this result and also shows what can go wrong even though we have convergence of attracting sets.

Example 1.1.1 Consider the two–dimensional ordinary differential equation given by

$$\dot{x} = \begin{pmatrix} 0 & 1 \\ -1 & 0 \end{pmatrix} x - \max\{\|x\| - 1, 0\}x$$

for $x = (x_1, x_2)^T \in \mathbb{R}^2$. □

Figure 1.1 shows, from left to right, two solutions of the original system, of its (explicit) Euler discretization (with time step $h = 1/2$) and of its implicit

Euler discretization (with time step $h = 1/2$), respectively. The initial values for these solutions are $x'_0 = (0, 2)$ and $x''_0 = (0, 1/2)$ and the solutions are computed for $t \in [0, 20]$. In addition, in the first two figures the shaded regions show the minimal attracting sets A and A_h.

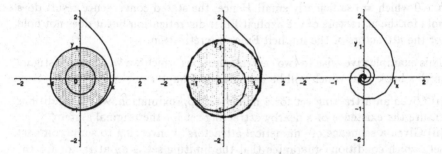

Fig. 1.1. Exact, explicit and implicit Euler solutions of Example 1.1.1

It is easily seen that for the original system each disc

$$D_a := \{x \in \mathbb{R}^2 \,|\, \|x\| \leq a\}$$

with $a \geq 1$ is an attracting set, while for the Euler discretization each set D_a with $a \in [(1 + h - \sqrt{1 - h^2})/h, c(h)]$ is an attracting set, where $c(h)$ is a constant tending to infinity as h tends to 0. For the implicit Euler discretization it turns out that each set D_a with $a \in [0, c(h)]$ is an attracting set, with $c(h)$ as above. Hence, indeed, for both discretizations there exist attracting sets approaching D_a for each $a \geq 1$.

It is now tempting to try the converse implication: Given a family of attracting sets A_h for the numerical systems which for $h \to 0$ converge to some compact set $\tilde{A}$ in the Hausdorff metric, can we say that this set $\tilde{A}$ is an attracting set for the original system? For the explicit Euler discretization this seems to be true, because each sequence of sets $A_h = D_{a_h}$—if convergent— must converge to some D_a with $a \geq 1$ (in fact, this property is true, but a formal proof is more complicated since there exist attracting sets which are not discs). In contrast to this, for the implicit Euler scheme this implication is easily seen to be false, since for instance $A_h = \{0\}$ is a sequence of attracting sets which converges to $\tilde{A} = \{0\}$ which is not an attracting set for the original system.

Another way to look at this problem emerges if we consider *attractors* instead of *attracting sets*. Here we define an attractor to be a compact attracting set which in addition is invariant, i.e., which is mapped exactly onto itself under the respective solution map (this implies that the attractor is the minimal closed attracting set with the given attracted neighborhood B, for details see Section 6.2). Now one might ask whether the convergence result of Kloeden and Lorenz remains valid if we replace *attracting sets* by *attractors*. For the attracted neighborhood $B = D_2$ it is easily seen that the original system has the attractor $A = D_1$, the explicit Euler scheme for time step $h > 0$ sufficiently small has the attractor $A_h = D_{a_h}$ with $a_h = (1 + h - \sqrt{1 - h^2})/h$ and the implicit Euler scheme has the attractor $A_h = \{0\}$ for all time steps $h > 0$ which are sufficiently small. Hence, the stated convergence result does hold for the attractors of the explicit Euler discretization but it does not hold for the attractors of the implicit Euler discretization.

This example gives rise to two central questions which we want to investigate in this book, and which will be answered in Chapter 6:

(i) Given an attracting set for a numerical approximation, which conditions ensure the existence of a nearby attracting set for the original system?
(ii) Given a sequence of "numerical attractors" converging to some compact set, which conditions guarantee that the limiting set is an attractor for the original system?

1.2 Different Approaches

There are basically three ways to obtain statements that tell us about the validity of numerical findings of long time behavior; all of them are used in this monograph. The first approach is to impose suitable conditions on the approximated system (or on the asymptotic object we are interested in), which ensure a faithful numerical approximation and exclude the appearance of numerical artifacts. This approach is closely related to the concept of *structural stability* in the theory of dynamical systems, which, roughly speaking, describes properties of dynamical systems which are robust against small perturbations. Typical examples of this approach are, for instance, the results on the numerical approximation of Morse–Smale systems by Garay [38, 39, 40, 41] and the investigation of gradient systems under discretization as presented in Section 7.7 of the monograph by Stuart and Humphries [113]. We will utilize a condition of this type for the approximation of domains of attraction in Chapter 7.

The second approach is to design algorithms which can be shown to converge to the right objects under no or under very mild conditions on the approximated system. An example for this approach is the subdivision algorithm for the computation of attractors based on a *rigorous discretization* as proposed by Junge [68, 69]. We will investigate this algorithm in Chapter 6 and present a related technique for the computation of domains of attraction in Chapter 7.

Most of the results we will develop here, however, follow a third approach. Instead of imposing conditions on the approximated system or designing clever—but expensive—algorithms we will consider standard methods (like one–step approximations of ordinary differential equations) and formulate *conditions on the behavior of the numerical systems* under which we can ensure convergence of the respective sets or the existence of respective nearby sets for the approximated system. A typical example for this approach in the literature is the study of the behavior of attracting sets in the Galerkin approximation to Navier–Stokes equations by Kloeden [74]. Here we will be able to give a number of conditions on the dynamical behavior of the numerical systems for the existence and convergence of attracting sets, attractors and domains of attraction. A typical statement of this type for one–step approximations is a robustness condition for numerical attracting sets which ensures the existence of a nearby attracting set for the approximated system, cf. Theorem 6.1.3. An example for a convergence result is Theorem 6.2.8, which—among other criteria—shows that a sequence of numerical attractors for vanishing time step converges to some real attractor if and only if we find nearby attracting sets for the numerical system which attract with a rate which is independent of the time step. Due to the fact that convergence occurs for $h \to 0$ we believe that in a general setting this is the strongest result one can obtain, i.e., we do not expect that there is a condition which can be verified using a finite number of time steps only. Of course, we are aware of the fact that a condition for an infinite sequence of vanishing time steps is impossible to check rigorously in practice. Nevertheless, apart from the fact that these results precisely show what we consider to be the principles of convergence of numerically approximated asymptotic objects, there are indeed ways to derive justified heuristic criteria for numerical approximations, cf. Remark 6.1.5.

While all of these approaches have their own advantages and disadvantages, there are a number of reasons why we believe this last approach to be particularly useful. For example, it applies to standard schemes which are implemented in most scientific software packages. Even though sophisticated algorithms are now available for many problems in numerical dynamics, it is a common practice to use standard tools for numerical simulations when one wants to obtain a first impression about what is going on in a system, and clearly it is important to have criteria at hand which facilitate the interpretation of these simulation results. Another reason for which we consider these results to be helpful is the fact that often structural stability conditions are difficult to verify (like, e.g., hyperbolicity) or do not hold for systems coming from real applications. Nevertheless, one might expect that—like in the explicit Euler discretization in Example 1.1.1 and in contrast to Murphy's law—numerical simulations yield reasonable results even for "fragile" objects, and our results allow a precise description of the cases where this is true. Finally, the use of numerical criteria does not exclude the use of structural stability conditions, on the contrary, sometimes these concepts can be efficiently combined as indicated in Remark 6.2.10.

Many of the results which are formulated in this book do not only give qualitative existence or convergence results but also quantitative information about the discretization error. The question behind this is the following: Given a numerical approximation (obtained from a time and/or space discretization) with some local discretization error, what can we say about the global discretizations error, i.e., the distance between the "real" and the "numerical" attracting sets, attractors or domains of attraction? While for finite time approximations the local error (plus some stability condition) directly implies a corresponding global error, the situation is more complicated for objects which are defined via the asymptotic behavior. Nevertheless, it turns out that the local error still determines the global error, however, not directly but in connection with a suitably defined *robustness gain* for the set which we want to approximate. Although in general these gains are not available explicitly from the systems equation, we can show that they always exist and that for an attracting set they are strongly related to the rate of attraction to this set, cf. Theorems 3.4.6 and 4.4.5.

1.3 Basic Idea

The basic idea we will use for the development of our results is to interpret the numerical approximation as a perturbation of the approximated system, and, vice versa, to interpret the approximated system as a perturbation of the numerical approximation. This classical technique from numerical analysis allows us to use abstract results about perturbed dynamical systems which we will develop for this purpose.

The main principle we are going to use for the treatment of perturbed system is adopted from mathematical control theory, namely we will work with a variant of the *input–to–state stability* property. The concept of input–to–state stability was introduced by Sontag [102] and provides a way to characterize the asymptotic behavior of nonlinear systems in the presence of perturbations. It can be considered as a nonlinear generalization of the "finite energy gain" property for linear control systems, where, actually, the term "nonlinear generalization" can be made mathematically precise using suitable nonlinear coordinate transformations, see [57]. Several variants and modifications of this property have been introduced in order to describe various aspects of the asymptotical behavior of nonlinear systems. Here we are going to introduce yet another variant, which is qualitatively equivalent to input–to–state stability (i.e., it describes the same dynamical behavior) but turns out to be more convenient when we want to deduce quantitative estimates, cf. Section 3.2 and Proposition 3.4.4.

In order to use this abstract concept for analyzing the behavior of numerical systems we will then investigate ways to embed numerical systems into suitably perturbed systems. While for internally perturbed systems the internal and the "numerical" perturbation act in the same way, for control systems the control and the perturbation can be considered as opponents. For instance, the control might want to achieve attraction to some set while the perturbation wants to keep the system away from it. This leads to the adoption of ideas from dynamical game theory, namely the use of nonanticipating strategies for modeling the "numerical" perturbations. It turns out that one has to be careful with the definition of "nonanticipation" in order to cover all possible numerical errors, cf. Examples 4.2.4 and 5.3.9.

In all cases we will use a very rich set of perturbation values, which allows the perturbation to act in any possible direction, with the only restriction being on its amplitude. This concept of an *inflated system* enables us to capture all possible numerical errors of a given magnitude without using any further information about their structure. Since most of the abstract perturbation results are formulated for more specific perturbations (i.e., not only for inflated systems) one could well include additional information about the numerical error in its modeling via perturbations. Here, however, we do not follow this idea because we want to consider general purpose numerical schemes without any additional structure. A typical objection against this type of "worst case analysis" is that it usually leads to very conservative results. While this criticism is justified in our case as long as the *quantitative* results are concerned (certainly, a numerical system can by chance or by good reasons perform much better than an inflated system), this does not apply to our *qualitative* results. The reason for this is the basic idea indicated above: Embedding the numerical system into the inflated original system *and* the original system into the inflated numerical system, we are able to use results for inflated systems in order to obtain necessary and sufficient conditions (i.e., equivalence statements), e.g. for the convergence of numerical attractors.

1.4 Outline of the Results

The approach we have just sketched is reflected in the arrangement of the material in the following chapters. After fixing notation and defining the types of systems we are going to consider, we start with the development of a perturbation theory for attracting sets. This part is split into the Chapters 3 and 4, where Chapter 3 is devoted to internally perturbed systems (in which case we speak of *strongly attracting sets*) while Chapter 4 contains the results for control systems (where we speak of *weakly attracting sets*). These two chapters have identical structure, at least as far as the differences between weak and strong attraction permit. We first define the respective concepts of attraction (along with other dynamical properties which will be needed) and then introduce a number of robustness concepts for these sets, i.e., methods to measure how much external perturbations affect the respective attraction properties. For the strongest of these concepts, which we call *input–to–state dynamical stability* we will then give alternative characterizations by means of a geometric criterion and using Lyapunov functions. On the one hand, these characterization are important tools for the application of this abstract concept, on the other hand they allow an exact description of the relation between the different robustness concepts we have introduced. In addition, we can use these characterizations to show that input–to–state dynamical stability is in fact an inherent property of asymptotically stable attracting sets, at least for sufficiently small compact perturbation ranges. We will further provide a stability analysis of these robustness properties, which includes the definition of the important concept of *embedding* systems into each other, and then state a number of results which are valid for inflated systems and go beyond what we could prove for general perturbations. Finally, we investigate the relation between these robustness concepts for continuous and discrete time systems, which will be needed for the interpretation of numerical results, since, in practice, a continuous time system can only be approximated by a discrete time system.

In the next Chapter 5 we will study the relation between numerical discretization and the perturbation concepts from the previous chapters. We present abstract frameworks first for time and then for space discretizations and show how the resulting numerical systems can be embedded into the perturbed systems considered in Chapter 3 and 4. In addition, we discuss a number of numerical schemes for systems without inputs, for internally perturbed systems and for control systems. A great part of this chapter is devoted to the presentation of a systematic development of high–order one–step schemes for systems affine in the input, which were recently proposed by Kloeden and the author in [55], and show how they fit into the abstract framework. Similarly, we discuss space discretization techniques, where particular attention is payed to *rigorous discretization* techniques as developed by Junge [68, Section 2.2] and [69]. In both presentations we will not go into too much implementational details, but restrict ourselves to a description of those main ideas, which we believe to be necessary in order to understand how these schemes work and to show that they are indeed implementable schemes

After all these preparatory investigations, in Chapter 6 we finally come to the presentation of the results on the discretization of attracting sets and attractors. In terms of the robustness properties from Chapter 3 and 4 we give conditions under which the existence of such a set in the approximated system implies the existence of a nearby set in the numerical scheme, and vice versa, which also include quantitative estimates for the distances between these sets. Furthermore, we give several conditions, which ensure that the limit of a sequence of numerical attracting sets (or attractors) is an attracting set (or an attractor) for the approximated system. For attractors we also formulate a sufficient condition on the behavior of the numerical scheme which not only implies convergence but also allows an estimate for the convergence rate. In addition to these results, which apply to general time and space discretizations, we also provide a convergence analysis for the rigorous subdivision algorithm for the computation of attractors from [68, 69], which turns out to be very straightforward using the "right" robustness concept for attractors. Preliminary versions of some of the results in this chapter for systems without inputs (i.e., without control or internal perturbation) have appeared in the papers [50, 51, 54], which were written during the research for this monograph. However, thanks to the systematic development of the abstract perturbation theory the results given here considerably improve these preliminary versions, even for systems without inputs.

The final Chapter 7 then focuses on domains of attraction and reachable sets. It turns out that essentially the same concepts which are used for attracting sets can be used here, because the complement of a domain of attraction is nothing but an attracting set for the time reversed system, provided that the system is reversible in time. Since for discrete time systems this is not necessarily the case we will not directly use this observation but formulate a *dynamical robustness* property for domains of attraction in forward time which is equivalent to the input–to–state dynamical stability for their complements under time reversal. Since we do not want to rephrase all the statements from Chapter 3 and 4 we use a "shortcut" and define this property directly in terms of Lyapunov functions. In order to prove that—similar to attracting sets—this dynamical robustness is an inherent property of domains of attraction we then have to show the existence of a suitable Lyapunov function. For this purpose we use generalizations of what is called Zubov's method for perturbed and for controlled systems, which were recently obtained by Camilli, Wirth and the author [16, 17] and Wirth and the author [58]. After summarizing (and slightly extending) the results from these references we show that the Lyapunov functions obtained by Zubov's method can be used to construct Lyapunov functions characterizing the dynamical robustness property. Having established this result we turn to the analysis of domains of attractions under discretization. Just as limits of numerical attracting sets do not need to be "real" attracting sets, limits of numerical domains of attraction do not need to be "real" domains of attraction. Hence we end up with similar results as those for attracting sets and attractors in Chapter 6 based on conditions for the behavior of the numerical system. In addition, we introduce a structural stability condition for domains of attractions (via the forward invariance of the complement of the domain of attraction) which allows to conclude convergence without imposing conditions on the numerical systems. After these results for general schemes we formulate a subdivision algorithm for the computation of domains of attraction, and show its convergence both without the structural stability condition (provided that the underlying space discretization is rigorous) and with this condition (in this case we can obtain an estimate also for non–rigorous discretizations). Finally, we discuss reachable sets, and show how the results for domains of attraction can be transferred to these sets. In this context we re–investigate the structural stability condition introduced before and show that for reachable sets this condition can be reformulated via chain reachable sets. Hence this condition turns out to be equivalent to a robustness condition well known in the geometric analysis of nonlinear control systems as presented, e.g., in the monograph by Colonius and Kliemann [22].

Two concepts which are used extensively throughout this book are *viscosity solutions* and *comparison functions*. Viscosity solutions are a generalized notion of solutions to partial differential equations and play a vital role in the Lyapunov function characterization of robustness properties in Chapter 3 and 4, as well as for the generalization of Zubov's method for controlled and perturbed systems in Chapter 7. Comparison functions provide an elegant way to formulate robustness, attraction and asymptotic stability properties without using ε–δ formalisms and in addition lead to a natural definition of robustness gains and rates of attraction, for which reason we use them throughout all the Chapters in this monograph. Since these notions might not be well known to all readers we have compiled some elementary background information in the two Appendices A and B. In addition, we use these appendices to formulate and prove several statements about viscosity solutions and comparison functions which did not fit into the other chapters, but are nevertheless needed for the formulation or proofs of some results.

1.5 Open Questions and Future Research

Although this monograph tries to give a self contained treatment of the mentioned problems regarding the asymptotic behavior of systems under perturbation and discretization, it is clear that not all questions arising in this context can be ultimately answered here. Before starting with the development of our results in the next chapter, we therefore want to summarize some open questions and some ideas for future research.

First of all, we believe that the characterization of the robustness of attracting sets by means of Lyapunov functions has not yet reached its final form. Both for internally perturbed and for control systems we are able to prove the existence of *discontinuous* Lyapunov functions, which exactly represent the attraction rate and the robustness gain. While these functions are sufficient for our applications in this book, from a theoretical point of view it is nevertheless interesting to know whether one can find *continuous* Lyapunov functions with the same properties. In the perturbed case we were at least able to show the existence of such functions which *approximately* represent these rates and gains, while in the case of control systems we could not even achieve this result.

Concerning the results on numerical approximations, the probably most important case which is not covered here is the analysis of schemes with adaptive timestepping. Our results only apply to one–step discretizations with fixed time step $h > 0$, but we conjecture that the principles used for these schemes can also be used for the analysis of schemes with step–size control. The reason why we did not include results for this case is that adaptive schemes in general do not induce a standard discrete time dynamical (or control) system. Recently, Kloeden and Schmalfuß [80] and Lamba [85] have proposed different techniques to overcome this difficulty, and we believe that based on these ideas results can be obtained, cf. the Discussion after Definition 5.1.5 in Chapter 5.

Another important issue is the development of an efficient implementation of the subdivision algorithm for the computation of domains of attraction. An first straightforward implementation of this algorithm shows very promising results, cf. Appendix C. We plan to develop such an efficient implementation and hope to be able to present results for more complex systems in the near future.

We have intentionally formulated the perturbation theory in Chapter 3 and 4 in much more generality than needed for our applications to numerical error analysis. Due to this fact we believe that these results are of independent interest and can be used in various different contexts. Certainly, since these results emerge from mathematical control theory, there should be a number of control theoretic applications, in particular for those problems where quantitative results are of interest. An example is the analysis of coupled systems by Jiang, Teel and Praly [67] and Teel [119], where the particular form of the robustness gain decides about stability or instability of the coupled system. In fact, our perturbation analysis is based on conceptionally similar ideas as used in this reference and we believe that many of the results in these references can be recovered and even refined using our approach, see [53, Section 4] for first steps in this direction. Another example is the relation between dynamical and control systems as investigated by Colonius and Kliemann [22]. One result in this area states that under suitable conditions on a control system one can conclude the existence of control sets (i.e., regions of complete controllability) around chain recurrent attractors of the corresponding uncontrolled system. Our results immediately lead to an estimate about the distance between the attractor and the control set, cf. [51, Section 8.3] for a first result in this direction.

A special type of perturbation occurs if we consider the effect of sampling on the performance of a control system. This effect can be investigated by similar techniques as used for numerical approximations, see, e.g., Nešić, Teel and Kokotović [93]. A particular question in this context is whether asymptotic stability can be achieved by sampled controls with some positive lower bound on the sampling rate, a property which by now could only be verified for homogeneous systems [49, 52]. It seems reasonable to expect that the perturbation theory used in this book can provide some new insights into this class of problems.

Apart from control theoretic applications, also for the analysis of numerical errors one could used more sophisticated perturbation models than simple inflated systems. For instance, it might be possible to figure out more restrictive classes of perturbations which still suffice to capture the error caused by specialized numerical schemes like, e.g., energy preserving schemes or symplectic Runge–Kutta methods.

Finally, discretization effects do not only occur in the approximation of finite dimensional systems. It would be a challenging project to investigate which parts of the perturbation theory can be carried over to infinite dimensional systems in order to analyze the behavior of partial differential equations under finite difference, finite element or other kinds of discretizations.

2 Setup and Preliminaries

In this chapter we will introduce the basic models we shall consider, fix the notation, formulate frequently used assumptions and introduce several basic concepts we will work with. For a summary of the notation see also the list on p. 219.

2.1 Continuous Time Systems

In this section we will introduce the systems of differential equations which we want to investigate. We consider finite dimensional ordinary differential equations in euclidean space of the type

$$\dot{x}(t) = f^0(x(t), u(t)) \tag{2.1}$$

where $f^0 : \mathbb{R}^n \times \mathbb{R}^m \to \mathbb{R}^n$ is a continuous map and the input function u satisfies $u \in \mathcal{U} := \{u : \mathbb{R}_0^+ \to U \mid u \text{ measurable}\}$, where $U \subset \mathbb{R}^m$ is some compact set. For functions $u_1, u_2 \in \mathcal{U}$ we define the concatenation at time $s \geq 0$ by

$$u_1 \&_s u_2(t) = \begin{cases} u_1(t), & t \leq s \\ u_2(t-s), & t > s \end{cases} \tag{2.2}$$

On $\mathcal{U}$ we use the L_∞-norm

$$\|u\|_\infty := \operatorname{ess\,sup}_{t \geq 0} \|u(t)\|$$

and its restriction to intervals $[a, b] \subset \mathbb{R}_0^+$

$$\|u\|_{[a,b]} := \operatorname{ess\,sup}_{t \in [a,b]} \|u(t)\|,$$

where $\|\cdot\|$ denotes the usual euclidean norm on $\mathbb{R}^m$. When we consider several different sets of control values we will denote them, e.g., by $\tilde{U}$ or U^*. In this case we use the convention that $\tilde{\mathcal{U}}$ or $\mathcal{U}^*$ denote the corresponding spaces of measurable functions with values in the respective set.

The input functions $u \in \mathcal{U}$ will have two different interpretations: First, they can be internal perturbations of the model, in which case we are interested in the behavior of (2.1) under all possible u. Second, they may be control functions which we are free to choose depending on the initial value. In this case we will consider the behavior of (2.1) under certain $u \in \mathcal{U}$ which are chosen to guarantee a specific performance of the system.

In both cases (i.e., also if $\mathcal{U}$ models internal perturbations) we refer to (2.1) as the *unperturbed system* since we will introduce further external perturbation to (2.1) below. To systems of the type $\dot{x}(t) = f(x(t))$, which we will occasionally consider, we refer as *systems without input*.

When defining properties characterizing the dynamical behavior of (2.1), and the $u \in \mathcal{U}$ are considered as perturbations we will speak about *strong* concepts, while when the functions $u \in \mathcal{U}$ are interpreted as control functions we will use the term *weak* concepts. For instance, a point x^* will be called *strongly asymptotically stable*, if there exists a neighborhood $\mathcal{N}$ of x^* such that each trajectory starting in $\mathcal{N}$ stays close to x^* and eventually converges to x^* *for all* $u \in \mathcal{U}$ (in some uniform sense to be specified later), whereas we will call x^* *weakly asymptotically stable*, if for each x in some neighborhood $\mathcal{N}$ of x^* *there exists* $u \in \mathcal{U}$ such that the corresponding trajectory stays close and eventually converges to x^* (again, in some uniform sense that we will make precise later).

Note that a number of different terminologies exist for these properties: In mathematical control theory, e.g., instead of strong asymptotic stability one also finds the term "robust (asymptotic) stability", while weak asymptotic stability is also termed "asymptotic controllability" or "open loop asymptotic stability".

The explicit definition of "weak" and "strong" properties apparently goes back to the study of stability properties of so called general control systems by Roxin [98], and since has been used by a number of authors, like e.g. Szegö and Treccani [114] or Kloeden [73], mostly in the context of general (i.e., set valued) dynamical systems or differential inclusions. Despite the fact that usually this terminology is used for differential inclusions rather that for control systems of type (2.1), we decided to adopt it here since to a great extent it allows a unified analysis and presentation of dynamical properties both "for all u" and "for some u".

For some purposes it will be convenient to consider sets of control functions $u \in \mathcal{U}$ depending on the initial value. For this purpose for any subset $B \subseteq \mathbb{R}^n$ we define the set

$$\mathcal{U}(B) := \{\bar{u} : B \times \mathbb{R}_0^+ \to U \mid \bar{u}(x, \cdot) \in \mathcal{U} \text{ for all } x \in B\}. \tag{2.3}$$

Our main object of interest is the effect of external perturbations on (2.1). For this purpose we consider a set of perturbation values $W \subseteq \mathbb{R}^l$ and a continuous map $f : \mathbb{R}^n \times U \times W \to \mathbb{R}^n$ satisfying

$$f(x, u, 0) = f^0(x, u) \text{ for all } x \in \mathbb{R}^n, u \in U. \tag{2.4}$$

Here $W \subseteq R^l$ can be an arbitrary set, in particular it can be compact, bounded or unbounded.

The actual system associated to f now depends on the interpretation of u. If u is modeling an internal perturbation of the system we consider the perturbed equation

$$\dot{x}(t) = f(x(t), u(t), w(t)) \qquad (2.5)$$

where $w \in \mathcal{W} := \{w : \mathbb{R} \to W \,|\, w \text{ measurable and locally essentially bounded}\}$. As on $\mathcal{U}$ we use the L_∞-norm $\|\cdot\|_\infty$ on $\mathcal{W}$, and for $\alpha \geq 0$ we define $\mathcal{W}_\alpha$ as the set of all $w \in \mathcal{W}$ with $\|w\|_\infty \leq \alpha$.

If u is a control function then we have to take care about how the control u and the perturbation w interact. For this purpose we use a concept from differential game theory (see, e.g., [8, Chapter VIII]), which we slightly change to cover all of the "numerical perturbations" we will encounter in what follows.

Definition 2.1.1 (nonanticipating functions and strategies)

(i) For each $\delta > 0$ we define the set of sequences

$$\mathcal{T}^\delta := \{\mathbf{t} = (t_i)_{i \in \mathbb{N}_0} \,|\, t_0 = 0, \, t_i \in \mathbb{R}, \, t_{i+1} - t_i \in (0, \delta], \, t_i \to \infty \text{ as } i \to \infty\}.$$

(ii) Let C be an arbitrary set. For each $\delta > 0$ we call a function $b : \mathcal{U} \times \mathbb{R}_0^+ \to C$ δ-nonanticipating if there exists a sequence $\mathbf{t}(p) = (t_i)_{i \in \mathbb{N}_0} \in \mathcal{T}^\delta$ such that the implication

$$u_1(s) = u_2(s) \text{ for all } s \leq t_i \;\Rightarrow\; b(u_1, s) = b(u_2, s) \text{ for all } s \leq t_i$$

holds for all $i \in \mathbb{N}$, and we call b 0-nonanticipating (or simply nonanticipating) if the implication

$$u_1(s) = u_2(s) \text{ for all } s \leq t \;\Rightarrow\; b(u_1, t) = b(u_2, t) \text{ for all } s \leq t$$

holds for all $t > 0$.

(iii) For each $\delta \geq 0$ we define the set of δ-nonanticipating strategies $\mathcal{P}^\delta$ as the set of maps $p : \mathcal{U} \to \mathcal{W}$ with the property that each $p \in \mathcal{P}^\delta$ the map $b(u, t) := p[u](t)$ is δ-nonanticipating. A sequence $(t_i)_{i \in \mathbb{N}_0} \in \mathcal{T}^\delta$ for which the nonanticipativity holds will be denoted by $\mathbf{t}(p)$, where for $p \in \mathcal{P}^0$ we also allow $\mathbf{t}(p) = \mathbb{R}_0^+$. Note that $\mathbf{t}(p)$ is not unique.   □

Note that the inclusion $\mathcal{P}^{\delta'} \subset \mathcal{P}^\delta$ is immediate from the definition for all $\delta > \delta' \geq 0$.

For the same f as in (2.5) we now consider the perturbed equation

$$\dot{x}(t) = f(x(t), u(t), p[u](t)) \qquad (2.6)$$

for $p \in \mathcal{P}$, where $\mathcal{P}$ is a subset of $\mathcal{P}^\delta$ for some $\delta > 0$ which is closed under shifts, i.e. for each $p \in \mathcal{P}$, each $u^* \in \mathcal{U}$ and each $s > 0$ we require the existence of $\tilde{p} \in \mathcal{P}$ such that

$$\tilde{p}[u](t - s) = p[u^* \&_s u](t)$$

holds for all $u \in \mathcal{U}$ and almost all $t \geq s$ (recall the definition of the concatenation operator $\&_s$ from (2.2)). For $\alpha \geq 0$ we denote by $\mathcal{P}_\alpha$ the subset of strategies from $\mathcal{P}$ satisfying $\|p[u]\|_\infty \leq \alpha$ for all $u \in \mathcal{U}$. In particular, if $\mathcal{P} = \mathcal{P}^\delta$ for some $\delta \geq 0$ we write $\mathcal{P}_\alpha^\delta$ for this set.

The reason for using this concept will be discussed in detail when we investigate the relation between numerical discretization and perturbation in Chapter 5, see in particular Example 5.3.9.

We will now state the main assumptions we will make on the right hand side of the respective equations.

We assume that for each $R > 0$ there exist constants M_R and $L_R > 0$ such that

$$\|f(x, u, w)\| \leq M_R \text{ and } \|f(x, u, w) - f(y, u, w)\| \leq L_R \|x - y\| \qquad (2.7)$$

for all $x, y \in \mathbb{R}^n$ and $w \in W$ with $\|x\| \leq R$, $\|y\| \leq R$ and $\|w\| \leq R$ and all $u \in U$.

In our numerical studies we will frequently use the stronger assumption

$$\|f(x, u, w)\| \leq M + \rho(\|w\|) \text{ and } \|f(x, u, w) - f(y, u, w)\| \leq L \|x - y\| \quad (2.8)$$

for all $x, y \in \mathbb{R}^n$, all $w \in W$, all $u \in U$, some function ρ of class $\mathcal{K}_\infty$ (cf. Appendix B) and suitable constants $L, M > 0$. This is mainly done for convenience since numerically we only consider the behavior of (2.5) and (2.6) on compact subsets of $\mathbb{R}^n$ and for compact perturbation range W, in which situation we can always obtain (2.8) from (2.7) using standard cutoff techniques.

Another assumption we will need for certain statements is an a priori bound on the effect of the perturbation on the system. Here we assume that

$$\|f^0(x, u) - f(x, u, w)\| \leq \rho(\|w\|) \text{ for all } x \in \mathbb{R}^n, u \in \mathcal{U}, w \in W \qquad (2.9)$$

for some function ρ of class $\mathcal{K}_\infty$ (cf. Appendix B). Our main motivation for assumption (2.9) is the so called *inflated system*, which is obtained when f and W are given by

$$f(x, u, w) := f^0(x, u) + w, \quad W = \{w \in \mathbb{R}^n \,|\, \|w\| \leq \alpha_0\} \qquad (2.10)$$

for some $\alpha_0 \in \mathbb{R}^+ \cup \{\infty\}$. For a given α_0 we also call (2.10) the α_0-*inflated system*. For systems without inputs these inflated systems were introduced in [76] using differential inclusions and in [51] using control systems. Clearly, the inflated f from (2.10) satisfies (2.9), and if $f = f^0$ satisfies (2.8) then also the inflated f from (2.10) does so. We will also consider inflated systems where the inflation is state dependent, i.e. where f has the form

$$f(x, u, w) := f^0(x, u) + b(x)w \qquad (2.11)$$

with W as in (2.10) and some bounded nonnegative function $b : \mathbb{R}^n \to \mathbb{R}$ which is globally Lipschitz. Also this f satisfies (2.9), and it satisfies (2.8) if f^0 does so and $\alpha_0 < \infty$.

We will denote the trajectories of (2.5) for initial value $x \in \mathbb{R}^n$, functions $u \in \mathcal{U}$, $w \in \mathcal{W}$ and initial time $t = 0$ by $\varphi(t, x, u, w)$ and those of (2.6) by $\varphi(t, x, u, p[u])$, and assume that the system is forward complete, i.e. that each trajectory exists for all positive times for all initial values and all inputs $u \in \mathcal{U}$ and w or $p[u] \in \mathcal{W}$, respectively. Note that the first condition of (2.8) implies this property. With a slight abuse of terminology we will frequently identify a system with its trajectories, because it will often be convenient to refer to "a system φ of type (2.5)" without explicitly mentioning the underlying vector field f.

Whenever it is clear from the context that we consider the unperturbed system (2.1) we also write $\varphi(t, x, u)$ instead of $\varphi(t, x, u, 0)$. Often, for system (2.5) we will use the set valued mappings

$$\varphi_\alpha(t, x) := \bigcup_{u \in \mathcal{U},\, w \in \mathcal{W}_\alpha} \{\varphi(t, x, u, w)\}$$

and, for a function $\alpha : \mathbb{R} \to \mathbb{R}_0^+$,

$$\varphi_{\alpha(t)}(t, x) := \bigcup_{\substack{u \in \mathcal{U},\, w \in \mathcal{W} \\ \|w(t)\| \leq \alpha(t) \text{ for a.a. } t > 0}} \{\varphi(t, x, u, w)\}.$$

For subsets $B \subset \mathbb{R}^n$ we define

$$\varphi_\alpha(t, B) := \bigcup_{x \in B} \varphi_\alpha(t, x)$$

and analogously $\varphi_{\alpha(t)}(t, B)$. For $\alpha = 0$ we will use the more intuitive notation φ_u instead of φ_0. With

$$\mathcal{R}(B) := \bigcup_{\alpha > 0,\, t \geq 0} \varphi_\alpha(t, B)$$

we denote the *reachable set* of B for all perturbations.

By [88, Proposition 5.1] forward completeness implies that the sets $\varphi_\alpha(t, B)$ are bounded for all $t, \alpha > 0$ and all bounded sets $B \subset \mathbb{R}^n$.

In the case that u represents a control function we will sometimes need certain continuity assumption on the system trajectories. For general functions $b : \mathbb{R}_0^+ \times \mathcal{U} \to \mathbb{R}^n$ this assumption is as follows.

$$
\text{For all sequences } u_n \in \mathcal{U} \text{ there exists } u \in \mathcal{U} \text{ and a sub-} \atop \text{sequence } n_k \to \infty \text{ such that } b(t, u_{n_k}) \to b(t, u) \text{ for each } t \geq 0. \tag{2.12}
$$

For functions $b : I \times \mathcal{U} \to \mathbb{R}^n$ where I is some subset $I \subset \mathbb{R}$ we will require (2.12) only for $t \in I \cap [0, \infty)$, in which case we say that b satisfies (2.12) *with respect to* I.

Using (2.12) we can now define suitable continuity assumptions on our control systems. The first is on the unperturbed system (2.1).

$$
\text{For all } x \in \mathbb{R}^n \text{ the trajectory } \varphi(t, x, u, 0) \text{ satisfies (2.12)} \tag{2.13}
$$

By Filippov's Lemma (see [87, p. 267]) this is satisfied, e.g., if $f^0(x, U)$ is a convex set for all $x \in \mathbb{R}^n$.

The second continuity assumption is on both the perturbed system (2.6) and the set of strategies $\mathcal{P}$.

$$
\text{For all } p \in \mathcal{P} \text{ and all } x \in \mathbb{R}^n \text{ the trajectory } \varphi(t, x, u, p[u]) \text{ satisfies (2.12)} \tag{2.14}
$$

This condition is satisfied, e.g., if the system has the structure

$$
f(x, u, p) = g_0(x) + \sum_{i=1}^{m} u_i g_i(x) + \sum_{j=1}^{l} p_j g_{j+m}(x) \tag{2.15}
$$

and the set of perturbation strategies $\mathcal{P}$ is such that all elements $p \in \mathcal{P}$ are continuous with respect to the weak*–topology on $\mathcal{U}$ and $\mathcal{W}$. In this case for each sequence of control functions u_n we find a weak*–convergent subsequence $u_{n_k} \to u$. Then the continuity of p implies that the pair $(u_{n_k}, p[u_{n_k}])$ weak*–converges to $(u, p[u]) \in \mathcal{U} \times \mathcal{W}$ and the control affine structure of (2.15) gives the convergence of the trajectories, cf. [22, Lemma 4.3.2]. For further reference we will denote the strategies p satisfying this continuity property by $\mathcal{P}^{\delta, c}$, i.e.,

$$
\mathcal{P}^{\delta, c} := \left\{ p \in \mathcal{P}^\delta \; \middle| \; \begin{array}{l} p : \mathcal{U} \to \mathcal{W} \text{ is continuous with respect} \\ \text{to the weak*–topology on } \mathcal{U} \text{ and } \mathcal{W} \end{array} \right\}. \tag{2.16}
$$

2.2 Discrete Time Systems

Since we are going to investigate numerical discretizations it will be necessary
to look at discrete time systems. Apart from this motivation, discrete time
systems form an interesting class of systems on its own right, and many
statements in the following chapters make sense even without any numerical
interpretation.

Our discrete time analogue of system (2.1) is

$$x(t + h) = \Phi_h^0(x(t), u(t + \cdot)) \tag{2.17}$$

where $\Phi_h^0 : \mathbb{R}^n \times \mathcal{U} \to \mathbb{R}^n$ is continuous in x, $h > 0$ is some positive time step
and $\mathcal{U}$ is as in the continuous time case. Observe that we deviate from the
common practice to use sequences $u_i \in U, i \in \mathbb{N}$ as inputs to our discrete time
system; instead we use continuous time functions $u \in \mathcal{U}$. In order to avoid
non-causal dependence on u we assume that Φ_0^h satisfies the implication

$$u_1(t) = u_2(t) \text{ for almost all } t \in [0, h]$$
$$\Rightarrow \ \Phi_h^0(x, u_1(\cdot)) = \Phi_h^0(x, u_2(\cdot)) \text{ for all } x \in \mathbb{R}^n. \tag{2.18}$$

The reason to allow continuous time control functions is that we want the
time-h map of (2.1) given by

$$\Phi_h^0(x, u) = \varphi^h(x, u, 0) \text{ with } \varphi^h(x, u, w) := \varphi(h, x, u, w) \tag{2.19}$$

to be in our class of discrete time systems. Also, this general setup allows an
elegant formalism for numerical approximations of continuous time systems.
Note, however, that the case of input sequences $u_i \in U, i \in \mathbb{N}$ is included in
our setup since we can—and will occasionally—identify these sequences with
the piecewise constant control functions $u(t) \equiv u_i, t \in [ih, (i + 1)h)$.

We will not only consider perturbations of continuous time systems but also
of discrete time systems. For this purpose we consider discrete time systems
whose right hand side is given by a map $\Phi_h : \mathbb{R}^n \times \mathcal{U} \times \mathcal{W} \to \mathbb{R}^n$ which is
continuous in x and satisfies

$$\Phi_h(x, u, 0) = \Phi_h^0(x, u) \text{ for all } x \in \mathbb{R}^n, u \in \mathcal{U}. \tag{2.20}$$

Also for Φ_h we assume the causality condition (2.18), now in both arguments
u and w.

As in the continuous time case the actual system defined by Φ_h depends on
the interpretation of u. When u is interpreted as perturbation we consider
the system

$$x(t+h) = \Phi_h(x, u(t+\cdot), w(t+\cdot)) \tag{2.21}$$

with $w \in \mathcal{W}$ and $\mathcal{W}$ is the same function space as defined for (2.5), above. When u is interpreted as a control function we set

$$x(t+h) = \Phi_h(x, u(t+\cdot), p[u](t+\cdot)) \tag{2.22}$$

where $p \in \mathcal{P}$ and $\mathcal{P}$ is again a set of δ-nonanticipating strategies from $\mathcal{U}$ to $\mathcal{W}$ as defined for (2.6). Note that we do not impose any a priori conditions on the relation between the anticipation times $\mathbf{t}(p) = (t_i)_{i \in \mathbb{N}_0}$ and the discrete time steps hi, $i \in \mathbb{N}$, at which our system is evaluated (although occasionally we will need them).

We will make similar assumptions on (2.20) as in the continuous time case: We assume that for each $R > 0$ there exist constants M_R and $L_R > 0$ such that

$$\|\Phi_h(x, u, w) - x\| \le h M_R$$
$$\text{and} \quad \|(\Phi_h(x, u, w) - x) - (\Phi_h(y, u, w) - y)\| \le h L_R \|x - y\| \tag{2.23}$$

for all $x, y \in \mathbb{R}^n$ and $w \in W$ with $\|x\| \le R$, $\|y\| \le R$ and $\|w\| \le R$ and all $u \in U$. This specific form of the Lipschitz condition will turn out to be useful when we consider families of discrete time systems approximating a continuous time system. For a single map $\Phi_h(x, u, w)$, which is Lipschitz in x for $\|x\| \le R$ with constant $\tilde{L}_R$, this condition is satisfied for $L_R = L_R(h) = 1/h + \tilde{L}_R/h$. However, when considering a family of systems with different time steps $h > 0$ we usually want the estimates in (2.23) to by valid for constants which are independent of $h > 0$. Thus for families of systems (2.23) is stronger that the usual Lipschitz condition. Note that if (2.7) holds then the time-h map φ^h satisfies (2.23) for slightly enlarged constants for all $h > 0$ sufficiently small.

Sometimes we will require the more restrictive global version

$$\|\Phi_h(x, u, w) - x\| \le h(M + \rho(\|w\|))$$
$$\text{and} \quad \|(\Phi_h(x, u, w) - x) - (\Phi_h(y, u, w) - y)\| \le h L \|x - y\| \tag{2.24}$$

for all $x, y \in \mathbb{R}^n$, all $w \in W$, all $u \in U$, some function ρ of class $\mathcal{K}_\infty$ and suitable constants $M, L > 0$, and occasionally we will need the a priori bound

$$\|\Phi_h^0(x, u) - \Phi_h(x, u, w)\| \le h\rho(\|w\|) \text{ for all } x \in \mathbb{R}^n, u \in \mathcal{U}, w \in W \tag{2.25}$$

on the effect of the perturbation for some class $\mathcal{K}_\infty$ function ρ.

Again analogously to the continuous time case we consider the α_0-inflated system for $\alpha_0 \in \mathbb{R}^+ \cup \{\infty\}$

$$\Phi_h(x, u, w) := \Phi_h^0(x, u) + \int_0^h w(t)dt, \quad w \in \mathcal{W} \tag{2.26}$$

with $W = \{w \in \mathbb{R}^n \,|\, \|w\| \le \alpha_0\}$

and its state dependent version

$$\Phi_h(x, u, w) := \Phi_h^0(x, u) + b(x)\int_0^h w(t)dt, \quad \text{with } \mathcal{W} \text{ as in (2.26)} \tag{2.27}$$

for some bounded nonnegative function $b : \mathbb{R}^n \to \mathbb{R}$. Note that here in the discrete case we do not require continuity of b, which implies that the system might not be of type (2.21). We will, however, always be able to circumvent this technical difficulty by using that any trajectory of (2.27) for some w or p can be interpreted as a trajectory of (2.26) for some suitable $\tilde{w}$ or $\tilde{p}$.

Observe that these inflated systems can be identified with systems with discrete perturbations, namely with $\Phi_h^0(x, u) + hw$ and $\Phi_h^0(x, u) + b(x)hw$, respectively, where $w \in W$ with W from (2.26).

It is easily seen that the time-h map $\varphi^h(x, u, w)$ for the continuous inflated system (2.5) with f from (2.10) and the inflated time-h map $\Phi_h(x, u, w) = \varphi_h(x, u, 0) + \int_0^h w(t)dt$ do not coincide in general. Under suitable conditions, however, they are closely related as the Lemmata 3.8.4 and 4.8.4, below, will show.

The trajectories of (2.21) and (2.22) for any time $t \in h\mathbb{N}$ will be denoted by $\Phi_h(t, x, u, w)$ or $\Phi_h(t, x, u, p[u])$, respectively, and are defined inductively by

$$\Phi_h(h, x, u, w) := \Phi_h(x, u, w)$$

and

$$\Phi_h(t + h, x, u, w) := \Phi_h(\Phi_h(t, x, u, w), u(t + \cdot), w(t + \cdot))$$

for $t \ge h$, and analogously for $\Phi_h(t, x, u, p[u])$. Note that we do not assume the trajectories to exist for negative times.

Analogous to the continuous time case we define the set valued mappings $\Phi_{h,\alpha}$ and $\Phi_{h,\alpha(t)}$, where for $\alpha = 0$ we will use $\Phi_{h,\mathcal{U}}$ instead of $\Phi_{h,0}$. We assume boundedness of these sets, i.e., we require that $\Phi_{h,\alpha}(t, B)$ is bounded for all $t, \alpha > 0$ and all bounded sets $B \subset \mathbb{R}^n$. Note that forward completeness is trivially satisfied for our discrete time systems, but, since we did not make any assumptions on the u– and w–dependence of Φ_h, it does not necessarily imply boundedness of $\Phi_{h,\alpha}(t, B)$.

The reachable set $\mathcal{R}(B)$ is defined analogously to the continuous time case, now for all $t \ge 0$ with $t \in h\mathbb{Z}$.

As in the continuous time case for some statements we will require the continuity properties (2.13) and (2.14), where (2.12), of course, only needs to be satisfied with respect to $I = h\mathbb{N}_0$.

In what follows we will frequently make statements simultaneously for discrete and continuous time systems. For this purpose we use the symbol Φ, which either denotes the trajectory φ of the continuous time system or the trajectory Φ_h of the discrete time system. Similarly, we will use Φ_α and $\Phi_{\mathcal{U}}$. For the set of times we will use the symbol $\mathbb{T}$ which is either $\mathbb{R}$ or $h\mathbb{Z}$, $\mathbb{T}^+ := \{t \in \mathbb{T} \mid t > 0\}$ and $\mathbb{T}_0^+ := \mathbb{T}^+ \cup \{0\}$.

2.3 Sets, Distances and Limits

The problems, definitions and results in this book center around the behavior of dynamical and control systems on or relative to certain sets. In particular, distances between sets play an important role in the formulation of robustness gains and numerical errors. In this section we compile the necessary definitions as well as some basic properties.

Throughout this text we will denote the usual euclidean norm for $x \in \mathbb{R}^n$ by $\|x\|$. The closure of a set $C \subset \mathbb{R}^n$ will be denoted by $\operatorname{cl} C$, its interior by $\operatorname{int} C$ and its complement by C^c, i.e.,

$$C^c := \{x \in \mathbb{R}^n \mid x \notin C\}.$$

For points and sets we will use the following distances.

Definition 2.3.1 (distances between sets)
Consider sets $C, D \subseteq \mathbb{R}^n$ and a point $x \in \mathbb{R}^n$. Then we define

the point–set (euclidean) distance by

$$\|x\|_D := \inf_{y \in D} \|x - y\|,$$

the set–set *Hausdorff semidistance* by

$$\operatorname{dist}(C, D) := \sup_{x \in C} \|x\|_D,$$

the set–set *Hausdorff distance* by

$$d_H(C, D) := \max\{\operatorname{dist}(C, D), \operatorname{dist}(D, C)\}$$

and the set–set *minimal distance* by

$$d_{\min}(C, D) := \inf_{x \notin C} \|x\|_D.$$

$\square$

Note that we define these distances for arbitrary sets, i.e., we do not assume closedness or boundedness.

The closed ball around some set $D \subset \mathbb{R}^n$ for some $r > 0$ will be denoted by

$$\mathcal{B}(r, D) := \{x \in \mathbb{R}^n \,|\, \|x\|_D \le r\}.$$

If $D = \{x\}$ then we also write $\mathcal{B}(r, x)$ instead of $\mathcal{B}(r, \{x\})$.

The following Lemma summarizes several properties of the distances.

Lemma 2.3.2 Consider sets $C, D \subseteq \mathbb{R}^n$ and a point $x \in \mathbb{R}^n$. Then the following properties hold.

(i) $\|x\|_D = 0 \Leftrightarrow x \in \operatorname{cl} D$

(ii) $\operatorname{dist}(C, D) = 0 \Leftrightarrow \operatorname{cl} C \subseteq \operatorname{cl} D$

(iii) $d_H(C, D) = 0 \Leftrightarrow \operatorname{cl} C = \operatorname{cl} D$

(iv) $d_{\min}(C, D) \ge r > 0 \Leftrightarrow \mathcal{B}(r, D) \subset \operatorname{cl} C$

(v)   If C is open and C_k, $k \in \mathbb{N}$ is a sequence of sets satisfying $\operatorname{dist}(C_k^c, C^c) \to 0$ as $k \to \infty$ then for each $x \in C$ there exists $K \in \mathbb{N}$ such that $x \in C_k$ for all $k \ge K$. If, in addition, either C or C^c is bounded then $\operatorname{dist}(C, C_k) \to 0$ as $k \to \infty$.

Proof: Properties (i) to (iv) are immediate from the definitions.

(v) Let $x \in C$. Since C is open there exists $\varepsilon > 0$ such that $\|x\|_{C^c} = \varepsilon$. So $x \notin C_k$, i.e., $x \in C_k^c$ implies $\operatorname{dist}(C_k^c, C^c) \ge \varepsilon$. Hence $x \in C_k$ for all $k \in \mathbb{N}$ with $\operatorname{dist}(C_k^c, C^c) < \varepsilon$.

If C is bounded then for each $\varepsilon > 0$ we find a finite number of points $y_l \in C$, $l = 1, \ldots, m$ such that $\operatorname{dist}(C, \{y_1, \ldots, y_l\}) \le \varepsilon$. Since by the first assertion we obtain $\{y_1, \ldots, y_l\} \subset C_k$ for all k sufficiently large the desired convergence follows.

If C^c is bounded then also $\mathcal{B}(\varepsilon, C^c)$ is bounded for each $\varepsilon > 0$. Fix $\varepsilon > 0$ and define $C_\varepsilon := C \cap \mathcal{B}(\varepsilon, C^c)$. From the assumption we know that $\operatorname{dist}(C_k^c, C^c) < \varepsilon$ for k sufficiently large, hence $C_k^c \subset \mathcal{B}(\varepsilon, C^c)$ and consequently $C_k \supseteq \mathcal{B}(\varepsilon, C^c)^c = C \setminus C_\varepsilon$ implying

$$\operatorname{dist}(C, C_k) = \max\{\underbrace{\operatorname{dist}(C \setminus C_\varepsilon, C_k)}_{=0}, \operatorname{dist}(C_\varepsilon, C_k)\} = \operatorname{dist}(C_\varepsilon, C_k).$$

Now since $C_\varepsilon^c \supseteq C^c$ we know that $\operatorname{dist}(C_k^c, C_\varepsilon^c) \le \operatorname{dist}(C_k^c, C^c) \to 0$ as $k \to \infty$ and since C_ε is bounded we can proceed as in the case where C is bounded to obtain the assertion. $\square$

Remark 2.3.3 In general it is not possible to find a direct relation between $\mathrm{dist}(C, D)$ and $\mathrm{dist}(D^c, C^c)$, no matter whether these sets are open or closed. As an example, consider the triangular set $C \subset \mathbb{R}^2$ given by

$$C := \{(x, y)^T \in \mathbb{R}^2 \,|\, x \in [0, 1], y \in [0, sx]\}$$

for some $s > 0$ and, for $\varepsilon < s/2$, the smaller triangular set

$$D_\varepsilon := \{(x, y)^T \in \mathbb{R}^2 \,|\, x \in [2\varepsilon/s, 1 - \varepsilon], y \in [\varepsilon, sx - \varepsilon]\}.$$

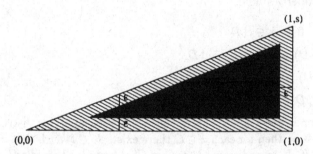

(1,s)

(0,0)   (1,0)

Fig. 2.1. The sets C (black and hatched) and D_ε (black)

In order to estimate $\mathrm{dist}(D_\varepsilon^c, C^c)$ observe that each point $(x, y)^T \in D_\varepsilon^c \setminus C^c$ (which is the interior of the hatched region in Figure 2.1) satisfies either $y < \varepsilon$ or $x > 1 - \varepsilon$ or $y < sx - \varepsilon$. Hence picking either $(x, y - \varepsilon)$ or $(x + \varepsilon, y)$ or $(x, y + \varepsilon)$ we always find points in C^c with distance less or equal to ε from $(x, y)^T$ and consequently $\mathrm{dist}(D_\varepsilon^c, C^c) \leq \varepsilon$.

For $\mathrm{dist}(C, D_\varepsilon)$ pick the point $(0, 0)^T \in C$. Obviously, the closest point in D_ε is $(2\varepsilon/s, \varepsilon)^T$ which has a distance bigger than $2\varepsilon/s$ from $(0, 0)^T$. From this we obtain the estimate $\mathrm{dist}(C, D_\varepsilon) \geq 2\varepsilon/s$ and hence

$$\mathrm{dist}(C, D_\varepsilon) \geq \frac{2}{s}\mathrm{dist}(D_\varepsilon^c, C^c)$$

which shows that—depending on s—the relative difference between these two distances can be arbitrary large.   □

For some constructions we will need the following definition of the limes superior of sets.

Definition 2.3.4 (limes superior for sets)

For a sequence of sets $C_k \subset \mathbb{R}^n$, $k \in \mathbb{N}$ we define the *limes superior* by

$$\text{Lim}\sup\nolimits_{k\to\infty} C_k := \bigcap_{K>0} \text{cl} \bigcup_{k\geq K} C_k.$$

□

Observe that we use the notation "Lim sup" with a capital "L" in order to distinguish this limit from the usual "lim sup" for real sequences. We will need the following lemma, which follows from [6, Proposition 1.1.5] and shows that the C_k uniformly approach their limes superior.

Lemma 2.3.5 Consider a sequence of sets $C_k \subset \mathbb{R}^n$ and a compact set $M \subset \mathbb{R}^n$. Let $C := \text{Lim}\sup_{k\to\infty} C_k$. Then for each $\varepsilon > 0$ there exists $K > 0$ such that

$$C_k \cap M \subset B(\varepsilon, C \cap M)$$

for all $k \geq K$.

The next lemma shows some relations between distances and the Lim sup for sets.

Lemma 2.3.6 Consider a sequence of sets $C_k \subset \mathbb{R}^n$ and a set $C \subset \mathbb{R}^n$. Then the following implications hold.

(i) If C is closed and $\lim_{k\to\infty} \text{dist}(C, C_k) = 0$, then $C \subseteq \text{Lim}\sup_{k\to\infty} C_k$.

(ii) If C is open and $\lim_{k\to\infty} \text{dist}(C_k^c, C^c) = 0$, then $C \subseteq \text{int} \, \text{Lim}\sup_{k\to\infty} C_k$.

(iii) If C is closed and $\lim_{k\to\infty} \text{dist}(C_k, C) = 0$, then $C \supseteq \text{Lim}\sup_{k\to\infty} C_k$.

(iv) If C is closed and $\lim_{k\to\infty} d_H(C_k, C) = 0$, then $C = \text{Lim}\sup_{k\to\infty} C_k$.

Proof: (i) Let $x \in C$. Then the dist assumption yields the existence of $x_k \in C_k$ such that $x_k \to x$. Hence $x \in \text{cl} \bigcup_{k\geq K} C_k$ for all $K > 0$ which implies the assertion.

(ii) By Lemma 2.3.2 (v) we obtain that each $x \in C$ is contained in C_k for k sufficiently large. This shows the assertion.

(iii) Let $x \in \text{Lim}\sup_{k\to\infty} C_k$. Then we know the existence of a sequence of points $x_k \in C_k$ with $x_k \to x$. Assume $x \notin C$. Then—since C is closed— we obtain that $\|x\|_C =: \varepsilon > 0$. Hence for all k sufficiently large we obtain $\|x_k\|_C \geq \varepsilon/2$, thus $\text{dist}(C_k, C) \geq \varepsilon/2$ contradicting the assumption.

(iv) follows from (i) and (iii). □

3 Strongly Attracting Sets

In this chapter we will investigate continuous and discrete time systems of type (2.5) and (2.21), along with their unperturbed counterparts (2.1) and (2.17). Here $u \in \mathcal{U}$ shall always be interpreted as an internal perturbation, i.e., we are going to investigate strong dynamical properties.

We will start by defining a number of these properties of sets like forward invariance, attraction, asymptotic stability, and then turn to the formulation of different robustness concepts which measure how much the presence of the external perturbation $w \in \mathcal{W}$ influences the behavior of the unperturbed systems (2.1) and (2.17).

Once these concepts are defined we will proceed to give several characterizations of these robustness properties by means of geometric criteria and Lyapunov functions. Along the way we will clarify the relation between the different types of robustness and also to other concepts in the literature.

Finally, we will investigate the "stability of robustness", i.e. we will see what happens to limits of systems with certain robustness properties and to systems which are close to systems with these properties. We will do this, consecutively, for general systems, for inflated systems, and the for the special case where continuous time systems are approximated by their time-h map.

Throughout this chapter Φ, $\Phi_{\mathcal{U}}$ and Φ_α denote either solutions of the continuous time system (2.5) or of the discrete time system (2.21), with $\mathbb{T} = \mathbb{R}$ or $\mathbb{T} = h\mathbb{Z}$, respectively.

3.1 Strong Attraction

In this section we give the basic definitions of the strong dynamical properties we will investigate in this chapter and discuss the relation between them.

Definition 3.1.1 (strong forward invariance)

(i) A set A is called strongly forward invariant if $\Phi_{\mathcal{U}}(t, A) \subseteq A$ for all $t \in \mathbb{T}^+$.

(ii) A set A_α is called strongly α-forward invariant if $\Phi_\alpha(t, A_\alpha) \subseteq A_\alpha$ for all $t \in \mathbb{T}^+$. □

Let us now define attracting sets. In many references in the literature these sets are supposed to be compact. Here we allow a bit more generality since we assume that A is closed and either A or A^c is bounded. The second case will turn out to be useful later when considering domains of attraction and reachable sets in Chapter 7. In fact, most of the subsequent statements remain true for arbitrary closed sets, if we are willing to assume (2.8) or, at least, uniform boundedness and Lipschitz continuity in a neighborhood of the sets under consideration, as used, e.g., in [70]. In general, for arbitrary closed sets things can go wrong if we do not have these properties.

Definition 3.1.2 (c-boundedness and strong attraction)

(i) A closed set A is called complementarily bounded or c-bounded if either A or A^c is bounded.

(ii) A closed c-bounded set A is called strongly attracting with attracted open neighborhood B if A is strongly forward invariant and satisfies

$$\mathrm{dist}(\Phi_{\mathcal{U}}(t,\widetilde{B}), A) \to 0 \text{ as } t \to \infty$$

for all subsets $\widetilde{B} \subset \mathrm{cl}\, B$ with $d_H(\widetilde{B}, A) < \infty$.

(iii) Let $\alpha > 0$. A closed c-bounded set A_α is called strongly α-attracting with attracted open neighborhood B if A_α is strongly α-forward invariant and satisfies

$$\mathrm{dist}(\Phi_\alpha(t,\widetilde{B}), A_\alpha) \to 0 \text{ as } t \to \infty$$

for all subsets $\widetilde{B} \subset \mathrm{cl}\, B$ with $d_H(\widetilde{B}, A_\alpha) < \infty$.   □

A closely related concept is that of asymptotic stability. Here we formulate it using the very convenient notation of comparison functions. For readers not familiar with these functions we refer to Appendix B for more information.

Definition 3.1.3 (strong asymptotic stability)

(i) A closed c-bounded set A is called strongly asymptotically stable if there exists a class $\mathcal{KL}$ function β and an open neighborhood B of A such that

$$\mathrm{dist}(\Phi_{\mathcal{U}}(t,x), A) \leq \beta(\|x\|_A, t) \text{ for all } x \in B.$$

(ii) A closed c-bounded set A is called strongly α-asymptotically stable if there exists a class $\mathcal{KL}$ function β and an open neighborhood B of A such that

$$\mathrm{dist}(\Phi_\alpha(t,x), A) \leq \beta(\|x\|_A, t) \text{ for all } x \in B.$$

□

Lemma 3.1.4 Every strongly (α-)attracting set A is strongly (α-)asymptotically stable.

Proof: We show the assertion for some $\alpha > 0$ and for continuous time systems. For $\alpha = 0$ and for discrete time systems it follows similarly. Since A is c-bounded its boundary ∂A is compact. Hence on any neighborhood $\mathcal{B}(\delta, \partial A)$, $\delta > 0$, the system is uniformly Lipschitz with some Lipschitz constant L_δ. Using Gronwall's Lemma and the α–forward invariance of A we obtain that $\text{dist}(\Phi_\alpha(t, x), A) \le e^{L_\delta t}\text{dist}(x, A)$ for all $x \in \mathbb{R}^n$ and all $t > 0$ satisfying $\Phi_\alpha(\tau, x) \subset \mathcal{B}(\delta, A)$ for all $\tau \in [0, t]$. Now define $B_r := \mathcal{B}(r, A) \cap B$. We claim that

$$a(r, t) := \sup_{s \ge t} \text{dist}(\Phi_\alpha(s, B_r), A)$$

can be bounded from above by some function of class $\mathcal{KL}$ which proves the desired asymptotic stability. For this purpose we verify the assumptions of Remark B.1.5 in Appendix B.

The properties $a(r, t) \to 0$ as $t \to \infty$ and all $r > 0$ and $a(r', t) \le a(r, t)$ for all $0 < r' \le r$ and all $t \ge 0$ are immediate, and imply Assumption (ii) of Remark B.1.5. Also, the property $a(r, t') \le a(r, t)$ for all $r > 0$ and all $t' \ge t \ge 0$ is obvious. In order to verify Assumption (i) of Remark B.1.5 it remains to show that $a(r, 0) \to 0$ as $r \to 0$. In order to see this fix some $r_0 > 0$ and let $\delta := a(r_0, 0)$. Fix some $\varepsilon > 0$ and let $t_\varepsilon > 0$ be the infimal time with $a(r_0, t_\varepsilon) \le \varepsilon$. Then we obtain that $a(r, 0) \le \max\{e^{L_\delta t_\varepsilon} r, \varepsilon\}$, and hence $a(r, 0) \le \varepsilon$ for all $r \in (0, r_0]$ with $r \le e^{-L_\delta t_\varepsilon}\varepsilon$. This shows the desired property. $\qquad\square$

Note that this lemma also holds for unbounded attracting sets A provided the system is uniformly Lipschitz and bounded on a neighborhood with positive minimal distance from A.

3.2 Robustness Concepts

In this section we will define various robustness concepts for the perturbed systems φ_α and $\Phi_{h,\alpha}$. Loosely speaking, these are methods to estimate how much the dynamical behavior of the perturbed systems (2.5) or (2.21) deviates from that of the unperturbed systems (2.1) or (2.17), respectively.

Definition 3.2.1 (γ-robust strong forward invariance)

A strongly forward invariant set C for the system (2.5) is called *directly γ-robust* for some γ of class $\mathcal{K}_\infty$ if for each $\alpha > 0$ there exists a strongly α-forward invariant set C_α with $d_H(C_\alpha, C) \le \gamma(\alpha)$.

It is called *inversely γ-robust* for some γ of class $\mathcal{K}_\infty$ if for each $\alpha > 0$ there exists a strongly α-forward invariant set C_α with $d_H(C_\alpha^c, C^c) \le \gamma(\alpha)$. $\qquad\square$

Both direct and inverse robustness will turn out to be useful and "natural" conditions, depending on the situation and on certain additional properties of C, cf. also Remark 7.4.6.

Definition 3.2.2 (γ-robust strong attraction)

A strongly attracting set A for the system (2.5) is called γ-*robust* with an attracted open neighborhood $B \subset \mathbb{R}^n$ with $A \subset B$, if for each $\alpha \geq 0$ there exists a strongly α-attracting set $A_\alpha \supseteq A$ with an attracted neighborhood B, which satisfies

$$d_H(A_\alpha, A) \leq \gamma(\alpha).$$

$\square$

Observe that this definition includes the assumption $A_\alpha \supseteq A$, which will be a crucial technical property in many proofs in this chapter. If A is just one point then γ-robust strong attraction is closely related to the so called a-$\mathcal{L}_\infty$ stability bound property introduced by Teel in [119].

Let us give a sufficient condition for γ-robustness.

Lemma 3.2.3 Consider a strongly attracting set A, which is directly γ-robustly strongly forward invariant for some γ of class $\mathcal{K}_\infty$ with $A \subseteq C_\alpha$ for C_α from Definition 3.2.1. Let B be an attracted neighborhood of A such that $\mathrm{cl}\,\Phi_\alpha(t, B) \subset B$ for all $\alpha > 0$ and all $t > T_\alpha$ for some $T_\alpha > 0$. Assume furthermore that

$$\limsup_{t \to \infty} d_H(\Phi_\alpha(t, \widetilde{B}), A) \leq \gamma(\alpha)$$

for all $\widetilde{B} \subseteq \mathrm{cl}\, B$ with $d_H(\widetilde{B}, A) < \infty$. Then the set A is a γ-robust strongly attracting set.

Proof: Consider a sequence $t_i \to \infty$ and define the sets

$$C_\alpha^i := \bigcup_{t \geq t_i} \Phi_\alpha(t, \mathcal{B}(2\gamma(\alpha), A) \cap B).$$

Then it is easily verified (recall the robust forward invariance assumption) that the sets

$$A_\alpha := \operatorname{Lim\,sup}_{i \to \infty} C_\alpha^i \cup \mathrm{cl} \bigcup_{t \geq 0} \Phi_\alpha(t, A)$$

are strongly α-forward invariant, contain A and have the desired distance from A. It remains to show that they are strongly α-attracting with attracted neighborhood B. In order to see this we use Lemma 2.3.5 (applied with $M = \mathcal{B}(2\gamma(\alpha), \partial A)$) in order to conclude that for each $\varepsilon > 0$ there exists $t_i > 0$ such that

$$\text{dist}\left(\text{cl} \bigcup_{t \geq t_i} \Phi_\alpha(t, \mathcal{B}(2\gamma(\alpha), A) \cap B), A_\alpha\right) \leq \varepsilon.$$

Since by the assumption we obtain that each $\widetilde{B} \subseteq \text{cl } B$ with $d_H(\widetilde{B}, A) < \infty$ is mapped by Φ_α into $\mathcal{B}(2\gamma(\alpha), A) \cap B$ for some $\widetilde{T} > 0$ we obtain that

$$\text{dist}(\Phi_\alpha(t, \widetilde{B}, A), A_\alpha) \leq \varepsilon$$

for all $t > \widetilde{T} + T_\varepsilon$, which shows the attractivity of A_α. □

Remark 3.2.4 Clearly, each γ-robust strongly attracting set A satisfies the assumption of Lemma 3.2.3. Furthermore, the sets A_α constructed in the proof of this Lemma satisfy $A_{\alpha'} \subseteq A_\alpha$ for all $\alpha' > \alpha > 0$. Thus for each γ-robust strongly attracting set A we can assume without loss of generality that the A_α in Definition 3.2.2 satisfy $A_{\alpha'} \subseteq A_\alpha$ for all $\alpha' > \alpha > 0$. □

Remark 3.2.5 Lemma 3.2.3 also shows that the function $\tilde{\gamma}$ defined by

$$\tilde{\gamma}(\alpha) := \max\{\sup_{t \in \mathbb{T}_0^+} \text{dist}(\Phi_\alpha(t, A), A), \sup_{\substack{\widetilde{B} \subseteq \text{cl } B \\ d_H(\widetilde{B}, A) < \infty}} \limsup_{t \to \infty} \text{dist}(\Phi_\alpha(t, \widetilde{B}), A)\}$$

forms a lower bound for the gain γ in the formulation of γ-robustness. In particular, if $\tilde{\gamma}$ is continuous then we can conclude the existence of an optimal robustness gain of class $\mathcal{K}_\infty$. □

The γ-robustness property only demands that the perturbed system has α-attracting sets nearby the unperturbed attracting set. By Lemma 3.1.4 we know that each of them is α-asymptotically stable; they can, however, have different rates of attraction, which in particular can be arbitrary slow. We will now introduce robustness properties with uniform rates of attraction for all $\alpha > 0$.

The following definition was introduced by Sontag [102] and has—along with various variations—by now become an essential property in the analysis of perturbed nonlinear systems. See, e.g., [20, 48, 57, 66, 67, 83, 95, 100, 105, 122] for just a small selection of references.

Definition 3.2.6 (input-to-state stability)

A strongly attracting set A for Φ is called input-to-state-stable (ISS) on a neighborhood B of A with rate of attraction β of class $\mathcal{KL}$ and robustness gain γ of class $\mathcal{K}_\infty$ if the inequality

$$\|\Phi(t, x, u, w)\|_A \leq \max\{\beta(\|x\|_A, t), \gamma(\|w\|_\infty)\}$$

holds for all $t \in \mathbb{T}^+$, $x \in B$, $u \in \mathcal{U}$ and $w \in \mathcal{W}$. □

The original definition in [102] was global, i.e., for $B = \mathbb{R}^n$ and perturbation values $W = \mathbb{R}^l$. Our version here includes this case, but it also includes the local version given by Sontag and Wang in [106], depending on the choice of W and B.

For our quantitative studies of the effect of perturbations it will be convenient to introduce yet another variation of the ISS property. Its main features are on the one hand that we require the rate of convergence to be determined by a one-dimensional dynamical system composed with a class $\mathcal{K}_\infty$ function to measure the overshoot. For this purpose we call a function $\mu : \mathbb{R}_0^+ \times \mathbb{R} \to \mathbb{R}_0^+$ of class $\mathcal{KLD}$ if its restriction to $\mathbb{R}_0^+ \times \mathbb{R}_0^+$ is of class $\mathcal{KL}$ and if additionally it defines a dynamical system on $\mathbb{R}_0^+$ (hence the "$\mathcal{D}$"), i.e., if it satisfies

$$\mu(r, 0) = r \text{ and } \mu(\mu(r,t), s) = \mu(r, t + s) \text{ for all } r \geq 0,\, s,\, t \in \mathbb{R},$$

cf. Definition B.1.2 in Appendix B. (Actually, this does not make any *qualitative* difference to ISS as any class $\mathcal{KL}$ function can be bounded from above by the composition of a class $\mathcal{KLD}$ function with a class $\mathcal{K}_\infty$ function, see Lemma B.1.4).

On the other hand, we also want this dynamical effect to be visible in the effect of the perturbation on the right hand side of the estimate, i.e. we want the influence of past perturbation to be decreasing with the same rate as the effect of large initial values. This "memory fading" idea was also used by Praly and Wang [95] in their notion of exp-ISS. There the perturbation is first fed into a one-dimensional dynamical system whose output then enters the right hand side of the ISS estimate. Here, instead, we will use the value of the perturbation at each time instance as an initial value of our dynamical system, which leads to the following definition.

Definition 3.2.7 (input-to-state dynamical stability)

A strongly attracting set A is called (strongly) *input-to-state dynamically stable* (ISDS), if there exists a function μ of class $\mathcal{KLD}$ and functions σ and γ of class $\mathcal{K}_\infty$ such that the inequality

$$\|\Phi(t, x, u, w)\|_A \leq \max\{\mu(\sigma(\|x\|_A), t),\, \nu(w, t)\}.$$

holds for all $t \in \mathbb{T}^+$, $x \in B$, $u \in \mathcal{U}$ and $w \in \mathcal{W}$, where $\nu = \nu_0$ for continuous time systems and $\nu = \nu_h$ for discrete time systems, and ν_0 and ν_h are defined by

$$\nu_0(w, t) := \text{ess sup}_{\tau \in [0,t]}\ \mu(\gamma(\|w(\tau)\|), t - \tau) \quad \text{and}$$

$$\nu_h(w, t) := \max_{\tau \in [h,t] \cap h\mathbb{Z}}\ \mu(\gamma(\|w\|_{[\tau - h, \tau]}), t - \tau).$$

(3.1)

Here we call the function μ the *rate of attraction*, the function σ the *overshoot gain* and the function γ the *robustness gain*.   □

There are several reasons why we have chosen this particular definition; substantial parts of the remainder of this chapter will center around the proof of the following properties:

- We obtain a "sharp" Lyapunov characterization. By this we mean that under the ISDS condition we can always find a Lyapunov function, which in turn implies the ISDS property with exactly the same rate and gains, see Theorem 3.5.3, below.

- The ISDS formulation explicitly takes into account that if $w(t) \to 0$ as $t \to \infty$, and we start inside B, then each trajectory will eventually converge to A. This allows the direct derivation of a "small gain" statement about the behavior of the system when the amplitude of the perturbation depends on the state x and is small relative to $\|x\|_A$ (like, e.g., for perturbations of type (2.11)). In particular, the robustness gain γ gives a precise upper bound for the amplitude of the perturbation in order to guarantee asymptotic stability of the perturbed system, see Theorem 3.7.4.

- ISDS is "almost" *quantitatively* equivalent (with respect to the robustness gain γ) to the apparently weaker γ-robustness and ISS properties, which is made precise in Proposition 3.4.4. It is furthermore *qualitatively* equivalent to ISS, which is also shown in Proposition 3.4.4.

- ISDS is an inherent property for strongly asymptotically stable sets, at least for sufficiently small bounded perturbation ranges W, see Theorem 3.4.6.

In fact, one could also define ISDS for discrete time systems using ν_0 instead of the (slightly larger) ν_h. The use of ν_h is motivated by a subtle detail regarding the definition of inflated systems, cf. Remark 3.8.6, below. Note that for piecewise constant perturbations (i.e., $w|_{[ih,(i+1)h)} \equiv w_i$ for all $i \in \mathbb{N}_0$) the values $\nu_0(w,t)$ and $\nu_h(w,t)$ coincide for all $t \geq 0$.

We will now state a few basic properties of ν_0 and ν_h.

Lemma 3.2.8 Let $\nu = \nu_0$ or $\nu = \nu_h$. Then the following properties hold.

(i) $\nu(w(\tau + \cdot), t) \leq \nu(\dot{w}, t + \tau)$ for all $t, \tau \in \mathbb{T}_0^+$.

(ii) If $\nu(w,t) \leq \mu(r,t)$ for some $r > 0$, $t \in \mathbb{T}^+$ then $\nu(w,\tau) \leq \mu(r,\tau)$ for all $\tau \in (0,t] \cap \mathbb{T}$.

(iii) $\limsup_{h \searrow 0} \nu_0(w,t-h) \leq \nu_0(w,t)$ and $\liminf_{h \searrow 0} \nu_0(w,t+h) \geq \nu_0(w,t)$.

Proof: The properties (i) and (ii) follow immediately from the definition of ν.

For (iii) observe that the inequalities

$$\nu_0(w, t_1 + t_2) \geq \mu(\nu_0(w, t_1), t_2) \text{ and } \mu(\nu_0(w, t_1 + t_2), -t_2) \geq \nu_0(w, t_1)$$

hold for all t_1, $t_2 > 0$. Hence setting $t_1 = t - h$ and $t_2 = h$ for the first assertion and $t_1 = t$, $t_2 = h$ for the second implies the assertion by the continuity of μ in t.   □

We will occasionally assume that the comparison functions σ, γ and μ in the ISDS definition are smooth on $\mathbb{R}^+$ or $\mathbb{R}^+ \times \mathbb{R}$, respectively, with non-vanishing derivatives. Proposition B.2.3 implies that if we have ISDS with non-smooth functions this assumption can be satisfied by slightly enlarging these functions.

3.3 Geometric Characterizations

In this section we will derive a geometric condition for the ISDS property of some attracting set A. Geometric in this context means that we are going to establish a criterion by means of a nested family of sets which shrink down to A.

Definition 3.3.1 (contracting family of neighborhoods)

Consider a strongly attracting set A with open attracted neighborhood B. A family $(B_\alpha)_{\alpha \in \mathbb{R}_0^+}$ of compact sets $B_\alpha \subset \mathbb{R}^n$, $\alpha \in \mathbb{R}_0^+$ together with a class $\mathcal{KLD}$ function ϑ is called a *contracting family of neighborhoods* for A w.r.t. B if

(i) $B_{\alpha'} \subseteq B_\alpha$ for all $\alpha' < \alpha$

(ii) $B_0 = A$

(iii) for each subset $\widetilde{B} \subseteq \text{cl } B$ with $d_H(\widetilde{B}, A) < \infty$ there exists an $\alpha^* \in \mathbb{R}_0^+$ with $\widetilde{B} \subseteq B_{\alpha^*}$.

(iv) $\Phi(t, B_\alpha) \subseteq B_{\vartheta(\alpha, t)}$ for all $t \in \mathbb{T}^+$

The family $(B_\alpha)_{\alpha \in \mathbb{R}_0^+}$ is called *strictly contracting* with respect to some open set O if, in addition,

(v) $B_{\alpha'} \cap O \subset \text{int } B_\alpha$ for all $\alpha' < \alpha$.

The family is called (strictly) α-contracting if (iv) can be sharpened to

(iv') $\Phi_{\alpha(t)}(t, B_\alpha) \subseteq B_{\vartheta(\alpha, t)}$ for all $t \in \mathbb{T}^+$, where $\alpha(t) = \vartheta(\alpha, t)$ for continuous time systems and $\alpha(t) = \vartheta(\alpha, (i + 1)h)$ for $t \in (ih, (i + 1)h]$ for discrete time systems.

□

Note that Property (iv') applies only to those $w \in W$ which are bounded by the given $\alpha(t)$. The following Lemma shows the implication of Property (iv') for arbitrary w.

Lemma 3.3.2 Consider an α-contracting family of neighborhoods B_α. Then for all $x \in B_\alpha$, all $u \in \mathcal{U}$, all $w \in W$ and all $t \in \mathbb{T}^+$ the relation

$$\Phi(t, x, u, w) \in B_{\alpha'}$$

holds for $\alpha' = \max\{\vartheta(\alpha, t), \bar{\nu}(w, t)\}$ with $\bar{\nu} = \nu_0$ or $\bar{\nu} = \nu_h$ from (3.1) with $\mu = \vartheta$ and $\gamma = \mathrm{id}_{\mathbb{R}}$.

Proof: Let $x \in B_\alpha$, $u \in \mathcal{U}$, $w \in W$ and $t \in \mathbb{T}^+$. Let $\tilde{\alpha} \geq \alpha$ minimal with $\vartheta(\tilde{\alpha}, t) \geq \bar{\nu}(w, t)$. Since $\tilde{\alpha} \geq \alpha$, property (i) of the α-contracting family yields $x \in B_{\tilde{\alpha}}$. By Lemma 3.2.8(ii) the choice of $\tilde{\alpha}$ implies $\vartheta(\tilde{\alpha}, \tau) \geq \bar{\nu}(w, \tau)$ for all $\tau \in (0, t] \cap \mathbb{T}$, hence in particular $\vartheta(\tilde{\alpha}, \tau) \geq \|w(\tau)\|$ for almost all $\tau \in [0, t]$ in the continuous time case and $\vartheta(\tilde{\alpha}, \tau) \geq \|w\|_{[\tau-h,\tau]}$ for all $\tau \in [0, t] \cap \mathbb{T}$ in the discrete time case. Thus we obtain $\Phi(t, x, u, w) \subset \Phi_{\tilde{\alpha}(t)}(t, B_{\tilde{\alpha}})$ with $\tilde{\alpha}(\tau) = \vartheta(\tilde{\alpha}, \tau)$ for continuous time systems and $\tilde{\alpha}(\tau) = \vartheta(\tilde{\alpha}, (i+1)h)$ for $\tau \in (ih, (i+1)h]$ for discrete time systems. Now the α-contracting property (iv') yields

$$\Phi_{\vartheta(\tilde{\alpha},t)}(t, B_{\tilde{\alpha}}) \subseteq B_{\vartheta(\tilde{\alpha},t)}$$

which implies the assertion since the choice of $\tilde{\alpha}$ implies $\alpha' = \vartheta(\tilde{\alpha}, t)$. $\quad\Box$

We now state the relation between the existence of an α-contracting family and the ISDS property.

Proposition 3.3.3 Consider a strongly attracting set A. Then A is ISDS with rate μ, overshoot gain σ and robustness gain γ if and only if there exists an α-contracting family of neighborhoods B_α with

$$d_H(B_\alpha, A) \leq \gamma(\alpha), \quad B(\sigma^{-1}(\gamma(\alpha)), A) \cap B \subseteq B_\alpha$$

and

$$\vartheta(\alpha, t) = \gamma^{-1}(\mu(\gamma(\alpha), \tau)).$$

Proof: Assume the existence of the α-contracting family, let $x \in B$, $u \in \mathcal{U}$, $w \in W$ and $t \in \mathbb{T}^+$. By the assumption on the B_α there exists $\alpha > 0$ such that $x \in B_\alpha$ and $d_H(B_\alpha, A) = \sigma(\|x\|_A)$, i.e. $\alpha \leq \gamma^{-1}(\sigma(\|x\|_A))$. By Lemma 3.3.2 we obtain $\Phi(t, x, u, w) \in B_{\alpha'}$, with $\alpha' = \max\{\vartheta(\alpha, t), \bar{\nu}(w, t)\}$. Here $\bar{\nu}$ from Lemma 3.3.2 satisfies $\nu = \gamma(\bar{\nu})$ for ν from the ISDS definition, hence

$$\|\Phi(t, x, u, w)\|_A \leq \max\{\gamma(\vartheta(\alpha, t)), \gamma(\bar{\nu}(w, t))\} \leq \max\{\mu(\sigma(\|x\|_A), t), \nu(w, t)\},$$

which implies ISDS.

Conversely, assume ISDS. Then we define the sets

$$B_\alpha := \left\{ x \in \mathbb{R}^n \;\middle|\; \begin{array}{l} \|\Phi(t,x,u,w)\|_A \leq \max\{\mu(\gamma(\alpha),t), \nu(w,t)\} \\ \text{for all } u \in \mathcal{U}, \, w \in \mathcal{W}, \, t \in \mathbb{T}_0^+ \end{array} \right\}$$

for all $\alpha \geq 0$. Obviously $B_{\alpha'} \subseteq B_\alpha$ for $\alpha' \leq \alpha$. The assertions on the distance are immediate and imply, in particular, that the sets shrink down to A. The fact that we can choose $\vartheta(\alpha,t) = \gamma^{-1}(\mu(\gamma(\alpha),\tau))$ follows directly from the construction. $\qquad \square$

It will turn out to be useful to have a characterization via a *strictly α-contracting* family of neighborhoods. We can obtain this at least with respect to $O = \mathcal{R}(B)$ by slightly relaxing the bounds in the previous proposition.

Proposition 3.3.4 Consider a strongly attracting set A. Then A is ISDS with attraction rate μ, overshoot gain σ and robustness gain γ if and only if for each $\varepsilon > 0$ there exists a strictly (with respect to $O = \mathcal{R}(B)$) α-contracting family of neighborhoods B_α with

$$d_H(B_\alpha, A) \leq (1+\varepsilon)\gamma(\alpha), \quad B(\sigma^{-1}(\gamma(\alpha)), A) \cap B \subset B_\alpha$$

and $\vartheta(\alpha,t) = \gamma^{-1}(\mu(\gamma(\alpha),(1-\varepsilon)t))$.

If, in addition, μ is C^1 with $\frac{\partial}{\partial r}\mu(r,t) > 0$ for all $r > 0$ and all $t > 0$, then the sets B_α can be chosen such that for each $x \in \mathcal{R}(B)$ there exists an open neighborhood $N(x)$ and a constant $C(x) > 0$ with

$$\|y - y'\| \geq C(x)|\gamma(\alpha) - \gamma(\alpha')| \text{ for all } y \in N(x) \cap \partial B_\alpha, \; y' \in N(x) \cap \partial B_{\alpha'} \quad (3.2)$$

for some $\alpha, \alpha' > 0$.

Proof: Assume the existence of the strictly α-contracting families from the assertion. Then straightforward calculations show that $d_H(B_\alpha, A) \leq \gamma_\varepsilon(\alpha)$, $B(\sigma_\varepsilon^{-1}(\gamma_\varepsilon(\alpha)), A) \cap B \subset B_\alpha$ and $\vartheta(\alpha,t) = \gamma_\varepsilon^{-1}(\mu_\varepsilon(\gamma_\varepsilon(\alpha),\tau))$ with $\gamma_\varepsilon(r) = (1+\varepsilon)\gamma(r)$, $\sigma_\varepsilon(r) = \sigma((1+\varepsilon)r)$ and $\mu_\varepsilon(r,t) = (1+\varepsilon)\mu((1-\varepsilon)r,(1-\varepsilon)t)$.

Thus for all $\varepsilon > 0$, all $x \in B$, all $u \in \mathcal{U}$, all $w \in \mathcal{W}$ and all $t \in \mathbb{T}^+$ Proposition 3.3.3 implies

$$\|\Phi(t,x,u,w)\|_A \leq \max\{\mu_\varepsilon(\sigma_\varepsilon(\|x\|_A),t), \nu_\varepsilon(u(\cdot),t)\}.$$

with $\nu_\varepsilon(u,t)$ from (3.1) with $\mu = \mu_\varepsilon$ and $\gamma = \gamma_\varepsilon$. Since $\varepsilon > 0$ was arbitrary this implies the desired ISDS estimate with robustness gain γ, overshoot gain σ and attraction rate μ.

Conversely, assume ISDS with robustness γ, overshoot σ and rate μ. Fix some $\varepsilon > 0$ and set $\rho_\varepsilon(r) := \varepsilon(1 - e^{-r}) + 1$. Then ρ_ε is strictly increasing for $r > 0$, $\rho_\varepsilon(0) = 1$ and $\rho_\varepsilon(r) \nearrow 1 + \varepsilon$ as $r \to \infty$. Using this function we define the following sets

$$B_\alpha := \left\{ x \in \mathbb{R}^n \;\middle|\; \begin{array}{l} \|\Phi(t,x,u,w)\|_A \leq \rho_\varepsilon(\mu(\gamma(\alpha),t)) \\ \qquad \max\{\mu(\gamma(\alpha),(1-\varepsilon)t),\,\nu(w,t)\} \\ \text{for all } u \in \mathcal{U},\, w \in \mathcal{W},\, t \in \mathbb{T}_0^+ \end{array} \right\}.$$

We show the following properties which imply the assertion:

(i) the sets B_α are closed sets satisfying $B_{\alpha'} \cap \mathcal{R}(B) \subset \mathrm{int}\, B_\alpha$ for all $\alpha' < \alpha$, and satisfying (3.2) under the additional assumption on μ

(ii) $\Phi(t,x,u,w) \in B_{\vartheta(\alpha,t)}$ for all $x \in B_\alpha$, all $u \in \mathcal{U}$, all $t \in \mathbb{T}^+$ and all $w \in \mathcal{W}$ with $\|w(\tau)\| \leq \vartheta(\alpha,\tau)$ for almost all $\tau \in [0,t]$

(iii) $d_H(B_\alpha, A) \leq \rho_\varepsilon(\gamma(\alpha))\gamma(\alpha) < (1+\varepsilon)\gamma(\alpha)$ and $B(\sigma^{-1}(\gamma(\alpha)), A) \subset B_\alpha$

(i) Closedness follows from the definition and from continuous dependence on the initial value. The inclusion $B_{\alpha'} \subseteq B_\alpha$ is immediate from the definition, hence it remains to show $(\partial B_\alpha \cap \mathcal{R}(B)) \cap B_{\alpha'} = \emptyset$. In order to accomplish this, observe that for all $x \in \mathcal{R}(B)$ there exists $t^* \in \mathbb{T}_0^+$ and an open neighborhood $N(x)$ such that $N(x) = \Phi(t^*, N^*, u^*, w^*)$ for some open and bounded set $N^* \subset B$, some $u^* \in \mathcal{U}$ and some essentially bounded $w^* \in \mathcal{W}$. Hence for all $\tilde{x} \in N(x)$ the ISDS estimate implies

$$\Phi(t, \tilde{x}, u, w) \leq \max\{\mu(R, t), \nu(w, t)\} \tag{3.3}$$

for $R = \max\{\sup_{y \in N^*} \mu(\sigma(\|y\|, t^*)), \|w^*\|_\infty\}$. Now let $x \in \partial B_\alpha \cap \mathcal{R}(B)$. Then there exist points $x_n \to x$, functions $u_n \in \mathcal{U}$ and $w_n \in \mathcal{W}$ and times $t_n \in \mathbb{T}^+$ such that

$$\|\Phi(t_n, x_n, u_n, w_n)\|_A > \rho_\varepsilon(\mu(\gamma(\alpha), t_n)) \max\{\mu(\gamma(\alpha), (1-\varepsilon)t_n), \nu(w_n, t_n)\}. \tag{3.4}$$

Now observe that for all $R > 0$ there exists $T > 0$ such that $\mu(R, t) < \mu(\gamma(\alpha), (1-\varepsilon)t)$ for all $t > T$, hence (as we may assume without loss of generality $x_n \in N(x)$) by (3.3) the times t_n are bounded by some $T = T(x) > 0$. Also by (3.3) we can assume $\nu(w_n, t_n) \leq \mu(R, t_n)$, otherwise we obtain a contradiction to (3.4). Hence all w_n are essentially bounded on $[0, t_n]$ independent of n. Thus by continuous dependence on the initial value we obtain a sequence $\varepsilon_n \to 0$ such that

$$\|\Phi(t_n, x, u_n, w_n)\|_A > \rho_\varepsilon(\mu(\gamma(\alpha), t_n)) \max\{\mu(\gamma(\alpha), (1-\varepsilon)t_n), \nu(w_n, t_n)\} - \varepsilon_n.$$

Since the right hand side of this inequality is continuous in α, for each $\alpha' < \alpha$ there exists $n \in \mathbb{N}$ with

$$\|\Phi(t_n, x, u_n, w_n)\|_A > \rho_\varepsilon(\mu(\gamma(\alpha'), t_n)) \max\{\mu(\gamma(\alpha'), (1-\varepsilon)t_n), \nu(w_n, t_n)\}$$

which implies $x \notin B_{\alpha'}$. In order to see (3.2) recall that $|t_n|$ and $\|w_n\|_\infty$ are uniformly bounded for all n. Then the boundedness of $|t_n|$ and the smoothness assumption on μ imply the existence of a constant $C > 0$ such that

$$|\rho_\varepsilon(\mu(\gamma(\alpha), t_n)) - \rho_\varepsilon(\mu(\gamma(\alpha'), t_n))| \geq C|\gamma(\alpha) - \gamma(\alpha')|$$

for α and α' from some compact interval not containing 0. By Gronwall's inequality for continuous time systems or by induction for discrete time systems (observe that all trajectories under consideration stay inside some compact subset on which we have a uniform Lipschitz constant for our system) we furthermore obtain

$$\|\Phi(t_n, y, u_n, w_n)\|_A \geq \|\Phi(t_n, y', u_n, w_n)\|_A - C'\|y - y'\|$$

for some $C' > 0$. Thus for $y \in N(x) \cap \partial B_\alpha$ we obtain $y' \notin B_{\alpha'}$ for all $y' \in N(x)$ with $\|y' - y\| \leq C''|\gamma(\alpha) - \gamma(\alpha')|$ for some $C'' > 0$ depending on C, C' and $\max\{\mu(\gamma(\alpha'), (1 - \varepsilon)t_n), \nu(w_n, t_n)\}$, which shows (3.2).

(ii) Let $\alpha > 0$, $x \in B_\alpha$, consider $\alpha' = \vartheta(\alpha, t) = \gamma^{-1}(\mu(\gamma(\alpha), (1 - \varepsilon)t))$ and let $x \in B_\alpha$ and $w \in W$ with $\|w(\tau)\| \leq \vartheta(\alpha, \tau)$ for almost all $\tau \in [0, t]$. Then the definition of B_α implies

$$
\begin{aligned}
&\|\Phi(t + s, x, u, w)\| \\
&\leq \quad \rho_\varepsilon(\mu(\gamma(\alpha), t + s)) \max\{\mu(\gamma(\alpha), (1 - \varepsilon)(t + s)), \nu(w, t + s)\} \\
&\leq \quad \rho_\varepsilon(\mu(\gamma(\alpha'), s)) \max\{\mu(\gamma(\alpha'), (1 - \varepsilon)s), \mu(\nu(w, t), s), \nu(w(t + \cdot), s)\} \\
&\leq \quad \rho_\varepsilon(\mu(\gamma(\alpha'), s)) \max\{\mu(\gamma(\alpha'), (1 - \varepsilon)s), \nu(w(t + \cdot), s)\}
\end{aligned}
$$

for all $u \in \mathcal{U}$ since $\mu(\nu(w, t), s) \leq \mu(\gamma(\alpha'), s))$ by the choice of w. This shows that $\|\Phi(t, x, u, w)\| \in B_{\alpha'}$.

(iii) The first inequality follows from the definition of B_α setting $w \equiv 0$ and $t = 0$. The second follows from the fact that the ISDS estimate implies the inequality in the definition of B_α for all $x \in B$ with $\sigma(\|x\|) \leq \gamma(\alpha)$. $\qquad\square$

3.4 Relation between Robustness Concepts

In this section we show the relation between the various robustness concepts that we have introduced in this chapter. We start by showing that any strongly attracting set is γ-robust for some suitable robustness gain γ of class $\mathcal{K}_\infty$ and some suitable perturbation range W.

Proposition 3.4.1 Assume that system (2.5) or (2.21) satisfies (2.9) or (2.25), respectively, for some ρ of class $\mathcal{K}_\infty$. Then any strongly asymptotically stable set A with attraction rate β of class $\mathcal{KL}$ and some attracted neighborhood B satisfying $\text{dist}(B, A) < \infty$ is γ-robust for some γ of class $\mathcal{K}_\infty$ and some W with nonvoid interior. In particular, γ and W only depend on β, ρ, $r_0 := \text{dist}(B, A) < \infty$ and on the Lipschitz constant L of the system on the (compact) set $\text{cl}\,(B(\beta(2r_0, 0), A) \setminus A)$.

Proof: Assume strong asymptotic stability of A with some neighborhood B with $\text{dist}(B, A) < \infty$ and some β of class $\mathcal{KL}$. Set $r_0 := \text{dist}(B, A)$. Observe that by the assumption on an asymptotically stable set either A or A^c is bounded, hence $\text{cl}\,(\mathcal{B}(\beta(2r_0, 0), A)\backslash A)$ is compact and the system is uniformly Lipschitz with some constant L on this set. Now for all $r \in (0, r_0]$ we can define

$$T(r) = \min\left\{t \in \mathbb{T}_0^+ \,|\, \beta(r, t) \leq \frac{r}{4}\right\}.$$

Note that T is finite for all $r > 0$ and w.l.o.g. monotone decreasing, further- more we obtain $\beta(s, T(r) + t) \leq r/4$ for all $t \geq 0$ and all $s \in [0, r]$. Now for all $\alpha \leq \alpha_0 := \rho^{-1}(e^{-LT(r_0)} \min\{r_0, \beta(r_0, 0)\}/4)$ consider the sets

$$D_\alpha := \mathcal{B}(r(\alpha), A),$$

where $r(\alpha)$ is chosen such that $e^{LT(r(\alpha))}\rho(\alpha) \leq r(\alpha)/4$. Observe that both α_0 and $r(\cdot)$ only depend on β, ρ, r_0 and L, and $r(\alpha) \to 0$ as $\alpha \to 0$. We set $W = \mathcal{B}(\alpha_0, 0)$. Then by Gronwall's Lemma we obtain for $t \leq T(\|x\|_A)$

$$\|\Phi(t, x, u, w)\|_A \leq \beta(\|x\|_A, t) + e^{Lt}\rho(\alpha)$$

for all $u \in \mathcal{U}$ and all $w \in \mathcal{W}_\alpha$, which implies that for each point $x \in D_\alpha$ and all $u \in \mathcal{U}$ we obtain

$$\Phi(T(r(\alpha)), x, u, w) \in D_\alpha$$

and

$$\|\Phi(t, x, u, w)\|_A \leq \beta(r(\alpha), 0) + r(\alpha)/4 \text{ for all } t \in [0, T(r(\alpha))] \cap \mathbb{T}.$$

Furthermore, for any $w \in \mathcal{W}_\alpha$ and any $x \in B$ this inequality implies that any trajectory satisfies

$$\|\Phi(iT(r(\alpha)), x, u, w)\|_A \leq \max\{r_0/2^i, r(\alpha)\} \tag{3.5}$$

for $i \in \mathbb{N}$ and hence hits D_α in some uniformly bounded finite time. Now we set

$$A_\alpha := \bigcup_{t \in [0, T(r(\alpha))]} \text{cl}\,\Phi_\alpha(t, D_\alpha).$$

This set is strongly α-forward invariant by construction and hence by (3.5) it is strongly α-attracting. Furthermore it satisfies $\mathcal{B}(r(\alpha), A) \subseteq A_\alpha$ and $d_H(A_\alpha, A) \leq \gamma(\alpha) := \beta(r(\alpha), 0) + r(\alpha)/4$. This shows the desired robustness property. $\square$

Remark 3.4.2 The relation between β and γ is implicit in this proof, i.e. in general it does not give an explicit formula for γ involving β. In the special case of exponential attraction, i.e. when $\beta(r, t) = Ce^{-\lambda t}$ for constants C, $\lambda > 0$ we obtain that γ is linear in α (for the ISS property at fixed points this was already observed in [67, Lemma A.2]). We shall investigate this case in detail in Example 4.4.2 in the next chapter. $\square$

Observe that in this proposition in particular the allowed range of perturbations W depends on $\mathrm{dist}(B, A)$. The following example shows that this may indeed happen.

Example 3.4.3 Consider the system (2.5) with right hand side given by

$$f(x) = \begin{cases} \max\left\{-x, -\frac{1}{x}\right\} + w, & x > 0 \\ w, & x = 0 \\ \min\left\{-x, -\frac{1}{x}\right\} + w, & x < 0 \end{cases}$$

Clearly, $A = \{0\}$ is a strongly attracting set for the corresponding unperturbed system where each set $B = (-c, c)$, $c > 0$ is an attracted neighborhood. Fixing some $c > 0$ one sees that $A = \{0\}$ is γ-robust with attracted neighborhood B for each γ of class $\mathcal{K}_\infty$ with $\gamma(r) \leq \min\{r, 1/r\}$ for $r \in [0, c]$ and $W = [-d, d]$ where $d = \min\{c, 1/c\}$. Thus the bigger $\mathrm{dist}(B, A) = c$ becomes the smaller W must be.   □

We will now show that for compact perturbation ranges W the γ-robustness implies ISDS for some suitable function μ of class $\mathcal{KLD}$, if we slightly enlarge γ. The same is shown to be true for ISS, even without any assumptions on W.

Proposition 3.4.4 (i) A strongly attracting set A is γ-robust for some γ of class $\mathcal{K}_\infty$ if and only if for each compact subset $\widetilde{W}$ of W and each function $\tilde{\gamma}$ of class $\mathcal{K}_\infty$ with $\tilde{\gamma}(r) > \gamma(r)$ for all $r > 0$ it is ISDS for suitable functions $\tilde{\sigma}$ of class $\mathcal{K}_\infty$ and $\tilde{\mu}$ of class $\mathcal{KLD}$.

(ii) If a strongly attracting set is ISDS then it is ISS with the same robustness gain γ and class $\mathcal{KL}$ function $\beta(r, t) = \mu(\sigma(r), t)$. Conversely, if a strongly attracting set A is ISS for some γ of class $\mathcal{K}_\infty$ and some β of class $\mathcal{KL}$ then for any class $\mathcal{K}_\infty$ function $\tilde{\gamma}$ satisfying $\tilde{\gamma}(r) > \gamma(r)$ for all $r > 0$ the set A is ISDS with overshoot gain $\tilde{\sigma}(r) = \beta(r, 0)$ and some suitable decay rate $\tilde{\mu}$ of class $\mathcal{KLD}$.

Proof: The implications ISDS $\Rightarrow$ ISS $\Rightarrow$ γ-robustness are immediate using Lemma 3.2.3 for the second one.

In order to show the converse implications we construct an α-contracting family of neighborhoods B_α meeting the assumptions of Proposition 3.3.3. We show both (i) and (ii) in one proof. For this we construct a two sided sequence α_i, $i \in \mathbb{Z}$ as follows. We set $\alpha_2 = 1$, choose α_1 and α_0 with $\alpha_1 < \alpha_0$ arbitrary in $(\alpha_2, \gamma^{-1}(\tilde{\gamma}(\alpha_2)))$ and set $\alpha_{i+3} = \tilde{\gamma}^{-1}(\gamma(\alpha_i))$ for $i \geq 0$ and $\alpha_{i-3} = \gamma^{-1}(\tilde{\gamma}(\alpha_i))$ for $i \leq 2$. This sequence satisfies $\alpha_{i+1} < \alpha_i$, $\tilde{\gamma}(\alpha_i) = \gamma(\alpha_{i-3})$, $\alpha_i \to 0$ as $i \to +\infty$ and $\alpha_i \to \infty$ as $i \to -\infty$. We set $\delta_i = \gamma(\alpha_i)$.

Now for each $i \in \mathbb{Z}$ we define the set

$$B_i := \{x \in \mathrm{cl}\, B \mid \mathrm{dist}(\Phi_{\alpha_{j+1}}(t, x), A) \le \delta_j \text{ for all } j \le i, t \in \mathbb{T}_0^+\}.$$

From the construction we obtain $\mathrm{dist}(B_i, A) \le \delta_i$ (in particular compactness of $\mathrm{cl}\,(B_i \setminus A)$ for each B_i), $B_{i+1} \subseteq B_i$ and α_i-forward invariance of B_i. In case (i), observe that the compactness of $\widetilde{W}$ implies the existence of $\alpha_0 > 0$ such that $\Phi_\alpha = \Phi_{\alpha_0}$ for all $\alpha \ge \alpha_0$. Hence for each set $\widetilde{B} \subseteq \mathrm{cl}\, B$ with $d_H(\widetilde{B}, A) \le \infty$ we can consider

$$d(\widetilde{B}) := \sup_{t \ge 0} \mathrm{dist}(\Phi_{\alpha_0}(t, \widetilde{B} \setminus A), A).$$

This value is finite because $\Phi_{\alpha_0}(t, \widetilde{B} \setminus A)$ is bounded for each fixed $t > 0$ and by the robustness of A we know that $\limsup_{t \to \infty} \mathrm{dist}(\Phi_{\alpha_0}(t, \widetilde{B}), A) \le \gamma(\alpha_0)$. Thus we obtain $\widetilde{B} \subseteq B_i$ for all i with $\delta_i \ge d(\widetilde{B})$. In case (ii), this inclusion also holds for non–compact W since the ISS estimate implies $\widetilde{B} \subseteq B_i$ for $\delta_i > \beta(d_H(\widetilde{B}, A), 0)$, i.e. for all $i < 0$ sufficiently small. In both cases we can conclude that each set $\widetilde{B} \subseteq B$ with $d_H(\widetilde{B}, A) < \infty$ is contained in some B_i. Furthermore, since $B_i \subseteq \mathrm{cl}\, B$ in case (i) we obtain

$$\limsup_{t \to \infty} \mathrm{dist}(\Phi_{\alpha_{i+1}}(B_{i-1}, t), A) \le \gamma(\alpha_{i+1}) < \gamma(\alpha_i) = \delta_i$$

and in case (ii) we obtain

$$\mathrm{dist}(\Phi_{\alpha_{i+1}}(B_{i-1}, t), A) \le \max\{\beta(\delta_{i-1}, t), \gamma(\alpha_{i+1})\} \le \gamma(\alpha_i)$$

for all $t > 0$ sufficiently large. Thus in both cases there exists a $\Delta t_i \in \mathbb{T}^+$ such that $\Phi_{\alpha_{i+1}}(B_{i-1}, t) \subset B_i$ for all $t \ge \Delta t_i$. Without loss of generality we may assume $\sum_{i=k_1}^{k_2} \Delta t_i \to \infty$ if either $k_1 \to -\infty$ or $k_2 \to \infty$.

Now we define our contracting family of neighborhoods as follows: For each $\alpha \in [\alpha_{i+2}, \alpha_{i+1}]$ we set

$$B_\alpha := \Phi_{\alpha_{i+1}}\left(\frac{\alpha_{i+1} - \alpha}{\alpha_{i+1} - \alpha_{i+2}} \Delta t_i, B_{i-1}\right) \cup B_i$$

if Φ is a continuous time system, and

$$B_\alpha := \Phi_{\alpha_{i+1}}\left(\left[\frac{\alpha_{i+1} - \alpha}{\alpha_{i+1} - \alpha_{i+2}} \Delta t_i\right]_h, B_{i-1}\right) \cup B_i$$

if Φ is a discrete time system with time step h, where $[r]_h$ denotes the largest value $s \in h\mathbb{Z}$ with $s \le r$. This construction implies $B_{\alpha_i} = B_{i-2}$ and $B_\alpha \subseteq B_{\alpha'}$ for all $0 < \alpha \le \alpha'$.

We obtain the desired distance $d_H(B_\alpha, A) \le \tilde{\gamma}(\alpha)$ since for $\alpha \in [\alpha_{i+2}, \alpha_{i+1}]$ we have

$$\mathrm{dist}(B_\alpha, A) \le \mathrm{dist}(B_{i-1}, A) \le \delta_{i-1} = \tilde{\gamma}(\gamma(\delta_{i+2})) = \tilde{\gamma}(\alpha_{i+2}) \le \tilde{\gamma}(\alpha).$$

In case (ii), for any $x \in B$ with $\|x\|_A \leq \tilde{\sigma}^{-1}(\tilde{\gamma}(\alpha))$ for some $\alpha \in [\alpha_{i+2}, \alpha_{i+1}]$ we have $\beta(\|x\|_A, 0) = \tilde{\sigma}(\|x\|_A) \leq \tilde{\gamma}(\alpha_{i+1})$, hence

$$\beta(\|x\|_A, 0) \leq \gamma(\alpha_i) = \delta_i$$

which by the definition of the B_i implies $x \in B_i$, hence $x \in B_{\alpha_{i+2}} \subseteq B_\alpha$ which yields the desired inclusion $\mathcal{B}(\tilde{\sigma}^{-1}(\tilde{\gamma}(\alpha)), A) \subseteq B_\alpha$.

It remains to show the α-contraction. This follows by setting

$$\vartheta(\alpha_i, t) = \alpha_i - \frac{t(\alpha_i - \alpha_{i+1})}{\Delta t_{i-1}}$$

for $t \in [0, \Delta t_{i-1}]$ and extending this map for all $t \in \mathbb{R}$ via

$$\vartheta(\alpha_i, t) = \vartheta(\alpha_k, t - T_{i,k})$$

for all $t \in [T_{i,k}, T_{i,k+1}]$ with $T_{i,k}$ given inductively by $T_{i,i} = 0$, $T_{i,k+1} = T_{i,k} + \Delta t_{k-1}$ for $k \geq i$ and $T_{i,k-1} = T_{i,k} - \Delta t_{k-2}$ for $k \leq i$. Since for each $i \in \mathbb{Z}$ and each $\alpha \in [\alpha_{i+1}, \alpha_i]$ there exists a unique $t(\alpha) \in [0, \Delta t_i]$ such that $\vartheta(\alpha_i, t(\alpha)) = \alpha$ we can extend this map to a class $\mathcal{KLD}$ map by setting $\vartheta(\alpha, t) = \vartheta(\alpha_i, t(\alpha) + t)$. From the construction of the B_α we then obtain

$$\Phi_{\alpha(t)}(t, B_\alpha) \subseteq \Phi_{\alpha_{i+1}}(t, B_\alpha)$$
$$\subseteq \Phi_{\alpha_{i+1}}(t + t(\alpha), B_{i-1}) \cup B_i = B_{\vartheta(\alpha, t+t(\alpha))} = B_{\vartheta(\alpha, t)}$$

for $\alpha(t) = \vartheta(\alpha, t)$, $\alpha \in [\alpha_{i+2}, \alpha_{i+1}]$ and $t \in [t(\alpha), \Delta t_{i-1}]$. This shows property (iv'). $\qquad\square$

As already noted in Remark 3.2.5, the robustness gain γ in the formulation of γ-robustness can chosen "almost" optimal with the only restriction being the possible discontinuity of the optimal gain. In the ISDS formulation the situation is different. Here we have a tradeoff between μ and γ, which also appears when a continuous optimal gain γ for the γ-robustness property exists.

When passing from γ-robustness to ISDS this tradeoff is represented by the choice of $\tilde{\gamma}$: The smaller the difference $\tilde{\gamma} - \gamma$ becomes the slower the corresponding rate of attraction μ might become. The following examples illustrates this tradeoff between $\tilde{\gamma}$ and μ, where we choose $\tilde{\gamma} = \gamma/(1-\varepsilon)$ for $\varepsilon > 0$. In particular, it shows that even for very simple systems the parameter ε has to be chosen strictly positive when passing from one to another.

Example 3.4.5 Consider the system

$$\dot{x} = -x + w, \quad w \in \mathbb{R}.$$

Clearly, the set $A = \{0\}$ is attracting, furthermore for each $\alpha \geq 0$ the set $A_\alpha = [-\alpha, \alpha]$ is the smallest α-attracting set, which shows that $\gamma(r) = r$ is the optimal robustness gain. On the other hand, for $\gamma_\epsilon = \gamma/(1-\varepsilon)$ we obtain ISDS with (optimal) attraction rate $\mu(r,t) = e^{-\varepsilon t}r$. For $\gamma_0 = \gamma$, however, we obtain for $x > \alpha$ and $w(t) \equiv \alpha$ the inequality

$$\|\varphi(t,x,w)\|_A = e^{-t}(x-\alpha) + \alpha > \alpha \text{ for all } t \geq 0.$$

Since $\nu(w,t) = \alpha$ for this w and γ (no matter how μ is chosen), we obtain

$$\|\varphi(t,x,w)\|_A > \nu(w,t) \text{ for all } t \geq 0,$$

hence the system is not ISDS for the optimal robustness gain γ.   □

To conclude this section we will now combine the two preceding propositions of this section.

Theorem 3.4.6 Consider the system (2.5) or (2.21) with solutions denoted by Φ. Assume that (2.9) or (2.25), respectively, holds for some ρ of class $\mathcal{K}_\infty$. Then any strongly attracting set with attracted neighborhood B satisfying $d_H(B,A) < \infty$ is ISDS for some suitable attraction rate μ, suitable gains σ and γ and a suitable perturbation range W containing 0 in its interior. In particular, the gains γ and the perturbation range W only depend on the rate of attraction β, on ρ, on the Lipschitz continuity of the unperturbed system and on the distance $d_H(B,A) < \infty$.

Again, we refer to Example 4.4.2 in the next chapter for the dependence between β and the gains in the ISDS formulation in the case of exponential attraction.

3.5 Lyapunov Function Characterization

In this section we will characterize the ISDS property by means of a suitable Lyapunov function. Since their introduction by Lyapunov [92] more than one century ago these functions have played an important role in the analysis of dynamical and control systems.

The usefulness of these functions in our context lies in the fact that they provide a "nonlinear distance" to the asymptotically stable set A which replaces the euclidean distance $\|\cdot\|_A$ in the ISDS estimate in Definition 3.2.7, where this nonlinear distance is chosen in such a way that we can avoid the explicit use of the overshoot gain σ. This gain, in turn, can be recovered from bounds on the Lyapunov function. This way we end up with an ISDS estimate where we have immediate decay of the distance (provided the perturbation is small) instead of having a decay only after a certain transient time.

For robustness investigations Lyapunov functions have been used extensively for a long time. For instance, the result on convergence of attracting sets under one step discretizations from Kloeden and Lorenz [77] uses the fact that a Lipschitz continuous Lyapunov function has a certain built in robustness in the sense that small perturbations acting on the system only cause small deviations in the value of the Lyapunov function along the solution trajectories.

Here we will construct Lyapunov functions which have a built–in robustness. By this we mean that we do not have to rely on continuity properties in order to get information about the behavior of perturbed systems, but that the Lyapunov function is such that it already includes information about the behavior of perturbed systems. The following proposition makes this principle mathematically precise.

Proposition 3.5.1 Let $A \subset \mathbb{R}^n$ be a closed c-bounded strongly forward invariant set. Assume there exist functions σ_1, σ_2 and $\tilde{\gamma}$ of class $\mathcal{K}_\infty$, $\tilde{\mu}$ of class $\mathcal{KLD}$, a strongly forward invariant set $O \subseteq \mathbb{R}^n$, an open subset $P \subseteq O$ with $A \subset P$ and a function $V : O \to R_0^+$ with

$$\sigma_1(\|x\|_A) \leq V(x) \text{ for all } x \in O,$$

$$V(x) \leq \sigma_2(\|x\|_A) \text{ for all } x \in P$$

and

$$V(\Phi(t, x, u, w)) \leq \max\{\tilde{\mu}(V(x), t), \tilde{\nu}(w, t)\} \tag{3.6}$$

for all $x \in O$, all $u \in \mathcal{U}$, $w \in \mathcal{W}$ and all $t \in \mathbb{T}^+$, where $\tilde{\nu}$ is defined by (3.1) for $\tilde{\mu}$ and $\tilde{\gamma}$. Then the set A is ISDS with attracted neigborhood $B = P$, attraction rate $\mu(r, t) = \sigma_1^{-1}(\tilde{\mu}(\sigma_1(r), t))$, overshoot gain $\sigma(r) = \sigma_1^{-1}(\sigma_2(r))$ and robustness gain $\gamma(r) = \sigma_1^{-1}(\tilde{\gamma}(r))$.

Proof: Consider $\tilde{V}(x) := \sigma_1^{-1}(V(x))$. Then a straightforward calculation yields

$$\tilde{V}(\Phi(t, x, u, w)) \leq \max\{\mu(\tilde{V}(x), t), \nu(w, t)\} \text{ for all } x \in O,$$

which implies the assertion since $\|x\|_A \leq \tilde{V}(x) \leq \sigma(\|x\|_A)$.   □

For future reference we make the following definition.

Definition 3.5.2 (ISDS Lyapunov function)

A function V satisfying the assumptions of Proposition 3.5.1 will be called an ISDS *Lyapunov function*.   □

Next we show that the existence of an ISDS Lyapunov function V is necessary and sufficient for the ISDS property, even quantitatively, i.e., we can precisely characterize the rate and gains using V.

Theorem 3.5.3 Consider the system (2.5) or (2.21) with solutions Φ. Let A be a closed c-bounded strongly forward invariant set for the corresponding unperturbed system (2.1) or (2.17). Consider a class $\mathcal{KLD}$ function μ and class $\mathcal{K}_\infty$ functions γ and σ. Then the following properties are equivalent:

(i)   The set A is ISDS with gains γ and σ and rate μ.

(ii) For each $\varepsilon > 0$ there exists a continuous function $V_\varepsilon : \mathcal{R}(B) \to \mathbb{R}_0^+$ which satisfies

$$(1 + \varepsilon)V_\varepsilon(x) \geq \|x\|_A \text{ for all } x \in \mathcal{R}(B),$$

$$V_\varepsilon(x) \leq \sigma(\|x\|_A) \text{ for all } x \in B$$

and

$$V_\varepsilon(\Phi(t, x, u, w)) \leq \max\{\mu(V_\varepsilon(x), (1 - \varepsilon)t), \nu(w, t)\}$$

for all $x \in \mathcal{R}(B)$, all $u \in \mathcal{U}$, $w \in \mathcal{W}$ and all $t \in \mathbb{T}^+$ with ν from (3.1).

(iii) There exists a function $V : \mathcal{R}(B) \to \mathbb{R}_0^+$ which satisfies

$$V(x) \geq \|x\|_A \text{ for all } x \in \mathcal{R}(B),$$

$$V(x) \leq \sigma(\|x\|_A) \text{ for all } x \in B$$

and

$$V(\Phi(t, x, u, w)) \leq \max\{\mu(V(x), t), \nu(w, t)\}$$

for all $x \in \mathcal{R}(B)$, all $u \in \mathcal{U}$, $w \in \mathcal{W}$ and all $t \in \mathbb{T}^+$ with ν from (3.1).

If one of these properties holds and, in addition, μ is C^1 with $\frac{\partial}{\partial r}\mu(r, t) > 0$ for all $r > 0$ and all $t > 0$, then V_ε in (ii) can be chosen to be Lipschitz on $\mathcal{R}(B) \setminus A$.

Proof: (i)$\Rightarrow$(ii): Assume ISDS, let $\varepsilon > 0$ and consider the strictly α-contracting family of neighborhoods B_α given by Proposition 3.3.4. Since the family is strictly contracting for each point $x \in \mathcal{R}(B) \setminus A$ there exists a unique $\alpha(x) > 0$ such that $x \in B_\alpha$ for all $\alpha > \alpha(x)$ and $x \notin B_\alpha$ for all $\alpha < \alpha(x)$. Furthermore, the strict contraction and the closedness of the sets B_α implies continuity of $\alpha(x)$ in x and the shrinking property implies $\alpha(x) \to 0$ as $x \to A$.

Thus setting $V_\varepsilon(x) = \gamma(\alpha(x))$ for $x \in \mathcal{R}(B) \setminus A$ and $V_\varepsilon(x) = 0$ on A we obtain a continuous function. Then the properties of the sets B_α imply the desired upper and lower bounds for V, and the last inequality for V_ε follows from Lemma 3.3.2.

The claimed Lipschitz property follows immediately from estimate (3.2).

(ii)$\Rightarrow$(iii): Consider the function $V : \mathcal{R}(B) \to \mathbb{R}_0^+$ defined pointwise by $V(x) := \limsup_{\varepsilon \searrow 0} V_\varepsilon(x)$. Obviously the first two inequalities are satisfied. For the third, pick $x \in \mathcal{R}(B)$, $u \in \mathcal{U}$, $w \in \mathcal{W}$ and $t \in \mathbb{T}^+$, and consider the point $y = \Phi(t, x, u, w)$. Then for each $\delta > 0$ there exists $\varepsilon \in (0, \delta)$ such that $V(y) \leq V_\varepsilon(y) + \delta$ and $V_\varepsilon(x) \leq V(x) + \delta$. This yields

$$V(y) \leq \max\{\mu(V_\varepsilon(x), (1 + \varepsilon)t), \mu(w, t)\} + \delta$$
$$\leq \max\{\mu(V(x) + \delta, (1 + \delta)t), \mu(w, t)\} + \delta$$

which implies the assertion as $\delta \to 0$.

(iii)$\Rightarrow$(i): Follows immediately from Proposition 3.5.1 with $\sigma_1(r) = r$. $\square$

Remark 3.5.4 The construction of V via the contracting family of neighborhoods somewhat hides the principles we have used for their construction. For $W = 0$ (i.e., in the absence of external perturbations) an inspection of the proof reveals that V reduces to

$$V(x) := \sup_{u \in \mathcal{U}} \sup_{t \in \mathbf{T}_0^+} \{\mu(\||\Phi(t, x, u, 0)\||_A, -t)\}$$

while—when we neglect the ρ_ε term in the construction of the B_α, which is only needed in order to cope with the perturbation—the function V_ε becomes

$$V_\varepsilon(x) := \sup_{u \in \mathcal{U}} \sup_{t \in \mathbf{T}_0^+} \{\mu(\||\Phi(t, x, u, 0)\||_A, -(1 - \varepsilon)t)\}.$$

This type of Lyapunov functions and, in particular, the trick of slowing down the rate of attraction in the construction of V_ε is classical (see, e.g., Yoshizawa [127, Chapter 19]) and is frequently used in the stability analysis of dynamical and control systems, see, e.g., [77, 70, 113, 120] for just a few recent examples. A nice survey on Lyapunov functions characterizing strong stability can be found in Teel and Praly [121]. $\square$

We now turn to an infinitesimal characterization of ISDS Lyapunov functions. For continuous time systems we will characterize inequality (3.6) via a suitable first order partial differential inequality. In order to do this we impose the following assumptions on μ.

Assumption 3.5.5 The function μ solves the ordinary differential equation

$$\frac{d}{dt}\mu(r, t) = -g(\mu(r, t))$$

for some Lipschitz continuous function $g : \mathbb{R}^+ \to \mathbb{R}^+$. $\square$

Note that under this assumption a straightforward application of Gronwall's Lemma yields the existence of a function $C_\mu : \mathbb{R}^+ \times \mathbb{R}^+ \times \mathbb{R}_0^+ \to \mathbb{R}^+$ such that

$$|\mu(r, t) - \mu(r', t)| \leq C_\mu(a, b, T)|r - r'| \text{ for all } t \in [0, T], r, r' \in [a, a+b]. \quad (3.7)$$

Assumption 3.5.5 is satisfied for instance if μ is C^∞ on $\mathbb{R}_0^+ \times \mathbb{R}$, hence by Proposition B.2.3 it can always be assumed by slightly enlarging the ISDS rate and gains.

Let us now formulate the relation between the inequality for $V(\varphi(t, x, u, w))$ in Theorem 3.5.3 and a suitable partial differential inequality. At a first glance the partial differential inequality in the following proposition might seem to have an unnecessary reversal of signs, i.e., it would seem more natural to consider the equation multiplied by -1. However, this is needed in order to be consistent with the usual definition of viscosity supersolutions, cf. Appendix A.

Proposition 3.5.6 Assume Assumption 3.5.5. Then a continuous function $V : \mathcal{R}(B) \to \mathbb{R}_0^+$ satisfies the inequality

$$V(\varphi(t, x, u, w)) \leq \max\{\mu(V(x), t), \nu(w, t)\}$$

for all $u \in \mathcal{U}$, $w \in \mathcal{W}$ and all $t \geq 0$ with ν from (3.1) if and only if it is a viscosity supersolution of

$$\inf_{u \in U,\, w \in W:\, \|w\| < \gamma^{-1}(V(x))} \{-DV(x)f(x, u, w) - g(V(x))\} \geq 0.$$

Proof: Let V satisfy the inequality and fix $x \in \mathcal{R}(B)$, $u^0 \in U$ and $w^0 \in W$ with $\|w^0\| < V(x)$. Consider the constant perturbation functions $u(t) \equiv u^0$ and $w(t) \equiv w^0$. Then by continuity there exists $t > 0$ such that $\mu(V(x), \tau) \geq \gamma(\|w^0\|) = \nu(\tau, w)$ for all $\tau \in [0, t]$. Now let $\xi \in D^- V(x)$. Then for small $\tau > 0$ we obtain

$$\xi f(x, u^0, w^0) = \frac{\xi(\varphi(\tau, x, u, w) - x)}{\tau} + \frac{o(\tau)}{\tau}$$

and hence

$$\xi f(x, u^0, w^0) \leq \limsup_{\tau \to 0} \frac{V(\varphi(\tau, x, u, w)) - V(x)}{\tau}$$
$$\leq \limsup_{\tau \to 0} \frac{\mu(V(x), \tau) - V(x)}{\tau} = -g(V(x)).$$

This shows the claim.

Let conversely V be a viscosity supersolution of the given inequality and fix some $t > 0$. From Corollary A.2.4 applied with $b = V(x)$, $a = \mu(V(x), t)$ and $W = W_{\gamma^{-1}(\mu(V(x), t))}$ we obtain

$$V(\varphi(t, x, u, w)) \leq \mu(V(x), t) \text{ for all } u \in \mathcal{U} \text{ and all } w \in \mathcal{W}$$
$$\text{with } \gamma(\|w(\tau)\|) \leq \mu(V(x), t) \text{ for almost all } \tau \in [0, t]. \tag{3.8}$$

We now claim that (3.8) implies

$$V(\varphi(t, x, u, w)) \leq \mu(V(x), t) \text{ for all } u \in \mathcal{U} \text{ and all } w \in \mathcal{W}$$
$$\text{with } \gamma(\|w(\tau)\|) \leq \mu(V(x), \tau) \text{ for almost all } \tau \in [0, t] \tag{3.9}$$

(note that the difference to (3.8) is the "τ" in the argument of μ bounding $\gamma(\|w(\tau)\|)$).

In order to prove (3.9) fix some $t > 0$, let $w \in \mathcal{W}$ satisfy this constraint, and assume $V(\varphi(t, x, u, w)) > \mu(V(x), t)$. Then there exists $\delta > 0$ such that $V(\varphi(t, x, u, w)) > \mu(V(x), t) + \delta$. Now pick an arbitrary $\varepsilon < \delta$ and choose $t^* > 0$ such that $V(\varphi(t^*, x, u, w)) = \mu(V(x), t^*) + \varepsilon$ and $V(\varphi(\tau, x, u, w)) > \mu(V(x), \tau) + \varepsilon$ for all $\tau \in [t^*, t]$. From the assumption on w we obtain $\|w(\tau)\| \leq V(\varphi(\tau, x, u, w)) - \varepsilon$ for almost all $\tau \in [t^*, t]$. Using the continuity of $V(\varphi(\tau, x, u, w))$ in τ and the Lipschitz property of g we can now conclude the existence of times t_i, $i = 0, \ldots, k$ such that $t_0 = t^*$, $t_k = t$ and $\mu(V(\varphi(t_i, x, u, w), t_{i+1} - t_i) \geq V(\varphi(t_i, x, u, w)) - \varepsilon$, which implies $\|w(\tau)\| \leq \mu(V(\varphi(t_i, x, u, w))$ for almost all $\tau \in [t_i, t_{i+1}]$. Hence by (3.8) we can conclude

$$V(\varphi(t_{i+1}, x, u, w)) \leq \mu(V(\varphi(t_i, x, u, w)), t_{i+1} - t_i)$$

which by induction implies

$$V(\varphi(t, x, u, w)) \leq \mu(V(\varphi(t^*, x, u, w)), t - t^*).$$

From this inequality and (3.7) we obtain

$$V(\varphi(t, x, u, w)) \leq \mu(\mu(V(x), t^*) + \varepsilon, t - t^*)$$
$$\leq \mu(V(x), t) + C_\mu(\mu(V(x), t), V(x) + \delta - \mu(V(x), t), t) + \varepsilon$$

which contradicts the assumption as $\varepsilon \to 0$ and hence shows (3.9).

We finally use (3.9) to show the assertion. If $\nu(w, t) \leq \mu(V(x), t)$ then inequality (3.9) directly implies the assertion. Hence consider some $t_1 > 0$ such that

$$V(\varphi(t_1, x, u, w)) > \nu(w, t_1) > \mu(V(x), t_1). \tag{3.10}$$

We set $r = \mu(\nu(w, t_1), -t_1)$ and choose $t_0 > 0$ minimal such that the inequality $V(\varphi(t, x, u, w)) \geq \mu(r, t)$ holds for all $t \in [t_0, t_1]$. Since by the choice of r and by the second inequality in (3.10) we have $r > V(x)$ we obtain $V(\varphi(t_0, x, u, w)) \leq \mu(r, t_0)$. Now by Lemma 3.2.8 (i) and (ii) we obtain

$$\nu(w(t_0 + \cdot), t) \leq \mu(r, t_0 + t) \leq \mu(V(x_0), t)$$

for $x_0 = \varphi(t_0, x, u, w)$ which by (3.9) implies

$$V(\varphi(t_1, x, u, w)) = V(\varphi(t_1 - t_0, x_0, u, w))$$
$$\leq \mu(V(x_0), t_1 - t_0) \leq \mu(r, t_1) = \nu(w, t_1)$$

which contradicts (3.10) and hence shows the claim.   □

We can now state our main Theorem on the Lyapunov characterization of ISDS via viscosity solutions.

Theorem 3.5.7 Consider system (2.5) and let A be an attracting set for the corresponding unperturbed system (2.1). Consider a function μ of class $\mathcal{KLD}$ satisfying Assumption 3.5.5 and functions σ and γ of class $\mathcal{K}_\infty$. Then A is ISDS with attraction rate μ and gains σ and γ if and only if for each $\varepsilon > 0$ there exists a continuous function $V : \mathcal{R}(B) \to \mathbb{R}_0^+$ which satisfies

$$(1 + \varepsilon)V(x) \geq \|x\| \text{ for all } x \in \mathcal{R}(B),$$

$$V(x) \leq \sigma(\|x\|) \text{ for all } x \in B$$

and is a viscosity supersolution of the equation

$$\inf_{u \in U, w \in W : \|w\| < \gamma^{-1}(V(x))} \{-DV(x)f(x, u, w) - (1 - \varepsilon)g(V(x))\} \geq 0.$$

Proof: Follows immediately from Theorem 3.5.3 and Proposition 3.5.6.   □

As far as the applications in this book are concerned, the viscosity solution characterization is completely satisfying and provides everything we need. Nevertheless, for the sake of completeness we state the following theorem which shows that we can even pass to smooth Lyapunov functions. Using a smoothing result from Lin, Sontag and Wang [88, Theorem B.1] (which in turn is an adaptation of a result by Wilson [126]) we arrive at the following characterization.

Theorem 3.5.8 Consider system (2.5) with compact perturbation range W and let A be an attracting set for the corresponding unperturbed system (2.1). Consider a function μ of class $\mathcal{KLD}$ satisfying Assumption 3.5.5 and functions σ and γ of class $\mathcal{K}_\infty$. Then A is ISDS with attraction rate μ and gains σ and γ if and only if for each $\varepsilon > 0$ there exists a continuous function $V : \mathcal{R}(B) \to \mathbb{R}_0^+$ which is C^∞ on $\mathcal{R}(B) \setminus A$ and satisfies

$$V(x) \geq \|x\| \text{ for all } x \in \mathcal{R}(B),$$

$$V(x) \leq (1 + \varepsilon)\sigma(\|x\|) \text{ for all } x \in B$$

and

$$\sup_{u \in U, w \in W : \|w\| \leq \gamma^{-1}(V(x)/(1+\varepsilon))} DV(x)f(x, u, w) \leq -g(V(x)).$$

Proof: Assume the existence of V for each $\varepsilon > 0$. Since V is also a viscosity supersolution of the given equation multiplied by -1 Proposition 3.5.6 and Proposition 3.5.1 applied with $\sigma_{1,\varepsilon}(r) = r$, $\sigma_{2,\varepsilon}(r) = (1+\varepsilon)\sigma(r)$, $\gamma_\varepsilon(r) = (1+\varepsilon)\gamma(r)$ and $\mu_\varepsilon(r,t) = \mu(r,t)$ for each $\varepsilon > 0$ yield ISDS with ε-dependent rate and gains obtained from Proposition 3.5.1. Since for $\varepsilon \to 0$ these converge to the original rate and gains we obtain the assertion.

Conversely, assume ISDS, fix $\varepsilon > 0$ and let $\varepsilon_1 > \varepsilon_2 > 0$ be such that $(1 + \varepsilon_1)^3 \leq (1+\varepsilon)$ and $(1+\varepsilon_1)(1-\varepsilon_2) \geq 1$. Applying Theorem 3.5.7 with $\varepsilon = \varepsilon_2$ we can conclude the existence of a locally Lipschitz (away from 0) Lyapunov function V_1 satisfying

$$\frac{\|x\|}{1+\varepsilon_2} \leq V_1(x) \leq \sigma(\|x\|)$$

for all $x \in \mathbb{R}^n$. Since V_1 is Lipschitz, by Rademacher's theorem we can conclude that it satisfies

$$\gamma(\|u\|) \leq V_1(x)(1-\varepsilon_2) \quad \Rightarrow \quad DV_1(x)f(x,u) \leq -(1-\varepsilon_2)g(x)$$

for almost all $x \in \mathcal{R}(B) \setminus A$. Thus we can apply the smoothing result [88, Theorem B.1] in order to obtain a smooth function V_2 satisfying

$$\|V_1(x) - V_2(x)\| \leq \rho_1(x)$$

for all $x \in \mathcal{R}(B) \setminus A$ and

$$\gamma(\|w\|) \leq V_1(x)(1-\varepsilon_2) \quad \Rightarrow \quad DV_2(x)f(x,u,w) \leq DV_1(x)f(x,u,w) + \rho_2(x)$$

for arbitrary but fixed continuous functions $\rho_1, \rho_2 : \mathcal{R}(B) \setminus A \to (0, \infty)$. Note that the original formulation of this theorem requires compactness of the perturbation range $U \times W$, i.e., compactness of W. Inspection of the proof in [88], however, reveals that the construction also works if for any compact subset $K \subset \mathbb{R}^n$ we can restrict ourselves to a compact subset $W_K \subset W$. Since for any compact $K \subset \mathbb{R}^n$ we only need to consider the compact set W_K given by

$$W_K := \mathrm{cl}\left\{ w \in W \,\middle|\, \|w\| \leq \gamma^{-1}\left(\max_{x \in K} V_1(x)\right) \right\},$$

this is indeed possible in our setup. Setting $\rho_1(x) = \varepsilon_2 V_1(x)$ and $\rho_2(x) = \varepsilon_2 g(x)$ the function V_2 satisfies

$$\frac{\|x\|}{(1+\varepsilon_2)^2} \leq V_2(x) \leq (1+\varepsilon_2)\sigma(\|x\|)$$

and

$$\gamma(\|w\|) \leq V_2(x)(1-\varepsilon_2)^2 \quad \Rightarrow \quad DV_2(x) \cdot f(x,u,w) \leq -(1-\varepsilon_2)^2 g(x)$$

for all $x \in \mathbb{R}^n \setminus \{0\}$. Hence $V = (1+\varepsilon_1)^2 V_2$ is the desired function. $\qquad \square$

Remark 3.5.9 It seems reasonable to expect that the smoothing of V results in a slight loss of the gains. For the Lyapunov function characterization of optimal H_∞ gains this behavior could be rigorously verified by Rosier and Sontag [97]. What remains open at the moment is the question whether for a given ISDS set with robustness gain γ this gain can be represented by a continuous Lyapunov function.   □

Let us now return to viscosity solution characterizations. Since we know that γ-robustness implies the ISDS property we obtain the following characterization of γ-robustness.

Corollary 3.5.10 Consider system (2.5) and let A be an attracting set for the corresponding unperturbed system (2.1). Then A is a γ-robust attracting set for some function γ of class $\mathcal{K}_\infty$ if and only if for each $\varepsilon > 0$ and for each compact subset $\widetilde{W} \subset W$ there exist functions σ_ε, μ_ε, g_ε and $\widetilde{V}_\varepsilon$ such that

$$\gamma(\widetilde{V}_\varepsilon(x)) \geq \|x\|/(1+\varepsilon) \text{ for all } x \in \mathcal{R}(B),$$

$$\widetilde{V}_\varepsilon(x) \leq \sigma_\varepsilon(\|x\|) \text{ for all } x \in B$$

and $\widetilde{V}_\varepsilon$ is a viscosity supersolution of the equation

$$\inf_{u\in U,\, w\in W:\, \|w\|<\widetilde{V}_\varepsilon(x)} \{-D\widetilde{V}_\varepsilon(x)f(x,u,w) - g_\varepsilon(\widetilde{V}_\varepsilon(x))\} \geq 0.$$

Proof: Let A be a γ-robust strongly attracting set and fix $\varepsilon > 0$. By Proposition 3.4.4(a) we obtain that A is ISDS, and by Proposition B.2.3 we can assume that the gains are C^∞ by enlarging them by the factor $\sqrt{1+\varepsilon}$. Setting $\widetilde{V}_\varepsilon = \gamma^{-1}(V)$ for V from Theorem 3.5.7 applied with ε_1 such that $(1+\varepsilon_1) \leq \sqrt{1+\varepsilon}$, one easily verifies the assertion.

Conversely, if we have $\widetilde{V}_\varepsilon$ as in the assumption, then we set $V = \gamma(\widetilde{V}_\varepsilon)$. By Propositions 3.5.6 and 3.5.1 we obtain ISDS with robustness gain $(1+\varepsilon)\gamma$, and hence $(1+\varepsilon)\gamma$-robustness of A for each $\varepsilon > 0$. By Lemma 3.2.3 this implies γ-robustness of A.   □

In particular, we obtain the following sufficient condition for γ-robustness

Corollary 3.5.11 Consider system (2.5) and an open set O and a function $\widetilde{V} : \mathrm{cl}\, O \to \mathbb{R}_0^+$ which satisfies $\inf_{x\in\partial O} \widetilde{V}(x) =: \alpha_0 > 0$ and which is a viscosity supersolution of the equation

$$\inf_{u\in U,\, w\in W:\, \|w\|<\widetilde{V}(x)} \{-D\widetilde{V}(x)f(x,u,w) - g(\widetilde{V}(x))\} \geq 0$$

on O for some $g : \mathbb{R}_0^+ \to \mathbb{R}_0^+$ with $g(a) > 0$ for $a > 0$. Then for $W = W_{\alpha_0}$ the set $A := \{x \in O \,|\, \widetilde{V}(x) = 0\}$ is a γ-robust strongly attracting set with attracted neighborhood $B = O$ and $\gamma(\alpha) := \sup\{\|x\|_A \,|\, \widetilde{V}(x) \leq \alpha\}$.

Proof: Immediate from Corollary 3.5.10.   □

3.6 Stability of Robustness Concepts

In this section we will investigate the effect of additional external perturbations on strongly attracting, γ-robust and ISDS sets. We will show what happens if a sequence of systems possessing one of these sets converges to some limiting system, and we will see how far we can make statements for the existence of those sets for systems nearby some reference system.

For the investigation of limiting systems we first introduce the following definition.

Definition 3.6.1 (asymptotical boundedness)

(i) A sequence of functions $\beta_n : (0,\infty) \times (0,\infty) \rightarrow [0,\infty)$ is said to be *asymptotically bounded* by some class $\mathcal{KL}$ function β if for all $\varepsilon > 0$ and all $a, b \in \mathbb{R}$ with $0 < a < b$ there exists $N \in \mathbb{N}$ such that

$$\beta_n(r,t) \leq \beta(r,t) + \varepsilon \text{ for all } r, t \in [a,b] \text{ and all } n \geq N.$$

(ii) Sequences of functions $\gamma_n, \sigma_n : (0,\infty) \rightarrow (0,\infty)$ and $\mu_n : (0,\infty) \times (0,\infty) \rightarrow [0,\infty)$ are said to be (ISDS–) *asymptotically bounded* by class $\mathcal{K}_\infty$ functions σ and γ and a class $\mathcal{KLD}$ function μ if for all $\varepsilon > 0$ and all $a, b \in \mathbb{R}$ with $0 < a < b$ there exists $N \in \mathbb{N}$ such that

$$\sigma_n(r) \leq \sigma(r) + \varepsilon \text{ and } \gamma_n(r) \leq \gamma(r) + \varepsilon \text{ for all } r \in [a,b] \text{ and all } n \geq N$$

and

$$\mu_n(r,t) \leq \mu(r,t) + \varepsilon \text{ for all } r, t \in [a,b] \text{ and all } n \geq N.$$

□

Note that in this definition we do not demand that the function β_n, μ_n, σ_n and γ_n are of the classes $\mathcal{KL}$, $\mathcal{KLD}$ or $\mathcal{K}_\infty$, respectively.

Proposition 3.6.2 Consider a system (2.5) or (2.21) given by f or Φ_h, respectively, and a sequence of approximating systems f_n or $\Phi_{h,n}$ with

$$\|f(x,u,w) - f_n(x,u,w)\| \leq \varepsilon_n$$

or, respectively,

$$\|\Phi_h(x,u,w) - \Phi_{h,n}(x,u,w)\| \leq h\varepsilon_n$$

for all $u \in \mathcal{U}$, $w \in \mathcal{W}$ $x \in \mathbb{R}^n$ and some sequence ε_n, $n \in \mathbb{N}$ with $\varepsilon_n \rightarrow 0$ as $n \rightarrow \infty$. Denote the trajectories by Φ_n and Φ and consider closed c-bounded sets A and A_n such that $d_H(A, A_n) \rightarrow 0$ as $n \rightarrow \infty$ and an open set B with $A_n \subset B$, $A \subset B$. Then the following properties hold
(i) If $\beta_n : (0,\infty) \times (0,\infty) \rightarrow [0,\infty)$ are functions which are asymptotically

bounded by some class $\mathcal{KL}$ function β and for each $T > 0$ there exists $N \in \mathbb{N}$ with

$$\|\Phi_n(t, x, u)\|_{A_n} \leq \beta_n(\|x\|_{A_n}, t)$$

for all $x \in B$, $u \in \mathcal{U}$ and all $t \in \mathbb{T} \cap [0, T]$ then A is an asymptotically stable set for Φ with attraction rate β.

(ii) If μ_n, γ_n and σ_n are functions which are (ISDS–) asymptotically bounded by μ, ρ and σ and for each $T > 0$ there exists $N \in \mathbb{N}$ with

$$\|\Phi_n(t, x, u, w)\|_{A_n} \leq \max\{\mu_n(\sigma_n(\|x\|_{A_n}), t), \nu_n(w, t)\}$$

for all $x \in B$, $u \in \mathcal{U}$, $w \in \mathcal{W}$ and all $t \in \mathbb{T} \cap [0, T]$ then A is an ISDS set for Φ with attraction rate μ and gains σ and γ.

Proof: Immediate from the continuity of the trajectories. $\square$

Note that for γ-robust strongly attracting sets A_n this limiting property is not true, as the following example shows.

Example 3.6.3 Consider the family of $2d$ systems

$$\dot{x} = -x/2^n$$
$$\dot{y} = -y + w.$$

For each of these systems the set $A_n = \{0\}$ is a γ-robust attracting set with $\gamma(r) = r$. Nevertheless, for the limiting system

$$\dot{x} = 0$$
$$\dot{y} = -y + w$$

the set $A = \{0\}$ clearly is not attracting. $\square$

We can only recover the following partial version of Proposition 3.6.2 for γ-robustness.

Proposition 3.6.4 Consider a system of type (2.5) or (2.21), closed sets A and A_n such that $d_H(A, A_n) \to 0$ as $n \to \infty$ and an open set B with $A_n \subset B$, $A \subset B$. Let A be c-bounded and assume that there exist functions γ_n of class $\mathcal{K}_\infty$ which are asymptotically bounded by some class $\mathcal{K}_\infty$ function γ and assume that each A_n is a γ_n-robust strongly attracting set. Then A is a γ-robust strongly attracting set.

Proof: By the assumption for each $\alpha \geq 0$ and each $n \in \mathbb{N}$ there exists an α-attracting set $A_{n,\alpha}$ with $d_H(A_{n,\alpha}, A_n) \leq \gamma_n(\alpha)$. Now for each $\alpha \geq 0$ consider the set

$$A_\alpha = \operatorname{Lim\,sup}_{n \to \infty} A_{n,\alpha}.$$

Note that α-forward invariance of A_α is immediate, furthermore we obtain $A \subseteq A_\alpha$ for each $\alpha > 0$. By Lemma 2.3.5 (applied with $M = \mathcal{B}(2\gamma(\alpha), A)$ if A is compact and $M = \mathcal{B}(2\gamma(\alpha), A) \setminus \operatorname{int} A$ if A^c is bounded) we obtain that for each $\varepsilon > 0$ we find $N > 0$ such that $A_{\alpha,n} \subset \mathcal{B}(\varepsilon, A_\alpha)$, which implies that $\mathcal{B}(2\varepsilon, A_\alpha)$ contains an α-attracting set and since $\varepsilon > 0$ was arbitrary A_α is α-attracting. Since the distance $d_H(A_\alpha, A) \leq \gamma(\alpha)$ is immediate from the definition of the A_α we obtain the assumption. $\qquad \Box$

We will now investigate what happens if we do not go to the limit but just consider a single system which is nearby some "reference" system. In order to obtain strong results we will introduce the following concept of embedding systems into each other.

Definition 3.6.5 (strong (α, C)-embedded system)

Consider two perturbed systems, both either of type (2.5) or of type (2.21) with same perturbation range U and perturbations ranges W and W^*, respectively. Denote the trajectories of the systems by Φ and Ψ, respectively, and let $\alpha \geq 0$ and $C \geq 1$. Then we say that the second system Ψ is (α, C)-embedded in the first Φ on some set $B \subseteq \mathbb{R}^n$ if for each $x \in B$, each $u \in \mathcal{U}$ and each $w^* \in W^*$ there exist $w \in W$ with $\|w(t)\| \leq \alpha + C\|w^*(t)\|$ for almost all $t > 0$ and

$$\Psi(t, x, u, w^*) = \Phi(t, x, u, w)$$

for all $t \in \mathbb{T}_0^+$ with $\Psi(\tau, x, u, w) \in B$ for all $\tau \in [0, t] \cap \mathbb{T}$.

Here we call Ψ the *embedded system* and Φ the *embedding system*. $\qquad \Box$

This definition allows us to characterize the behavior of nearby embedded systems.

Proposition 3.6.6 Consider a system of type (2.5) or (2.21) with trajectories Ψ, which is (α, C)-embedded on some open set B in some other system of the same type with trajectories denoted by Φ for some $\alpha \geq 0, C \geq 1$. Assume that the embedding system Φ has a strongly attracting set A which is ISDS on B with rate μ and gains σ and γ. Then for each $D > 1$ the embedded system Ψ has a strongly attracting set $\tilde{A}$ which satisfies $d_H(\tilde{A}, A) \leq \gamma(D\alpha)$ and the "ISDS-like" estimate

$$\|\Psi(t, x, u, w^*)\|_{\tilde{A}} \leq \max\{\mu(\sigma(\|x\|_{\tilde{A}} + \gamma(D\alpha)), t), \nu(CDw^*/(D-1), t)\}$$

for each $w^* \in W^*$. If $\alpha = 0$ then the set $\tilde{A} = A$ satisfies the ISDS estimate

$$\|\Psi(t, x, u, w^*)\|_{\tilde{A}} \leq \max\{\mu(\sigma(\|x\|_{\tilde{A}}), t), \nu(Cw^*, t)\}.$$

Proof: Fix $x \in B$, $u \in \mathcal{U}$ and $w^* \in \mathcal{W}^*$ and let $w \in \mathcal{W}$ be the perturbation for which the embedding is obtained. Consider the ISDS Lyapunov function V from Theorem 3.5.3 (iii). We set $\tilde{A} := \text{cl}\,\{x \in \mathbb{R}^n \,|\, V(x) \leq \gamma(D\alpha)\}$. Then the bounds on V imply that $d_H(A_\alpha, A) \leq \gamma(D\alpha)$ implying $\|x\|_{\tilde{A}} \leq \|x\|_A \leq \|x\|_{\tilde{A}} + \gamma(D\alpha)$. Setting

$$\tilde{V}(x) := \begin{cases} V(x), & V(x) > \gamma(D\alpha) \\ 0, & V(x) \leq \gamma(D\alpha) \end{cases}$$

we obtain a function satisfying

$$\sigma(\|x\|_{\tilde{A}} + \gamma(D\alpha)) \geq \tilde{V}(x) \geq \|x\|_{\tilde{A}}.$$

Furthermore for $\tilde{V}$ we have the implication

$$\tilde{V}(x) \leq \gamma(D\alpha) \;\Rightarrow\; \tilde{V}(x) = 0. \tag{3.11}$$

Now observe that defining

$$w^{\geq}(t) := \begin{cases} w(t), & \|w(t)\| \geq D\alpha \\ 0, & \|w(t)\| < D\alpha \end{cases} \quad \text{and} \quad w^{<}(t) := \begin{cases} w(t), & \|w(t)\| < D\alpha \\ 0, & \|w(t)\| \geq D\alpha \end{cases}$$

we obtain $\nu(w,t) = \max\{\nu(w^{\geq},t), \nu(w^{<},t)\}$. Since $\nu(w^{<},t) \leq \gamma(D\alpha)$ for all $t > 0$ by (3.11) we obtain

$$\tilde{V}(\Psi(t,x,v,w^*)) \leq \max\{\mu(\tilde{V}(x),t), \nu(w^{\geq},t)\}.$$

For all $t > 0$ with $\|w(t)\| \geq D\alpha$ we obtain $\|w(t)\| - \alpha \geq (D-1)\alpha$, hence a simple computation shows $\|w^{\geq}(t)\| \leq D(\|w(t)\| - \alpha)/(D-1) \leq CD\|w^*(t)\|/(D-1)$ for almost all $t > 0$. This yields

$$\tilde{V}(\Psi(t,x,u,w^*)) \leq \max\{\mu(\tilde{V}(x),t), \nu(CDw^*/(D-1),t)\}$$

which gives the assertion by the bounds on $\tilde{V}$. $\qquad\qquad\square$

We can state a similar proposition for γ-robustness instead of ISDS.

Proposition 3.6.7 Consider a system of type (2.5) or (2.21) with trajectories Ψ, which is (α, C)-embedded on some open set B in some other system of the same type with trajectories denoted by Φ for some $\alpha \geq 0$, $C \geq 1$. Assume that the embedding system Φ has a strongly attracting set A which is γ-robust on B for some gain γ of class $\mathcal{K}_\infty$. Then for each $D > 1$ the embedded system Ψ has a strongly attracting set $\tilde{A}$, which is $\gamma(CD \cdot /(D-1))$-robust and satisfies $d_H(\tilde{A}, A) \leq \gamma(D\alpha)$. If $\alpha = 0$ then the set $\tilde{A} = A$ itself is a $\gamma(C\cdot)$-robust strongly attracting set for Ψ.

Proof: We set $\tilde{A} = A_{D\alpha}$. The assumption on the (α, C)-embedding implies the inclusions

$$\Psi_{\alpha'}(t, x) \subseteq \Phi_{D\alpha}(t, x) \text{ for all } \alpha' \in [0, (D-1)\alpha/C]$$

and

$$\Psi_{\alpha'}(t, x) \subseteq \Phi_{CD\alpha'/(D-1)}(t, x) \text{ for all } \alpha' \geq (D-1)\alpha/C.$$

Hence setting $\tilde{A}_{\alpha'} = A_{D\alpha}$ for $\alpha' \in [0, (D-1)\alpha/C]$ and $\tilde{A}_{\alpha'} = A_{CD\alpha'/(D-1)}$ for $\alpha' \geq D\alpha/C$ gives attracting sets $\tilde{A}_\alpha$ for $\Psi_{\alpha'}$ satisfying

$$d_H(\tilde{A}_{\alpha'}, \tilde{A}) \leq d_H(\tilde{A}_{\alpha'}, A) \leq CD\alpha'/(D-1) \text{ for all } \alpha' \geq 0.$$

This shows the claim. □

We can avoid the constant D if we do not require robustness of the resulting strongly attracting set for the embedded system.

Proposition 3.6.8 Consider a system of type (2.5) or (2.21) with trajectories Ψ, which is (α, C)-embedded on some open set B in some other system of the same type with trajectories denoted by Φ for some $\alpha \geq 0$, $C \geq 1$. Assume that the embedding system Φ has a strongly attracting set A which is γ-robust on B. Then the embedded system Ψ has a strongly attracting set $\tilde{A}$ with $A \subseteq \tilde{A}$ and $d_H(\tilde{A}, A) \leq \gamma(\alpha)$.

Proof: Consider the α-attracting sets A_α from the definition of γ-robustness. Then it is immediate from the definitions that $\tilde{A} = A_\alpha$ is the desired attracting set. □

3.7 Inflated Systems

We will now turn to perturbed systems (2.5) and (2.21) with right hand side of type (2.10) and (2.11) or (2.26) and (2.27), respectively. The main difference to the general situation is that here the perturbation is powerful enough to steer the system into a neighborhood of the unperturbed system, where we can even estimate the diameter of this neighborhood. This is done in the following Lemma.

Lemma 3.7.1 Consider a discrete or continuous time α_0-inflated system with right hand side given by (2.10) or (2.26) satisfying assumption (2.8) or (2.24), respectively. Let $\varepsilon > 0$ and $\alpha \geq 0$ such that $\alpha + \varepsilon \leq \alpha_0$. Then the following assertions hold.
(i) For each $w \in \mathcal{W}_{\alpha+\varepsilon}$ there exists $\tilde{w} \in \mathcal{W}_\alpha$ such that the estimate

$$\Phi(t, x, u, w) \in \mathcal{B}(\varepsilon(e^{Lt} - 1)/L, \Phi(t, x, u, \tilde{w}))$$

holds for all $x \in \mathbb{R}^n$, all $u \in \mathcal{U}$ and all $t \geq 0$.

(ii) Let $x \in \mathbb{R}^n$, $u \in \mathcal{U}$, $\tilde{w} \in \mathcal{W}_\alpha$ and $T \in \mathbb{T}^+$, and consider a function $x : [0, T] \to \mathbb{R}^n$ satisfying $x(0) = x$ and

$$\|x(t) - \Phi(t, x, v, \tilde{w})\| \leq t\varepsilon/(Lt + 1) \text{ for all } t \in \mathbb{T} \cap [0, T].$$

If $\Phi = \varphi$ is a continuous time system (2.5) then assume furthermore that x solves the differential equation $\dot{x}(t) = b(t)$ for some essentially bounded and measurable function $b(t)$ with

$$\|b(t) - f(\varphi(t, x, u, \tilde{w}), u(t), \tilde{w}(t))\| \leq \varepsilon/(Lt + 1) \text{ for almost all } t \in [0, T].$$

If $\Phi = \Phi_h$ is a discrete time system (2.21) for some $h > 0$ then assume that

$$\|(x(t + h) - x(t)) - (\Phi_h(t + h, x, u, \tilde{w}) - \Phi_h(t, x, u, \tilde{w}))\| \leq h\varepsilon/(Lt + 1)$$

for all $t \in \mathbb{T} \cap [0, T]$. Under these assumptions there exists $w \in \mathcal{W}_{\alpha + \varepsilon}$ such that

$$\Phi(t, x, v, w) = x(t)$$

for all $t \in \mathbb{T}^+ \cap [0, T]$.

(iii) For each $x \in \mathbb{R}^n$, each $u \in \mathcal{U}$, each $\tilde{w} \in \mathcal{W}_\alpha$, each $T \in \mathbb{T}^+$ and each $x_T \in \mathbb{R}^n$ with $\|x_T - \varphi(T, x, u, \tilde{w})\| \leq T\varepsilon/(LT + 1)$ there exists $w \in \mathcal{W}_{\alpha + \varepsilon}$ such that

$$\Phi(T, x, u, w) = x_T.$$

If $\Phi = \Phi_h$ is a discrete time system with time step $h > 0$ and $T = h$ then the assertion also holds if $\|x_T - \Phi_h(T, x, u, \tilde{w})\| \leq T\varepsilon$.

(iv) For each $T \in \mathbb{T}^+$, each $u \in \mathcal{U}$, each $\tilde{w} \in \mathcal{W}_\alpha$ and each two points x, $x^* \in \mathbb{R}^n$ satisfying $\|x - x^*\| \leq e^{-TL}T\varepsilon/(LT + 1)$ there exists $w \in \mathcal{W}_{\alpha + \varepsilon}$ such that

$$\Phi(T, x^*, u, w) = \Phi(T, x, u, \tilde{w}).$$

Proof: (i) This assertion follows immediately from Gronwall's Lemma for continuous time systems and by induction for discrete time systems.

(ii) We show the assertion for continuous time systems, for discrete time systems it follows with similar arguments. Fix T, x, u, $\tilde{w}$ and $x(t)$ as in the assumption. We claim that $w(t) := \tilde{w}(t) + b(t) - f(x(t), u(t), \tilde{w}(t))$ is the desired perturbation function. Indeed, we have that

$$\frac{d}{dt}x(t) = b(t) = f(x(t), u(t), \tilde{w}(t)) + w(t) - \tilde{w}(t) = f(x(t), u(t), w(t))$$

and

$$\frac{d}{dt}\varphi(t, x, u, w) = f(\varphi(t, x, u, w), u(t), w(t)).$$

Hence since $x(0) = \varphi(0, x, u, w)$, by uniqueness of the solution to this differential equation we can conclude

$$\varphi(t, x, u, w) = x(t)$$

for all $t \in [0, T]$. Since

$$\|w(t)\| \leq \|\tilde{w}(t)\| + \|b(t) - f(x(t), u(t), \tilde{w}(t))\|$$
$$\leq \alpha + \|b(t) - f(\varphi(t, x, u, \tilde{w}), u(t), \tilde{w}(t))\| + L\|\varphi(t, x, u, \tilde{w}) - x(t)\|$$
$$\leq \alpha + \varepsilon/(Lt + 1) + Lt\varepsilon/(Lt + 1) \leq \alpha + \varepsilon$$

for almost all $t \in [0, T]$ we obtain that $w \in \mathcal{W}_{\alpha + \varepsilon}$.

(iii) Set $\Delta x_T := x_T - \varphi(T, x, u, \tilde{w})$. Then it is easy to check that $x(t) = \varphi(t, x, u, \tilde{w}) + t\Delta x_T/T$ satisfies the assumptions of (ii) (by setting $b(t) = f(\varphi(t, x, v, \tilde{w}), v(t), \tilde{w}(t)) + \Delta x_T/T$ in the continuous time case). Hence the assertion follows.

(iv) From Gronwall's Lemma or by induction in the discrete time case we obtain

$$\|\varphi(T, x^*, u, \tilde{w}) - \varphi(T, x, u, \tilde{w})\| \leq T\alpha/(LT + 1).$$

Hence we can apply (iii) to $x_T = \varphi(T, x^*, u, \tilde{w})$ which shows the claim. $\square$

As shown in Example 3.6.3, for a sequence of systems Φ_n possessing a sequence of γ-robust strongly attracting sets A_n it is in general not true that the limiting set A (if existing) is also γ-robust for the limiting system. If these systems, however, are inflated, then this property is true and we can formulate the analogous statement of Proposition 3.6.2 also for γ-robustness.

Proposition 3.7.2 Consider an inflated system (2.5) or (2.21) given by f or Φ_h from (2.10) or (2.26), respectively, and a sequence of approximating inflated systems f_n or $\Phi_{h,n}$ with

$$\|f(x, u, w) - f_n(x, u, w)\| \leq \varepsilon_n$$

or, respectively,

$$\|\Phi_h(x, u, w) - \Phi_{h,n}(x, u, w)\| \leq h\varepsilon_n$$

for all $u \in \mathcal{U}$, $w \in \mathcal{W}$ $x \in \mathbb{R}^n$ and some sequence ε_n, $n \in \mathbb{N}$ with $\varepsilon_n \to 0$ as $n \to \infty$. Denote the trajectories by Φ_n and Φ. Consider furthermore closed sets A and A_n such that $d_H(A, A_n) \to 0$ as $n \to \infty$ and an open set B with $A_n \subset B$, $A \subset B$. Assume that A is c-bounded and that there exists a sequence of class $\mathcal{K}_\infty$ functions γ_n which is asymptotically bounded by some class $\mathcal{K}_\infty$ function γ such that for each n the set A_n is a γ_n-robust strongly attracting set for Φ_n. Then A is a γ-robust strongly attracting set for Φ.

Proof: Note that due to the fact that Φ_n is an inflated system, Φ is $(\varepsilon_n, 1)$-embedded in Φ_n. Hence by Proposition 3.6.7 for each $D > 1$ there exists a $\gamma_n(D \cdot /(D-1))$-robust attracting set $A_{n,D}$ for Φ with $d_H(A_{n,D}, A_n) \leq \gamma_n(D\varepsilon_n)$. Setting $D_n = \sqrt{\varepsilon_n}$ we thus obtain $\gamma_n(\cdot/(1-\sqrt{\varepsilon_n}))$-robust attracting sets $\tilde{A}_n$ with $d_H(\tilde{A}_n, A_n) \leq \gamma_n(\sqrt{\varepsilon_n}) \to 0$ as $n \to \infty$, hence also $d_H(\tilde{A}_n, A) \to 0$ as $n \to \infty$. Since $\gamma(\cdot/(1-\sqrt{\varepsilon_n}))$ converges to γ uniformly on compact subsets of $[0,\infty)$, the sequence $\tilde{\gamma}_n := \gamma_n(\cdot/(1-\sqrt{\varepsilon_n}))$ is asymptotically bounded by γ and we can apply Proposition 3.6.4 which gives the assertion. $\qquad\square$

The following corollary shows an alternative version of the partial differential inequality from Corollary 3.5.10 for inflated systems.

Corollary 3.7.3 Consider the inflated system (2.5) with right hand side (2.10) and let A be an attracting set for the corresponding unperturbed system (2.1). Then A is a γ-robust attracting set for some function γ of class $\mathcal{K}_\infty$ if and only if for each $\varepsilon > 0$ and for each compact subset $\widetilde{W} \subset W$ there exist continuous functions σ_ε, μ_ε, g_ε of class $\mathcal{K}$ and $\tilde{V}_\varepsilon : \mathbb{R}^n \to \mathbb{R}_0^+$ such that

$$\gamma(\tilde{V}_\varepsilon(x)) \geq \|x\|/(1+\varepsilon) \text{ for all } x \in \mathcal{R}(B),$$

$$\tilde{V}_\varepsilon(x) \leq \sigma_\varepsilon(\|x\|) \text{ for all } x \in B$$

and $\tilde{V}_\varepsilon$ is a viscosity supersolution of the equation

$$\inf_{u \in U}\{-D\tilde{V}_\varepsilon(x)f^0(x,u) - \|D\tilde{V}_\varepsilon(x)\|\tilde{V}_\varepsilon(x) - g_\varepsilon(\tilde{V}_\varepsilon(x))\} \geq 0.$$

Proof: Since for each vector $p \in \mathbb{R}^n$ and each $\alpha \geq 0$ the equality

$$\inf_{w \in \mathbb{R}^n:\, \|w\| \leq \alpha} pw = -\alpha\|p\|$$

holds, the partial differential inequality here is equivalent to that from Corollary 3.5.10 and the assertion follows. $\qquad\square$

Let us now turn to state dependent inflation. We will show a property, which is one of the basic reasons for choosing the particular definition of the ISDS property. It shows what happens to an ISDS set for an inflated system when considering the associated system with state dependent inflation amplitude, i.e. with right hand side of type (2.11) or (2.27), and impose a suitable bound on b.

Basically, this is a small gain type of theorem, which in particular shows that when the amplitude of the perturbation is smaller than $\gamma^{-1}(\|x\|_A)$ then we still have asymptotic stability of the ISDS set (for discrete time systems we will have to make this bound slightly smaller depending on h). In this sense, it gives a quantitative version of the corresponding statement for ISS systems by Sontag and Wang [105, Theorem 1], where under the ISS assumption the existence of such an upper bound for the perturbation amplitude ("stability margin") was shown.

The fact that the robustness gain γ determines the stability margin should come as no surprise as this was already used by Jiang, Teel and Praly [67] in an application for coupled systems based on the ISS property. The main advantage of the ISDS formulation is that here we also obtain an estimate for the corresponding rate of attraction.

Theorem 3.7.4 Consider an α_0-inflated system of type (2.5) or (2.21) with right hand side given by (2.10) or (2.26), and let A be an ISDS set with rate μ, gains γ and σ and attracted neighborhood B. Consider the corresponding state dependent inflated system (2.11) or (2.27) with trajectories denoted by Φ and with b satisfying the bounds $b(x) \leq \alpha_0$ and $b(x) \leq \max\{\gamma^{-1}(\|x\|_A), \rho\}$ for continuous time systems or $b(x) \leq \alpha_0$ and $b(x) \leq \max\{\gamma^{-1}(\mu(\|x\|_A, h)), \rho\}$ for discrete time systems, respectively, for all $x \in B$ and some $\rho > 0$. Then for each $x \in B$, all $u \in \mathcal{U}$ and each $w \in \mathcal{W}_1$ the trajectories of the state dependent inflated system satisfy the inequality

$$\|\Phi(t, x, u, w)\|_A \leq \max\{\mu(\sigma(\|x\|_A, t), \gamma(\rho)\}$$

for all $t \in \mathbb{T}^+$.

Proof: We first show the statement for continuous time systems. Let $x \in B$ and $u \in \mathcal{U}$, fix $\varepsilon > 0$, pick some $w \in \mathcal{W}_1$ and set $w_\varepsilon = (1 - \varepsilon)w$. Note that by the structure of (2.11) for each $t \in \mathbb{T}^+$ the value $\Phi(t, x, u, w_\varepsilon)$ depends continuously on ε. Define $\tilde{w}_\varepsilon(t) := b(\Phi(t, x, u, w_\varepsilon))w_\varepsilon(t)$. Then $\tilde{w}_\varepsilon$ satisfies

$$\nu(t, \tilde{w}_\varepsilon) \leq \sup_{\tau \in [0,t]} \mu(\gamma(b(\Phi(\tau, x, u, w_\varepsilon))(1 - \varepsilon)), t - \tau)$$
$$\leq C(t) \max\{ \sup_{\tau \in [0,t]} \mu(\|\Phi(\tau, x, u, \tilde{w}_\varepsilon)\|_A, t - \tau), \gamma(\rho)\}$$

for each $t \geq 0$ and constants $C(t) < 1$. Thus, by continuity of μ and $b(\Phi(t, x, u, w_\varepsilon))$ in t, for all $t > 0$ there exist times $t^*(t) > 0$ such that

$$\nu(t + t^*(t), \tilde{w}_\varepsilon) \leq \max\{ \sup_{\tau \in [0,t]} \mu(\|\Phi(\tau, x, u, \tilde{w}_\varepsilon)\|_A, t + t^*(t) - \tau), \gamma(\rho)\} \quad (3.12)$$

Now assume that the asserted inequality is not valid for some $t > 0$. Then by continuity of all expressions involved we can pick $T > 0$ and $\delta > 0$ such that

$$\|\Phi(t, x, u, \tilde{w}_\varepsilon)\|_A \leq \max\{\mu(\sigma(\|x\|_A), t), \gamma(\rho)\} \text{ for all } t \in [0, T] \quad (3.13)$$

and

$$\|\Phi(t, x, u, \tilde{w}_\varepsilon)\|_A > \max\{\mu(\sigma(\|x\|_A), t), \gamma(\rho)\} \text{ for all } t \in (T, T + \delta].$$

From (3.12) and (3.13) we obtain

$$\|\Phi(T + t^*(T), x, u, \tilde{w}_\varepsilon)\|_A$$
$$\leq \max\{\mu(\sigma(\|x\|_A), T), \ \nu(T + t^*(T), \tilde{w}_\varepsilon)\}$$
$$\leq \max\{\mu(\sigma(\|x\|_A), T + t^*(T)),$$
$$\sup_{\tau \in [0,T]} \mu(\|\Phi(\tau, x, u, \tilde{w}_\varepsilon)\|_A, T + t^*(T) - \tau), \gamma(\rho)\}$$
$$\leq \max\{\mu(\sigma(\|x\|_A), T + t^*(T)),$$
$$\sup_{\tau \in [0,T]} \mu(\max\{\mu(\sigma(\|x\|_A), \tau), \gamma(\rho)\}, T + t^*(T) - \tau), \gamma(\rho)\}$$
$$\leq \max\{\mu(\sigma(\|x\|_A), T + t^*(T)), \ \gamma(\rho)\},$$

where we can assume without loss of generality that $t^*(T) \leq \delta$ and hence this inequality contradicts the choice of T. Thus we obtain the desired inequality for $\tilde{w}_\varepsilon$ for each $\varepsilon > 0$, hence by continuity in ε also for w.

For discrete time systems pick $x \in B$, $u \in \mathcal{U}$, $w \in \mathcal{W}_1$, let V be the ISDS Lyapunov function from Theorem 3.5.3(iii) and abbreviate $x_i = \Phi(hi, x, u, w)$ for each $i \in \mathbb{N}_0$. Setting $\tilde{w}(t) := b(x_i)w(t)$ for all $t \in [hi, h(i+1))$, $i \in \mathbb{N}_0$ the assumption on b implies

$$\nu_h(h, \tilde{w}(hi + \cdot)) \leq \max\{\mu(\|x_i\|_A, h), \gamma(\rho)\} \leq \max\{\mu(V(x_i), h), \gamma(\rho)\}$$

for all $i \in \mathbb{N}_0$. Thus by the properties of V we obtain

$$V(x_{i+1}) \leq \max\{\mu(V(x_i), h), \ \nu_h(h(i+1), \tilde{w})\}$$
$$\leq \max\{\mu(V(x_i), h), \ \gamma(\rho)\}$$

for all $i \in \mathbb{N}_0$. From this estimate we obtain the desired inequality by induction: For $i = 0$ the inequality is trivially satisfied. Now assume $V(x_i) \leq \max\{\mu(V(x), hi), \gamma(\rho)\}$ for some $i \in \mathbb{N}_0$. Then we obtain

$$\mu(V(x_i), h) \leq \max\{\mu(V(x), h(i+1)), \ \mu(\gamma(\rho), h)\}$$
$$\leq \max\{\mu(V(x), h(i+1)), \ \gamma(\rho)\}$$

and hence

$$V(x_{i+1}) \leq \max\{\mu(V(x_i), h), \ \gamma(\rho)\} \leq \max\{\mu(V(x), h(i+1)), \ \gamma(\rho)\}$$

i.e., the desired inequality. $\qquad\Box$

Remark 3.7.5 If the perturbation range W is such that $(1-\varepsilon)W_1 \subseteq W$ and we have continuity of $\Phi(t, x, u, w_\varepsilon)$ in ε then exactly the same proof shows that the preceding theorem also holds for arbitrary perturbed systems $f(x, u, w)$ or $\Phi_h(x, u, w)$ of type (2.5) or (2.21), where $f(x, u, b(x)w)$ or $\Phi_h(x, u, b(x)w)$ with b as in (2.11) plays the role of the state dependent inflation. The restriction to inflated systems is motivated by our application: We will use this result for the analysis of space discretizations (cf. Theorem 6.1.6) where it will be sufficient to consider inflated systems. $\qquad\Box$

A similar property as for ISDS sets holds true for γ-robust strongly forward invariant sets.

Proposition 3.7.6 Consider an α_0-inflated system Φ with right hand side given by (2.10) or (2.26), and let C be a direct γ-robust strongly forward invariant set for some γ of class $\mathcal{K}_\infty$. Then the corresponding state dependent inflated system (2.11) or (2.27) with b satisfying $b(x) \leq \rho$ on $\mathcal{B}(\gamma(\rho), C)$ for some $\rho \in [0, \alpha_0]$ has a strongly forward invariant set $\tilde{C}$ which is strongly forward invariant for all $w \in \mathcal{W}_1$ and satisfies $d_H(\tilde{C}, C) \leq \gamma(\rho)$.

Proof: From the assumption we know the existence of a strongly ρ-forward invariant set C_ρ with $d_H(C_\rho, C) \leq \gamma(\rho)$. Now observe that by the assumption on b each trajectory of the state dependent inflated system in C_ρ for some $w \in \mathcal{W}_1$ coincides with some trajectory of the inflated system for some $w \in \mathcal{W}_\rho$. Hence $\tilde{C} = C_\rho$ satisfies the assumption.   □

At the end of this section we will briefly discuss the relation of our robustness concepts to a slightly different concept for inflated systems which was introduced by the author in [54]. We recall its definition.

Definition 3.7.7 ((γ, ρ)-attracting set)

Consider two class $\mathcal{K}_\infty$ functions γ and ρ, and a strongly attracting set A. We say that A is (γ, ρ)-attracting with (bounded) attracted neighborhood B if there exists a family of sets C_τ with $B \subseteq C_0$, which satisfies the following properties:

(i) $C_{\tau'} \subseteq C_\tau$ for all $\tau, \tau' \in \mathbb{R}_0^+$, $\tau' \geq \tau$

(ii) $A = \bigcap_{\tau \in \mathbb{R}_0^+} C_\tau$

(iii) $\Phi(t, C_\tau) \subseteq C_{\tau+t}$ for all $\tau \in \mathbb{R}_0^+$ and all $t \in \mathbb{T}^+$

(iv) $d_H(C_\tau, A) \leq \rho(1/\tau)$

(v)   for all $\tau \geq 0$ there exists $T_\tau \in \mathbb{T}^+$ such that for all $t \in [0, T_\tau] \cap \mathbb{T}$ the inequality

$$d_{\min}(\Phi(t, C_\tau), C_\tau) \geq t\gamma^{-1}(d_H(C_\tau, A))$$

is satisfied

□

Condition (v) implies α-forward invariance for each $\alpha = \gamma^{-1}(d_H(C_\tau, A))$, conversely by Lemma 3.7.1 (iii) α-forward invariance for $\alpha = \gamma^{-1}(d_H(C_\tau, A))$ implies condition (v) with $(1 - \varepsilon)\gamma^{-1}(d_H(C_\tau, A))$ for all $\varepsilon > 0$ for continuous time systems, and also for $\varepsilon = 0$ for discrete time systems.

The family C_τ bears several similarities to the α-contracting family B_α. Besides minor technical issues (e.g. B is supposed to be bounded, the C_τ are parameterized by time) the main difference between this concept and an α-contracting family lies in the fact that here contraction and robustness are decoupled, i.e. the contraction condition (iii) is only required for the unperturbed system while the robustness enters in form of the forward invariance condition (v).

We will now analyze how (γ, ρ)-attraction is related to γ-robust attraction and ISDS. The following propositions show that this concept lies in between these two concepts, i.e. for the same class $\mathcal{K}_\infty$ function γ and all $K > 1$ we obtain the implications

$$\text{ISDS with gain } \gamma \Rightarrow (\gamma(K \cdot), \rho)\text{-attraction}$$
$$(\gamma, \rho)\text{-attraction} \Rightarrow \gamma\text{-robustness}.$$

We prove these implications for continuous time systems, similar arguments show the same results for discrete time systems, cf. also [54].

Proposition 3.7.8 Assume that a set A admits an α-contracting family of neighborhoods B_α with some $\vartheta(\alpha, t)$ satisfying $d_H(B_\alpha, A) \leq \gamma(\alpha)$. Then for each open set B with $B \subset B_{\alpha_0}$ for some $\alpha_0 > 0$ and each constant $K > 1$ the set A is a $(\gamma(K \cdot), \rho(\cdot))$-attracting set with attracted neighborhood B and $\rho(1/\tau) = \gamma(\vartheta(\alpha_0, \tau))$.

In particular, if A is ISDS with rate μ and gain γ then for each $K > 1$ the set A is a $(\gamma(K \cdot), \rho(\cdot))$-attracting set with attracted neighborhood B and $\rho(1/\tau) = \mu(\gamma(\alpha_0), \tau)$.

Proof: Fix some α_0 and consider the sets $C_\tau := B_{\vartheta(\tau, \alpha_0)}$. Then the properties (i)–(iv) follow immediately, and (v) follows from Lemma 3.7.1 (iii). The assertion for ISDS sets follows easily from Proposition 3.3.3. □

Proposition 3.7.9 Assume that a set A is (γ, ρ)-attracting. Then A is a γ-robust attracting set for the α_0-inflated system with $\alpha_0 = \gamma^{-1}(d_H(C_0, A))$.

Proof: Fix some $\alpha \leq \gamma^{-1}(d_H(C_0, A))$. Let $\tau_0 > 0$ be minimal with $d_H(C_{\tau_0}, A) \leq \gamma(\alpha)$. We show that if $\delta\gamma^{-1}(d_H(C_\tau, A)) \geq \alpha$ for some $\delta \in (0, 1)$ then there exists $t > 0$ such that

$$\varphi_\alpha(t, C_\tau) \subset C_{\tau + (1-\delta)t/2}. \tag{3.14}$$

Since (ii) and (iii) imply that $\lim_{\tau \to \tau_0} d_H(C_\tau, C_{\tau_0}) = 0$, from (3.14) we obtain

$$\lim_{t \to \infty} d_H(\varphi_\alpha(t, C_\tau), C_{\tau_0}) = 0$$

which yields the assertion.

In order to show (3.14) we set $D = \sqrt{(1+\delta)/(2\delta)} > 1$ and observe that by Lemma 3.7.1 (i) and (iii) we obtain $\varphi_{\alpha'}(t_1, \varphi(t_2, C)) \supset \varphi_\alpha(t_1 + t_2, C)$ for all t_1,

t_2 sufficiently small and all $\alpha' \geq D\alpha(t_1+t_2)/t_1$. Now observe that from the assumption on the C_τ (recall that Property (v) implies $\gamma^{-1}(d_H(C_\tau, A))$-forward invariance of each C_τ) we obtain that $\varphi_{\alpha'}(t_1, \varphi(t_2, C_\tau)) \subset C_{\tau+t_2}$ if $\alpha' \leq \gamma^{-1}(d_H(C_{\tau+t_2}, A))$. For t_2 sufficiently small we obtain $D\gamma^{-1}(d_H(C_{\tau+t_2}, A)) \geq \gamma^{-1}(d_H(C_\tau, A))$. Hence for all t_1 and t_2 sufficiently small this yields

$$\varphi_{\tilde{\alpha}}(t_1 + t_2, C) \subset C_{\tau+t_2}$$

for all $\tilde{\alpha} \leq t_1 \gamma^{-1}(d_H(C_\tau, A))/(D^2(t_1 + t_2))$. Setting $t_2 = (1 - \delta)t/2$ and $t_1 = t - t_2 = (1 + \delta)t/2$ hence yields this inclusion for all $\tilde{\alpha} \leq (1 + \delta)\gamma^{-1}(d_H(C_\tau, A))/(2D^2)$. By choice of D the α from (3.14) satisfies this inequality, hence we obtain (3.14).   □

The main difference between γ-robustness and (γ, ρ)-attraction is condition (iv), which was introduced in [54] in order to ensure (γ, ρ)-attraction for a limiting set of a sequence of (γ, ρ)-attracting sets. Here we have already seen that both γ-robustness and ISDS enjoy this "stability for limits" without any additional requirements, cf. the Theorems 3.7.2 and 3.6.2. Furthermore, if we are only interested in how much an attracting set may "blow up" for an inflated system then γ-robustness is a more natural concept; if we also want to know about rates of convergence, then ISDS fits better into the classical stability theory, provides more information (it includes bounds on the overshoot and allows estimates for fading perturbations) and allows nice characterizations (like, e.g., the Lyapunov function characterization). Hence, in what follows we will use these two concepts instead of (γ, ρ)-attraction.

3.8 Discrete and Continuous Time Systems

So far we have developed all results for either discrete or continuous time systems. In particular, the stability results in the two preceding sections are formulated when either *all* of the systems under consideration are continuous time systems or *all* are discrete time systems.

In numerical approximations, however, when starting from some continuous time system, we will in general have to use an approximating discrete time system, i.e., the numerical approximation approximates the time-h map rather than the original continuous time system. Any numerical information obtained from this approximation then must be retranslated to the continuous time system. This is the reason why we have to investigate the relation between the robustness behavior of the continuous time system φ and its time-h map φ^h defined by (2.19), which will be done in this last section of this chapter.

Proposition 3.8.1 Consider a continuous time system (2.5) satisfying (2.8) and its time-h map φ^h for some $h > 0$. Let $\alpha > 0$ and let $A_{\alpha,h}$ be a strongly

α-attracting set for the time-h map with attracted neighborhood B satisfying $d_{\min}(A_{\alpha,h}, B) > (M + \rho(\alpha))h$. Then there exists a strongly α-attracting set A_α for the continuous time system with $d_H(A_\alpha, A_{\alpha,h}) \leq (M + \rho(\alpha))h$.

Proof: Since $A_{\alpha,h}$ is α-forward invariant for φ^h we obtain $\mathrm{cl}\, \varphi_\alpha(h, A_{\alpha,h}) \in A_{\alpha,h}$. We set

$$A_\alpha := \bigcup_{t \in [0,h]} \mathrm{cl}\, \varphi_\alpha(t, A_{\alpha,h}).$$

Then the α-forward invariance and the distance estimate $d_H(A_\alpha, A_{\alpha,h}) \leq (M + \rho(\alpha))h$ is immediate, implying in particular $A \subset B$. In order to see the α-attraction, observe that by continuous dependence on the initial value for each $\rho > 0$ there exists $\varepsilon > 0$ such that

$$\mathrm{dist}(\varphi_\alpha(t, \mathcal{B}(\varepsilon, A_\alpha)), A_\alpha) \leq \rho \text{ for all } t \in [0, h].$$

Hence from

$$\mathrm{dist}(\varphi_\alpha(ih, B), A_{\alpha,h}) \to 0 \text{ as } i \to \infty$$

we obtain the desired attraction property. $\square$

We can prove a similar result for γ-robust attracting sets.

Proposition 3.8.2 Consider a continuous time system (2.5) satisfying (2.8) and its time-h map φ^h for some $h > 0$. Let $\alpha > 0$ and let A_h be a γ-robust strongly attracting set for the time-h map with attracted neighborhood B satisfying $d_{\min}(A_h, B) > Mh$. Let $\alpha_0 > 0$ be such that $d_{\min}(A_h, B) > (M + \rho(\alpha_0))h$. Then there exists a γ-robust strongly asymptotically stable set A for the continuous time system φ with perturbations from W_{α_0}, satisfying

$$d_H(A, A_h) \leq Mh$$

and

$$\gamma(r) = e^{Lh}\gamma(r) + (e^{Lh} - 1)\rho(r)/L.$$

Proof: Let $A_{\alpha,h}$ be the α-attracting sets for the discrete time system satisfying the distance estimate $d_H(A_{\alpha,h}, A_h) \leq \gamma(\alpha)$. We set

$$A_\alpha = \bigcup_{t \in [0,h]} \mathrm{cl}\, \varphi_\alpha(t, A_{\alpha,h}).$$

By Proposition 3.8.1 we obtain that each A_α is α-attracting and the desired distance follows easily from Gronwall's Lemma. $\square$

We can state the analogous results of Proposition 3.6.2 and 3.6.4 for the case that the approximating systems are time-h_n maps for $h_n \to 0$.

Proposition 3.8.3 Consider a continuous time system of type (2.5) satisfying (2.8), a sequence of time steps $h_n \to 0$ and the time h_n-maps φ^{h_n}. Consider furthermore closed and c-bounded sets A and A_n such that $d_H(A, A_n) \to 0$ as $n \to \infty$ and an open set B with $A_n \subset B$, $A \subset B$. Then the following properties hold

(i) If $\beta_n : (0, \infty) \times (0, \infty) \to [0, \infty)$ are functions which are asymptotically bounded by some class $\mathcal{KL}$ function β and for each $T > 0$ there exists $N \in \mathbb{N}$ with

$$\|\varphi^{h_n}(t, x, u)\|_{A_n} \leq \beta_n(\|x\|_{A_n}, t)$$

for all $x \in B$, $u \in \mathcal{U}$, all $t \in h_n\mathbb{Z} \cap [0, T]$ and all $n \geq N$ then A is a strongly attracting set for the continuous time system φ with attraction rate β.

(ii) If γ_n are functions which are asymptotically bounded by some class $\mathcal{K}_\infty$ function γ and each A_n is a γ_n-robust attracting set for φ^{h_n} then A is a γ-robust attracting set for the continuous time system φ.

(iii) If μ_n, γ_n and σ_n are functions which are (ISDS–) asymptotically bounded by μ, ρ and σ and for each $T > 0$ there exists $N \in \mathbb{N}$ with

$$\|\varphi^{h_n}(t, x, u, w)\|_{A_n} \leq \max\{\mu_n(\sigma_n(\|x\|_{A_n}), t),\, \nu_{h_n, n}(w, t)\}$$

for all $x \in B$, $u \in \mathcal{U}$, $w \in \mathcal{W}$ and all $t \in h_n\mathbb{Z} \cap [0, T]$ then A is an ISDS set for the continuous time system φ with rate μ and gains σ and γ.

Proof: (i) This follows easily from the fact that

$$\|\varphi(t, x, u, 0) - \varphi^{h_n}(ih_n, x, u, 0)\| \leq Mh_n \text{ for all } t \in [ih_n, (i+1)h_n].$$

(ii) From Proposition 3.8.2 we obtain the existence of a sequence of $\tilde\gamma_n$-robust attracting sets $\tilde A_n$ for φ with $d_H(\tilde A_n, A_n) \leq Mh_n$ and $\tilde\gamma_n(r) = e^{Lh_n}\gamma_n(r) + (e^{Lh_n} - 1)\rho(r)/L$. Clearly, the $\tilde\gamma_n$ are asymptotically bounded by γ and the $\tilde A_n$ converge to A. Hence the assertion follows by Proposition 3.6.4.

(iii) Fix $t > 0$, $x \in B$, $u \in \mathcal{U}$ and $w \in \mathcal{W}$ and define the value

$$C := \operatorname{ess\,sup}_{\tau \in [0,t]} \|f(\varphi(t, x, u, w), u(t), w(t))\| < \infty.$$

Then we obtain

$$\|\varphi(t, x, u, w) - \varphi^{h_n}(ih_n, x, u, w)\| \leq Ch_n \text{ for all } t \in [ih_n, (i+1)h_n].$$

Now fix some $t > 0$. Then we find a sequence $t_n \to t$ such that $t_n = i_n h_n$ for some $i_n \in \mathbb{N}$ and $t_n \leq t \leq t_n + h_n$. Hence we obtain the estimate

$$\|\varphi(t, x, u, w)\|_A \leq \|\varphi^{h_n}(t_n, x, u, w)\|_{A_n} + \varepsilon_n$$
$$\leq \max\{\mu(\sigma(\|x\|_{A_n}), t_n), \nu_{h_n}(w, t_n)\} + 2\varepsilon_n$$

for some suitable sequence $\varepsilon_n \to 0$. Now the desired estimate follows from the continuity of μ and Lemma 3.2.8 (iii) observing that $\mu(\nu_0(w, t), -h) \geq \nu_h(w, t)$, which is immediate from (3.1).   □

We end this section by investigating a peculiarity in the definition of inflated systems. In general, the time-h map $\varphi^h(x, u, w)$ corresponding to the solution $\varphi(t, x, u, w)$ of the inflated system

$$\dot{x} = f^0(x, u) + w \tag{3.15}$$

does not coincide with the (discrete time) inflated system

$$\Phi_h(x, u, w) := \varphi^h(x, u, 0) + \int_0^h w(t)dt \tag{3.16}$$

based on the time-h map $\varphi^h(x, u, 0)$ corresponding to the solution $\varphi(t, x, u, 0)$ of the unperturbed system (3.15) with $w = 0$.

Since it is desirable in the analysis of numerical approximation to be able to jump between these two system, in the following Lemma we show their relation. For simplicity we set $W = \mathbb{R}^n$ here, the restriction to the case $W = \mathcal{B}(\alpha_0, 0) \subset \mathbb{R}^n$ is straightforward.

Lemma 3.8.4 Let $h > 0$ and consider the discrete time systems φ^h and Φ_h from (3.15) and (3.16). Assume that the continuous time system (3.15) satisfies (2.8). Then for each $w \in \mathcal{W}$, each $x \in \mathbb{R}^n$ and each $u \in \mathcal{U}$ there exists $\tilde{w} \in \mathcal{W}$ with $\|\tilde{w}(t)\| \leq \|w(t)\| + hL\|w\|_{[ih,t]}$ for almost all $t \in [ih, (i+1)h]$ and all $i \in \mathbb{N}_0$ such that

$$\varphi^h(t, x, u, \tilde{w}) = \Phi_h(t, x, u, w) \text{ for all } t \in \mathbb{T}^+. \tag{3.17}$$

Conversely, for each $\tilde{w} \in \mathcal{W}$, each $x \in \mathbb{R}^n$ and each $u \in \mathcal{U}$ there exists $w \in \mathcal{W}$ with $\|w(t)\| \leq \|\tilde{w}(t)\| + (e^{Lh} - 1)\|\tilde{w}\|_{[ih,t]}$ for almost all $t \in [ih, (i+1)h]$ such that (3.17) holds.

Proof: It is sufficient to show (3.17) for $t = h$, since from that we obtain the assertion for arbitrary $t \in \mathbb{T}^+$ by a simple induction.

Let $x \in \mathbb{R}^n$, $u \in \mathcal{U}$ and $w \in \mathcal{W}$. For $t \in [0, h]$ we set

$$\tilde{w}(t) = f^0(\varphi(t, x, u, 0), u(t)) + w(t) - f^0\left(\varphi(t, x, u, 0) + \int_0^t w(\tau)d\tau, u(t)\right).$$

Then we obtain

$$\frac{d}{dt}\left(\varphi(t, x, u, 0) + \int_0^t w(\tau)d\tau\right) = f^0(\varphi(t, x, u, 0), u(t)) + w(t)$$

$$= f^0\left(\varphi(t, x, u, 0) + \int_0^t w(\tau)d\tau, u(t)\right) + \tilde{w}(t)$$

and

$$\frac{d}{dt}\varphi(t,x,u,\tilde{w}) = f^0(\varphi(t,x,u,\tilde{w}),\, u(t)) + \tilde{w}(t),$$

which by the uniqueness of the solution to this differential equation implies

$$\varphi^h(x,u,\tilde{w}) = \varphi(h,x,u,\tilde{w}) = \varphi(h,x,u,0) + \int_0^h w(\tau)d\tau = \Phi_h(x,u,w).$$

The estimate on the bound follows easily from the Lipschitz estimate on f^0.

Conversely, let $x \in \mathbb{R}^n$, $u \in \mathcal{U}$ and $\tilde{w} \in \mathcal{W}$. Setting

$$w(t) = f^0(\varphi(t,x,u,\tilde{w}),u(t)) + \tilde{w}(t) - f^0(\varphi(t,x,u,0),u(t))$$

similar arguments as above yield the assertion using Lemma 3.7.1(i) to obtain the estimate $\|\varphi(t,x,u,\tilde{w}) - \varphi(t,x,u,0)\| \le \|\tilde{w}\|_{[0,t]}(e^{Lt} - 1)/L$. □

The following corollary is immediate from this lemma.

Corollary 3.8.5 Let $h > 0$ and consider the discrete time systems φ^h and Φ_h from (3.15) and (3.16). Assume that the continuous time system (3.15) satisfies (2.8). Then, if A is a strongly attracting set for φ^h which is γ-robust or ISDS for some robustness gain γ of class $\mathcal{K}_\infty$ then A is also a γ-robust or ISDS strongly attracting set for Φ_h with robustness gain $\gamma(e^{Lh} \cdot)$.

Conversely, if A is a strongly attracting set for Φ_h which is γ-robust or ISDS for some robustness gain γ of class $\mathcal{K}_\infty$ then A is also a γ-robust or ISDS strongly attracting set for φ^h with robustness gain $\gamma((1 + Lh) \cdot)$.

Proof: Immediate from Lemma 3.8.4 and the definitions of γ-robustness and ISDS for discrete time systems. □

Remark 3.8.6 In fact, this relation between the two different types of inflation is the main reason to define ISDS for discrete time systems using ν_h instead of ν_0. All other results in this chapter concerning the ISDS property for discrete time systems hold (with the obvious modifications) also when ISDS is defined using ν_0. □

4 Weakly Attracting Sets

In this chapter we will investigate discrete and continuous time systems of type (2.6) and (2.22), along with their unperturbed counterparts (2.1) and (2.17). Here $u \in \mathcal{U}$ will always be interpreted as a control function, i.e., we are going to investigate weak dynamical properties.

As far as the structural differences between weak and strong attractivity, stability and robustness concepts allow, we will develop the results in parallel to Chapter 3. This means that again we will start by defining a number of dynamical properties of sets like forward invariance, attraction, asymptotic stability, but now in the weak sense, and then turn to the formulation of several different robustness concepts

As in the strong case we will give characterizations of these robustness properties by geometrical methods and by Lyapunov functions, investigate the relation between these properties and investigate the stability of these robustness properties under limits and additional perturbations.

Similar to the previous chapter, Φ, $\Phi_{\mathcal{U}}$ and Φ_α denote either solutions of the continuous time system (2.6) or of the discrete time system (2.22), with $\mathbb{T} = \mathbb{R}$ or $\mathbb{T} = h\mathbb{Z}$, respectively.

4.1 Weak Attraction

It will turn out that when considering continuous time systems, for our applications it is convenient and sufficient to ask for certain estimates or properties only at a countable number of positive times t_i with $t_i \to \infty$ rather than for all times $t \geq 0$. We will use the following definition. In order to ensure a consistent terminology we formulate it for both discrete and continuous time systems although it is somewhat redundant for discrete time systems.

Definition 4.1.1 (weak inequality and inclusion)

Let $\mathcal{T}$ denote the set of strictly increasing positive real sequences, i.e.

$$\mathcal{T} := \{\mathbf{t} = (t_i)_{i \in \mathbb{N}_0} \,|\, t_i \in \mathbb{T} \text{ for all } i \in \mathbb{N}_0, \, 0 = t_0 < t_1 < t_2 < \ldots\}.$$

Then for functions $N : \mathbb{R}^n \to \mathbb{R}, \rho : \mathbb{R} \to \mathbb{R}$ and $x \in \mathbb{R}^n$, $p \in \mathcal{P}$ and $T > 0$ we say that the system (2.6) or (2.22) satisfies

$$N(\Phi(t, x, u, p[u])) \leq \rho(t) \text{ weakly for all } t \geq T$$

if for each $\mathbf{t} = (t_i)_{i \in \mathbb{N}_0} \in \mathcal{T}$ there exists a $u^{\mathbf{t}} \in \mathcal{U}$ such that

$$N(\Phi(t_i, x, u^{\mathbf{t}}, p[u^{\mathbf{t}}])) \leq \rho(t_i) \text{ for all } i \in \mathbb{N}_0 \text{ with } t_i \geq T,$$

where the $\mathbf{t}$–dependence of the $u^{\mathbf{t}}$ is such that

> for all $\mathbf{t}^1 = (t^1{}_i)_{i \in \mathbb{N}_0}$, $\mathbf{t}^2 = (t^2{}_i)_{i \in \mathbb{N}_0} \in \mathcal{T}$ with $t^1{}_i = t^2{}_i$;
> for all $i = 0, \ldots, N$ the control functions $u^{\mathbf{t}_1}$ and $u^{\mathbf{t}_2}$ (4.1)
> satisfy $u^{\mathbf{t}_1}(t) = u^{\mathbf{t}_2}(t)$ for almost all $t \in [0, t^1{}_N]$.

Similarly, for a family of sets $B_t \subset \mathbb{R}^n$, $t \in \mathbb{T}^+$, we say that for some $x \in \mathbb{R}^n$, $p \in \mathcal{P}$ and $T > 0$ that the system satisfies

$$\Phi(t, x, u, p[u]) \in B_t \text{ weakly for all } t \geq T$$

if for each $\mathbf{t} = (t_i)_{i \in \mathbb{N}_0} \in \mathcal{T}$ there exists a $u^{\mathbf{t}} \in \mathcal{U}$ such that

$$\Phi(t_i, x, u^{\mathbf{t}}, p[u^{\mathbf{t}}]) \in B_{t_i} \text{ for all } i \in \mathbb{N}_0 \text{ with } t_i \geq T,$$

where the $\mathbf{t}$–dependence of the $u^{\mathbf{t}}$ again is such that (4.1) holds.   □

Note that if a continuous time system φ satisfies the weak inclusion

$$\varphi(t, x, u, p[u]) \in B \text{ weakly for all } t \geq T$$

for some fixed set $B \subset \mathbb{R}^n$ and the vector field is uniformly bounded in some neighborhood of B, then by setting $t_i = \sum_{k=1}^{i} 1/k$ we easily obtain the existence of a $u \in \mathcal{U}$ such that

$$\limsup_{t \to \infty} \|\Phi(t, x, u, p[u])\|_B = 0.$$

Definition 4.1.2 (weak forward invariance)

(i) A set $A \subset \mathbb{R}^n$ is called weakly forward invariant if for all $x \in A$ it satisfies

$$\Phi(t, x, u, 0) \in A \text{ weakly for all } t \geq 0.$$

(ii) Consider a set of perturbation strategies $\mathcal{P}$ and let $\alpha > 0$. A set $A_\alpha \subset \mathbb{R}^n$ is called weakly α-forward invariant if for all $x \in A_\alpha$ and all $p \in \mathcal{P}_\alpha$ it satisfies

$$\varphi(t, x, u, p[u]) \in A_\alpha \text{ weakly for all } t \geq 0.$$

□

Weak forward invariance is also known under the names of *controlled forward invariance* or *viability*, cf. [5].

Let us now define weakly attracting sets. Since here attraction depends on the choice of u the formulation looks different than in the strong case, however, it is the natural adaptation.

Definition 4.1.3 (weak attraction)

(i) A c-bounded set $A \subset \mathbb{R}^n$ is called called weakly attracting with open attracted neighborhood B if it is weakly forward invariant and there exists a class $\mathcal{KL}$ function β and a constant $C \geq 0$ such that for each $x \in B$ the inequality

$$\|\Phi(t, x, u, 0)\|_A \leq \beta(\|x\|_A + C, t) \text{ weakly for all } t \geq 0$$

holds.

(ii) Let $\alpha > 0$. A c-bounded set $A_\alpha \subset \mathbb{R}^n$ is called called weakly α-attracting with open attracted neighborhood B if it is weakly α-forward invariant and there exists a class $\mathcal{KL}$ function β and a constant $C \geq 0$ such that for each $x \in B$ and each $p \in \mathcal{P}_\alpha$ the inequality

$$\|\Phi(t, x, u, p[u])\|_{A_\alpha} \leq \beta(\|x\|_{A_\alpha} + C, t) \text{ weakly for all } t \geq 0$$

holds.   □

Weak asymptotic stability in addition demands that the trajectories remain close to A before eventually converging.

Definition 4.1.4 (weak asymptotic stability)

(i) A closed c-bounded set $A \subset \mathbb{R}^n$ is called called weakly asymptotically stable with open attracted neighborhood B if it is weakly forward invariant and there exists a class $\mathcal{KL}$ function β such that for each $x \in B$ the inequality

$$\|\Phi(t, x, u, 0)\|_A \leq \beta(\|x\|_A, t) \text{ weakly for all } t \geq 0$$

holds.

(ii) Let $\alpha > 0$. A closed c-bounded set $A_\alpha \subset \mathbb{R}^n$ is called called weakly α-asymptotically stable with open attracted neighborhood B if it is weakly α-forward invariant and there exists a class $\mathcal{KL}$ function β such that for each $x \in B$ and each $p \in \mathcal{P}_\alpha$ the inequality

$$\|\Phi(t, x, u, p[u])\|_{A_\alpha} \leq \beta(\|x\|_{A_\alpha}, t) \text{ weakly for all } t \geq 0$$

holds.   □

Unlike the strong concept, a weakly attracting set is not automatically weakly asymptotically stable, as the following example illustrates.

Example 4.1.5 Consider the two dimensional system

$$\dot{x} = \rho(y)xu_1$$
$$\dot{y} = -u_2 y + (u_2 - 1)(y - 3)$$

for $U = [-1, 1] \times [0, 1]$, where ρ is some monotone increasing Lipschitz continuous function with $\rho(y) = 0$, $y \in (-\infty, 1]$ and $\rho(y) = 1$, $y \in [2, \infty)$.

We claim that the set $A = \{0\}$ is weakly attracting (with $B = \mathbb{R}^2$) but not weakly asymptotically stable. Consider any point $(x, y)^T \in \mathbb{R}^2$. If $y < 2$ we can first use the constant control $u^1 \equiv (0, 0)$ to control the system to $(x, 2)$. From this point we may then use the constant control $u^2 \equiv (1, 0)$ or $u^2 \equiv (-1, 0)$ to steer the trajectory to some point $(0, y_1)$ with $y_1 \in [2, 3]$. Finally, the control $u^3 \equiv (0, 1)$ steers the trajectory to the origin exponentially fast. For initial values $(x, y)^T \in \mathbb{R}^2$ with $y \geq 2$ we only use u^2 and u^3. It is easily seen from the construction (cf. Remark B.1.5 in Appendix B), that the distance of the trajectory from A can be bounded by a term of the form $\beta(\|\|(x, y)\|\| + C, t)$ for some suitable class $\mathcal{KL}$ function and some suitable $C > 0$. Hence A is weakly attracting.

On the other hand, any point $x = (x, y)^T \in \mathbb{R}^2$ with $x \neq 0$, has to be controlled to some point (x_1, y_1) with $y_1 > 1$ before we can steer the x-component to 0. This contradicts the weak asymptotic stability.   □

Let us give criteria for the existence of weakly α-attracting and α-asymptotically stable sets.

Proposition 4.1.6 Consider a closed set $C \subset \mathbb{R}^n$ and an open set $B \subseteq \mathbb{R}^n$ with $C \subset B$. Assume that for each $r > 0$ there exists a time $T(r) < \infty$ such that for each $x \in B$, with $\|x\|_C \leq r$ and each $p \in \mathcal{P}_\alpha$ the inclusion $\Phi(t, x, u, p[u]) \in C$ holds weakly for all $t \geq T(r)$. Then there exists a weakly α-attracting set $A_\alpha \subseteq C$ with attracted neighborhood B. If, in addition, C is weakly α-forward invariant then we can choose $A_\alpha = C$.

Proof: Consider the set

$$A_\alpha := \{x \in C \,|\, \forall p \in \mathcal{P}_\alpha : \Phi(t, x, u, p[u]) \in C \text{ weakly for all } t \in \mathbb{T}^+\}.$$

Note that if C is weakly α-forward invariant then we obtain $A_\alpha = C$. We set

$$\tilde{\beta}(r, t) := \sup_{\|x\|_{A_\alpha} \leq r} \sup_{p \in \mathcal{P}_\alpha} \sup_{t \in \mathcal{T}} \inf_{u \in \mathcal{U}} \sup_{t_i \geq t} \|\Phi(t_i, x, u, p[u])\|_{A_\alpha}.$$

Obviously, $\tilde{\beta}(r, t)$ is monotone increasing in r for each t and $\tilde{\beta}(r, t) = 0$ for each $r > 0$ and $t \geq T(r)$. Hence fixing some $C > 0$ we find a class $\mathcal{KL}$ function β with

$$\beta(r + C, t) \geq \tilde{\beta}(r, t)$$

which shows the weak α-attractivity of A_α. $\qquad\qquad\qquad\qquad$ □

Proposition 4.1.7 A closed c-bounded set A is weakly α-asymptotically stable with attracted neighborhood B if there exists $\varepsilon_0 > 0$, maps T_1 : $[\varepsilon_0, \infty) \to \mathbb{T}$, $T_2 : (0, \varepsilon_0] \to \mathbb{T}$, and a continuous map $\delta : [0, \varepsilon_0] \to \mathbb{R}_0^+$ with $\delta(0) = 0$, such that for all $x \in B$ with $r := \|x\|_A \geq \varepsilon_0$ and all $p \in \mathcal{P}_\alpha$ there exists a $u' \in \mathcal{U}$ such that

$$\|\Phi(T_1(r), x, u', p[u'])\|_A < \varepsilon_0$$

and for all $x \in B$ with $r := \|x\|_A \leq \varepsilon_0$, all $p \in \mathcal{P}_\alpha$ and all $(t_i)_{i \in \mathbb{N}_0} \in \mathcal{T}$ there exists a $u^* \in \mathcal{U}$ such that

$$\|\Phi(T_2(r), x, u^*, p[u^*])\|_A < r/2$$

and

$$\|\Phi(t_i, x, u^*, p[u^*])\|_A < \delta(r) \text{ for all } i \in \mathbb{N}.$$

Proof: For $r \leq \varepsilon_0$ we set $\tau_0(r) = 0$, $\tau_k(r) = \sum_{i=1}^{k} T_2(r/2^{i-1})$ and define $\tilde{\beta}(r, t) := \delta(r/(2^{k-1}))$, $t \in [\tau_{k-1}, \tau_k]$. For $r > \varepsilon_0$ we set

$$\tilde{\beta}(r, t) = \sup_{t \in [0, T_1(r)] \cap \mathbb{T}, \, x \in \mathcal{B}(r, A) \setminus A} \text{dist}(\Phi_\alpha(t, x), A) < \infty \text{ for all } t \in [0, T_1(r))$$

and $\tilde{\beta}(r, t) = \tilde{\beta}(\varepsilon_0, t - T_1(r))$ for $t \geq T_1(r)$. Then it follows from the assumption that for all $x \in B$ and all $p \in \mathcal{P}_\alpha$ the inequality

$$\|\Phi(t, x, u, p[u])\|_A \leq \tilde{\beta}(\|x\|_A, t)$$

is weakly satisfied. Since we easily find a class $\mathcal{KL}$ function β majorizing $\tilde{\beta}$ this shows the claim. $\qquad\qquad\qquad\qquad$ □

Proposition 4.1.8 Let A_α be a weakly α-attracting set satisfying the following condition: There exists $\varepsilon > 0$ and $C > 0$ such that for all $x \in \mathbb{R}^n$ with $\text{dist}(x, A_\alpha) \leq \varepsilon$ and all $p \in \mathcal{P}_\alpha$ there exists $u^* \in \mathcal{U}$ and $t_x \in \mathbb{T}_0^+$ with $t_x \leq C\text{dist}(x, A_\alpha)$ such that

$$\varphi(t_x, x, u^*, p[u^*]) \in A_\alpha$$

for continuous time systems or

$$\Phi_h(\max\{t_x, h\}, x, u^*, p[u^*]) \in A_\alpha$$

for discrete time systems. Then A_α is weakly α-asymptotically stable. In addition, the class $\mathcal{KL}$ function β characterizing weak α-asymptotic stability can be chosen such that $\beta(r, t) \leq \tilde{\beta}(r + \tilde{C}, t)$ for all $r \geq 0$, where $\tilde{\beta}$ and $\tilde{C}$ characterize the α-attraction of A_α.

Proof: We show the statement for continuous time systems, for discrete time systems it follows similarly. Let $\tilde\beta$ and $\tilde C$ characterize the weak attraction of A. Let M be a bound on $\|f\|$ on the (compact) set $\mathrm{cl}\,\bigcup_{t\in[0,C\varepsilon]}\varphi_\alpha(\mathcal{B}(\varepsilon,\partial A),t)$. Using the weak forward invariance of A_α, for each x as in the assumption, each sequence $\mathbf{t}=(t_i)_{i\in\mathbb{N}_0}\in\mathcal{T}$ and each $p\in\mathcal{P}_\alpha$ we can extend the u^* from the assumption to some u' such that $\varphi(t_i,x,u',p[u'])\in A$ for all $i\in\mathbb{N}$ with $t_i\geq t_x$. Hence we obtain

$$\|\varphi(t_i,x,u',p[u'])\|_A\leq Mt_x\leq MC\|x\|_A \text{ for all } t_i\in[0,t_x)$$

and

$$\|\varphi(t_i,x,u',p[u'])\|=0 \text{ for all } t_i\geq t_x.$$

Hence for all $\delta\in(0,\varepsilon]$, any class $\mathcal{KL}$ function β with $\beta(r,t)\geq\tilde\beta(r+\tilde C,t)$, $r\geq\delta$ and $\beta(r,t)\geq MCr$, $r\leq\delta$, $t\leq MC\mathrm{dist}(x,A)$ satisfies the desired estimate. In particular, we can choose $\tilde\beta(r,t)\leq\tilde\beta(r+\tilde C,t)$ if $\delta>0$ is so small that $MC\delta<\tilde\beta(\delta+\tilde C,MC\delta)$.   □

We can drop the assumption of A being weakly α-attracting if we do not want an upper bound for β and assume $\mathrm{dist}(B,A)<\infty$.

Proposition 4.1.9 Let A_α be a weakly α-forward invariant set. Consider some open set B containing A_α with $\mathrm{dist}(B,A_\alpha)<\infty$ and assume that there exists $\varepsilon>0$ and $T>0$ such that for all $x\in B$ and all $p\in\mathcal{P}_\alpha$ there exists a $u^*\in\mathcal{U}$ such that $\Phi(T_{x,p},x,u^*,p[u^*])\in\mathcal{B}(\varepsilon,A_\alpha)$ for some $T_{x,p}\in[0,T]$. Assume furthermore that there exists $C>0$ such that for all $x\in\mathbb{R}^n$ with $\mathrm{dist}(x,A_\alpha)\leq\varepsilon$ and all $p\in\mathcal{P}_\alpha$ there exists a $u'\in\mathcal{U}$ and $t_x\in\mathbb{T}_0^+$ with $t_x\leq C\mathrm{dist}(x,A_\alpha)$ such that

$$\varphi(t_x,x,u',p[u'])\in A_\alpha$$

for continuous time systems and

$$\Phi_h(\max\{t_x,h\},x,u',p[u'])\in A_\alpha$$

for discrete time systems. Then A_α is weakly α-asymptotically stable with attracted neighborhood B.

Proof: Again we show the proof for continuous time systems. As in the proof of Proposition 4.1.8 for all $x\in\mathcal{B}(\varepsilon,A_\alpha)$ and all $p\in\mathcal{P}_\alpha$ we obtain

$$\|\varphi(t,x,u,p[u])\|_A\leq Mt_x\leq MC\|x\|_A \text{ weakly for all } t\leq t_x$$

and

$$\|\varphi(t,x,u,p[u])\|_A=0 \text{ weakly for all } t\geq t_x.$$

Hence for each $x\in B$ and each $p\in\mathcal{P}_\alpha$ we find a control $u_{x,p}$ which steers the trajectory to $\mathcal{B}(\varepsilon,A)$ in a time $T_{x,p}\in[0,T]$ and then satisfies the bounds from above. Thus for $r \leq \varepsilon$ we define $\tilde{\beta}(r,t) = MC\mathrm{dist}(x,A)$, $t \leq t_x$, $\tilde{\beta}(r,t) = 0$, $t \geq t_x$ and for $r > \varepsilon$ we set

$$\tilde{\beta}(r,t) = \max_{x \in B, \|x\|_A \leq r, p \in \mathcal{P}_\alpha, \tau \in [t, T_{x,p}]} \max\{\|\varphi(\tau, x, u_{x,p}, p[u_{x,p}])\|_A, \tilde{\beta}(\varepsilon, \tau - T_{x,p})\}$$

(using the convention $\tilde{\beta}(\varepsilon, t) = \varepsilon$ for $t < 0$). Since this function can be bounded from above by some class $\mathcal{KL}$ function we obtain the assertion.   □

In our robustness and stability analysis we will consider both weakly attracting and weakly asymptotically stable sets. As we will see, the latter have a certain built-in robustness which makes them preferable, cf. Corollary 4.4.4, below.

4.2 Robustness Concepts

Let us now define robustness concepts for weakly invariant, attracting and asymptotically stable sets.

Definition 4.2.1 (γ-robust weak forward invariance)

A weakly forward invariant set C for the system (2.5) is called *directly γ-robust* for some γ of class $\mathcal{K}_\infty$ if for each $\alpha > 0$ there exists a weakly α-forward invariant set C_α with $d_H(C_\alpha, C) \leq \gamma(\alpha)$.

It is called *inversely γ-robust* for some γ of class $\mathcal{K}_\infty$ if for each $\alpha > 0$ there exists a weakly α-forward invariant set C_α with $d_H(C_\alpha^c, C^c) \leq \gamma(\alpha)$.   □

Definition 4.2.2 (γ-robust weak attraction)

Let γ be a function of class $\mathcal{K}_\infty$. A weakly attracting set A is called *γ-robust* with attracted open neighborhood $B \subset \mathbb{R}^n$ with $A \subset B$ if there exist weakly α-attracting sets $A_\alpha \supseteq A$ with attracted neighborhood B for all $\alpha \geq 0$ satisfying $d_H(A_\alpha, A) \leq \gamma(\alpha)$ for each $\alpha > 0$ and $A_{\alpha'} \subseteq A_\alpha$ for all $\alpha \geq \alpha' \geq 0$.   □

Note that here we have explicitly assumed that the A_α are contained in each other, a property which we have obtained for free in the strong case, cf. Remark 3.2.4.

Let us state a sufficient condition for γ-robustness, which can be considered as the weak analogue of Lemma 3.2.3.

Lemma 4.2.3 Consider some γ of class $\mathcal{K}_\infty$, a weakly attracting set A with attracted neighborhood B, and assume that A is γ-robustly weakly forward invariant. Assume furthermore that for each $\alpha > 0$, each $\varepsilon > 0$ and each

$r > 0$ there exists a time $T(r, \alpha, \varepsilon) < \infty$ such that for each $x \in B$, with $\|x\|_A \leq r$ and each $p \in \mathcal{P}_\alpha$ the inequality $\|\Phi(t, x, u, p[u])\|_A \leq \gamma(\alpha) + \varepsilon$ is weakly satisfied for all $t \geq T$. Then for each $c > 1$ the set A is a $c\gamma$-robust weakly attracting set.

Proof: Fix $c > 1$. Then by the assumptions and Proposition 4.1.6 the sets

$$A_\alpha := \{x \in B \mid \|\Phi(t, x, u, p[u])\|_A \leq c\gamma(\alpha) \text{ weakly for all } p \in \mathcal{P}_\alpha, \, t \geq 0\}$$

have the necessary properties.   □

The following example shows the subtle dependence of the γ-robustness from the choice of the strategies $\mathcal{P}$.

Example 4.2.4 Consider the $1d$ system

$$\dot{x}(t) = u(t) + w(t)$$

with $U = [-1, 1]$ and $W = [-1/2, 1/2]$. Clearly, the set $A = \{0\}$ is weakly attracting for the unperturbed system with $B = \mathbb{R}$. Now consider the perturbation strategies $\mathcal{P} = \mathcal{P}^\delta$ for some $\delta > 0$. For some given initial value x_0 we define the strategy $p \in \mathcal{P}_\alpha^\delta$ with $\mathbf{t}(p) = \{0, \delta, 2\delta, \ldots\}$ inductively for $i \in \mathbb{N}_0$ by

$$p[u]|_{[i\delta,(i+1)\delta)} = \begin{cases} -\alpha, & \text{if } \varphi(\delta, \varphi(i\delta, x_0, u, p[u]), u, 0) \leq 0 \\ +\alpha, & \text{if } \varphi(\delta, \varphi(i\delta, x_0, u, p[u]), u, 0) > 0 \end{cases}$$

Then for each $u \in \mathcal{U}$ we obtain

$$|\varphi((i+1)\delta, x, u, p[u])| \geq \delta\alpha.$$

for all $i \in \mathbb{N}_0$, hence the gain γ in the γ-robustness property must satisfy $\gamma(\alpha) \geq \delta\alpha$. On the other hand, for all $x_0 \in \mathbb{R}$ we can use the control defined inductively

$$u|_{[i\delta,(i+1)\delta)} = \begin{cases} -\min\{1, 2x_0/\alpha\}, & \text{if } \varphi(i\delta, x_0, u, p[u]) \geq 0 \\ +\min\{1, 2x_0/\alpha\}, & \text{if } \varphi(i\delta, x_0, u, p[u]) < 0 \end{cases}$$

which guarantees that—regardless of the choice of $p \in \mathcal{P}_\alpha^\delta$—we reach the set $[-\delta\alpha, \delta\alpha]$ and remain inside this set for all future times. Hence $\gamma(\alpha) = \delta\alpha$ is actually the optimal robustness gain, and obviously depends on the choice of δ.

Observe that in particular for $p \in \mathcal{P}_\alpha^0$ and arbitrary $\varepsilon > 0$ we are able to steer each initial value x_0 to the set $A_\varepsilon = [-\varepsilon, \varepsilon]$ and hold it there. Hence for $\delta = 0$ the sets which can be reached are *independent of* α, while for each $\delta > 0$ these sets *depend on* α (although the effect scales down with δ). Thus we do in fact have a change in the qualitative behavior of the perturbed system when passing from $\delta = 0$ to $\delta > 0$.

We will see in Chapter 5 (see in particular Example 5.3.9) that this seemingly undesirable change in the qualitative behavior is in fact exactly what we need to cover different types of numerical perturbations: Some will turn out to be representable by perturbations $p \in \mathcal{P}^0$ (which means that for this example the numerical error does not affect the weak asymptotic stability of $\{0\}$), for others we will need $p \in \mathcal{P}^\delta$ for some $\delta > 0$ (since they are able to destroy the weak asymptotic stability of $\{0\}$ in this example).   □

Due to the fact that weak attraction does not imply weak asymptotic stability, Definition 4.2.2 is too weak for some implications we would like to obtain. Hence we will also use the following stronger one.

Definition 4.2.5 (nested γ-robust weak attraction)

A weakly attracting set A is called *nested γ-robust* if it is γ-robust and, in addition, if there exists a function χ of class $\mathcal{K}$ such that the α-attracting sets A_α satisfy

(i) $d_{\min}(A_\alpha, A) \geq \chi(\alpha)$ for each $\alpha > 0$ with $W_\alpha \neq W$

(ii) for all $\alpha'' > \alpha' > \alpha > 0$ there exists $T = T(\alpha'', \alpha', \alpha) > 0$ such that for all $p \in \mathcal{P}_\alpha$ and all $x \in A_{\alpha''} \setminus A_{\alpha'}$ there exists a $u^* \in \mathcal{U}$ and $\tau \in \mathbb{T}, \tau \leq T$ with $\Phi(t, x, u^*, p[u^*]) \in A_{\alpha''}$ for all $t \in [0, \tau] \cap \mathbb{T}$ and $\Phi(\tau, x, u^*, p[u^*]) \in A_{\alpha'}$.

□

Note that here as well as in Definition 4.2.2 the "approximating sets" A_α are only assumed to be attracting, and not asymptotically stable. Property (i) states that the A_α grow strictly, at least for α close to 0, and Property (ii) essentially states that we do not have to leave $A_{\alpha''}$ in order to steer to $A_{\alpha'}$.

Next we introduce the weak analogue to the ISDS property.

Definition 4.2.6 (weak input-to-state dynamical stability)

A weakly attracting set A is called *weakly input-to-state dynamically stable* (wISDS), if there exists a function μ of class $\mathcal{KLD}$ and functions σ and γ of class $\mathcal{K}_\infty$ such that for all $p \in \mathcal{P}$ and all $x \in B$ it satisfies the inequality

$$\|\Phi(t, x, u, p[u])\|_A \leq \max\{\mu(\sigma(\|x\|_A), t), \nu(p, t)\} \text{ weakly for all } t \geq 0,$$

where $\tilde{\nu} = \tilde{\nu}_0$ for continuous time systems and $\tilde{\nu} = \tilde{\nu}_h$ for discrete time systems, with

$$\tilde{\nu}_0(p, t) := \sup_{u \in \mathcal{U}} \nu_0(p[u], t) \quad \text{and} \quad \tilde{\nu}_h(p, t) := \sup_{u \in \mathcal{U}} \nu_h(p[u], t) \qquad (4.2)$$

for ν_0 and ν_h from (3.1).

Here we call the function μ the *rate of attraction*, the function σ the *overshoot gain* and the function γ the *robustness gain*.   □

Note that this definition implies weak asymptotic stability of A.

The motivation for wISDS is similar to that for ISDS which we have discussed after Definition 3.2.7, and indeed we will be able to prove most of the favorable features of ISDS also for the wISDS property. In this chapter we will not introduce a weak version of the ISS property since γ-robust weak attraction and wISDS will be sufficient for our purposes.

We give some properties of $\tilde{\nu}$.

Lemma 4.2.7 Let $\tilde{\nu} = \tilde{\nu}_0$ or $\tilde{\nu} = \tilde{\nu}_h$. Then the following properties hold.

(i) $\tilde{\nu}(p[\cdot](\tau + \cdot), t) \leq \tilde{\nu}(p, t + \tau)$ for all $t, \tau \in \mathbb{T}_0^+$.

(ii) If $\tilde{\nu}(p, t) \leq \mu(r, t)$ for some $r > 0$, $t \in \mathbb{T}^+$ then $\tilde{\nu}(p, \tau) \leq \mu(r, \tau)$ for all $\tau \in (0, t] \cap \mathbb{T}$.

(iii) $\limsup_{h \searrow 0} \tilde{\nu}_0(p, t - h) \leq \tilde{\nu}_0(p, t)$ and $\liminf_{h \searrow 0} \tilde{\nu}_0(p, t + h) \geq \tilde{\nu}_0(p, t)$.

Proof: These properties follow with the same arguments as the respective properties for ν_0 and ν_h in Lemma 3.2.8.   □

4.3 Geometric Characterizations

Analogously to what we did for ISDS we will now establish a geometric criterion for the wISDS property, again by means of a family of shrinking sets.

Definition 4.3.1 (contractible family of neighborhoods)

Consider a weakly attracting set A with open attracted neighborhood B. Then a family $(B_\alpha)_{\alpha \in \mathbb{R}_0^+}$ of sets $B_\alpha \subset \mathbb{R}^n$, $\alpha \in \mathbb{R}_0^+$ together with a class $\mathcal{KLD}$-function ϑ is called a *contractible family of neighborhoods* for A w.r.t. B if the following properties hold

(i) $B_{\alpha'} \subseteq B_\alpha$ for all $\alpha' < \alpha$

(ii) $B_0 = A$

(iii) for each subset $\tilde{B} \subseteq B$ with $d_H(\tilde{B}, A) < \infty$ there exists an $\alpha^* \in \mathbb{R}_0^+$ with $\tilde{B} \subseteq B_{\alpha^*}$.

(iv) for each $x \in B_\alpha$ and each $t \in \mathbb{T}^+$ there exists a $u^* \in \mathcal{U}$ such that $\Phi(t, x, u^*, 0) \in B_{\vartheta(\alpha, t)}$

The family is called α-*contractible* if (iv) can be sharpened to

(iv') for each $x \in B_\alpha$, each $t \in \mathbb{T}^+$ and all $p \in \mathcal{P}$, which for all $u \in \mathcal{U}$ satisfy $p[u](\tau) \leq \vartheta(\alpha, \tau)$ for almost all $\tau \in [0, t]$ for continuous time systems or $p[u](\tau) \leq \vartheta(\alpha, (i+1)h)$ for almost all $\tau \in [0, t]$ with $\tau \in (ih, (i+1)h]$ for discrete time systems, respectively, there exists $u^* \in \mathcal{U}$ such that $\Phi(t, x, u^*, p[u^*]) \in B_{\vartheta(\alpha, t)}$.

$\square$

The following Lemma shows analogous to Lemma 3.3.2 that we can generalize the α-contractibility to arbitrary w.

Lemma 4.3.2 Consider an α-contractible family of neighborhoods B_α. Then for all $\alpha > 0$, all $x \in B_\alpha$, all $t \in \mathbb{T}^+$ and all $p \in \mathcal{P}$ there exists a $u^* \in \mathcal{U}$ with

$$\Phi(t, x, u^*, p[u^*]) \in B_{\alpha'}$$

for $\alpha' = \max\{\vartheta(\alpha, \tau), \bar{\nu}(p, t)\}$ with $\bar{\nu} = \tilde{\nu}_0$ or $\bar{\nu} = \tilde{\nu}_h$ from (4.2) with $\mu = \vartheta$ and $\gamma = \mathrm{id}_\mathbb{R}$.

Proof: Let $x \in B_\alpha$, $p \in \mathcal{P}$ and $t \in \mathbb{T}^+$. Let $\tilde{\alpha} \geq \alpha$ minimal with $\vartheta(\tilde{\alpha}, t) \geq \bar{\nu}(w, t)$. Since $\tilde{\alpha} \geq \alpha$, property (i) of the α-contracting family yields $x \in B_{\tilde{\alpha}}$. By Lemma 4.2.7(ii) the choice of $\tilde{\alpha}$ implies $\vartheta(\tilde{\alpha}, \tau) \geq \bar{\nu}(p, \tau)$ for all $\tau \in (0, t] \cap \mathbb{T}^+$. This implies that for all $u \in \mathcal{U}$ we have the inequality $p[u](\tau) \leq \vartheta(\tilde{\alpha}, \tau)$ for almost all $\tau \in [0, t]$ for continuous time systems or $p[u](\tau) \leq \vartheta(\tilde{\alpha}, (i+1)h)$ for almost all $\tau \in [0, t]$ with $\tau \in (ih, (i+1)h]$ for discrete time systems, respectively. Thus property (iv') yields the existence of $u^* \in \mathcal{U}$ with

$$\Phi(t, x, u^*, p[u^*]) \subset B_{\vartheta(\tilde{\alpha}, t)}$$

which implies the assertion since the choice of $\tilde{\alpha}$ implies $\alpha' = \vartheta(\tilde{\alpha}, t)$. $\square$

Note that, in contrast to the analogous statement for the ISDS property in Lemma 3.3.2, here we only obtain an inclusion for one fixed time t, because the choice of u depends on t. A simple induction, however, generalizes this result to arbitrary sequences $(t_i)_{i \in \mathbb{N}_0} \in \mathcal{T}$.

Corollary 4.3.3 Consider an α-contractible family of neighborhoods B_α. Then for all $\alpha > 0$, all $x \in B_\alpha$, all $p \in \mathcal{P}$ the system satisfies

$$\Phi(t_i, x, u, p[u]) \in B_{\alpha'(t)} \text{ weakly for all } t \geq 0,$$

where the function $\alpha' : \mathbb{R}_0^+ \to \mathbb{R}_0^+$ is given by $\alpha'(\tau) = \max\{\vartheta(\alpha, \tau), \bar{\nu}(p, \tau)\}$ and $\bar{\nu}$ as in Lemma 4.3.2.

Proof: Consider $p \in \mathcal{P}$ and a sequence of times $(t_i)_{i \in \mathbb{N}_0} \in \mathcal{T}$. We define inductively $x_0 = x$, $x_{i+1} = \Phi(t_{i+1} - t_i, x_i, u_i, p_i[u_i])$, with u_i from Lemma 4.3.2 for $x = x_i$ and $p_i = p[u_1 \&_{t_1} u_2 \&_{t_2} \ldots \&_{t_i} u_i](t_i + \cdot)$. Then by induction we obtain $x_i \in B_{\alpha'_i(\tau)}$ where

$$\alpha'_i \leq \max\{\vartheta(\alpha'_{i-1}, t_i - t_{i-1}), \bar{\nu}(p_i, t_i - t_{i-1})\}.$$

Since $\bar{\nu}(p_i, t_i - t_{i-1}) \leq \bar{\nu}(p, t_i)$ we obtain by induction that

$$\alpha'_i \leq \max\{\vartheta(\alpha, t_i), \bar{\nu}(p, t_i)\},$$

i.e. the assertion. □

We can further improve this statement if the system under consideration satisfies the continuity property (2.14) and the B_α are closed.

Corollary 4.3.4 Consider a system (2.6) or (2.22) satisfying (2.14) and consider an α-contractible family of closed neighborhoods B_α. Then for all $x \in B_\alpha$ and all $p \in \mathcal{P}$ there exists $u^* \in \mathcal{U}$ such that

$$\Phi(t, x, u^*, p[u^*]) \in B_{\alpha'(t)}$$

holds for all $t \in \mathbb{T}^+$, $\alpha'(t) = \max\{\vartheta(\alpha, t), \bar{\nu}(p, t)\}$ and $\bar{\nu}$ as in Lemma 4.3.2.

Proof: For discrete time systems the assertion has already been shown in Corollary 4.3.3, even without assumption (2.14). For continuous time systems for each $j \in \mathbb{N}$ we consider the sequences $t^j = (t_i^j)_{i \in \mathbb{N}_0} \in \mathcal{T}$ with $t_i^j := i 2^{-j}$ for $i \in \mathbb{N}_0$. Fixing some $\alpha > 0$, some $x \in B_\alpha$ and some $p \in \mathcal{P}$ by Corollary 4.3.3 we find control functions $u^j \in \mathcal{U}$ such that

$$\varphi(t_i^j, x, u^j, p[u^j]) \in B_{\alpha'(t_i^j)},$$

where α' is independent of j. Let u^* denote the limiting control from (2.14). We claim that this u^* is the desired control. To this end fix some time $t > 0$ and for each $j \in \mathbb{N}$ denote by t_*^j the maximal time t_i^j with $t_i^j \leq t$. By the construction of the sequences $(t_i^j)_{i \in \mathbb{N}_0}$ we obtain $t_*^j \nearrow t$ and hence Lemma 4.2.7(iii) yields that $\alpha'(t) \geq \limsup_{j \to \infty} \alpha'(t_*^j)$. Observing that for continuous time systems property (iv') implies continuity of the B_α with respect to the Hausdorff distance and using the shrinking property of the B_α this implies for all $x \in \mathbb{R}^n$ the estimate

$$\|x\|_{B_{\alpha'(t)}} \leq \|x\|_{B_{\alpha'(t_*^j)}} + \varepsilon_1^j$$

for some $\varepsilon_1^j \to 0$ as $j \to \infty$. By continuity of the trajectory in t we thus obtain

$$\|\varphi(t, x, u^j, p[u^j])\|_{B_{\alpha'(t)}} \le \|\varphi(t, x, u^j, p[u^j])\|_{B_{\alpha'(t^j_*)}} + \varepsilon^j_1$$

$$\le \|\varphi(t^j_*, x, u^j, p[u^j])\|_{B_{\alpha'(t^j_*)}} + \varepsilon^j_1 + \varepsilon^j_2 = \varepsilon^j_1 + \varepsilon^j_2.$$

Hence (2.14) implies the equality $\|\varphi(t, x, u^*, p[u^*])\|_{B_{\alpha'(t)}} = 0$ which yields the desired inclusion $\varphi(t, x, u^*, p[u^*]) \in B_{\alpha'(t)}$ since the B_α are closed. $\Box$

We now state the relation between the existence of an α-contractible family and the wISDS property.

Proposition 4.3.5 Consider a system of type (2.6) or (2.22) and let A be a weakly asymptotically stable set for the corresponding unperturbed system (2.1) or (2.17), respectively, with attracted neighborhood B. Then A is wISDS with attraction rate μ, overshoot gain σ and robustness gain γ if and only if there exists an α-contractible family of neighborhoods B_α with

$$d_H(B_\alpha, A) \le \gamma(\alpha), \quad B(\sigma^{-1}(\gamma(\alpha)), A) \cap B \subset B_\alpha$$

and

$$\vartheta(\alpha, t) = \gamma^{-1}(\mu(\gamma(\alpha), \tau)).$$

If the system satisfies (2.14) then the sets B_α can be chosen as closed sets.

Proof: Assume the existence of the α-contractible family, let $x \in B$ and $p \in \mathcal{P}$. By the assumption on the B_α there exists $\alpha > 0$ such that $x \in B_\alpha$ and $d_H(B_\alpha, A) = \sigma(\|x\|_A)$, i.e., $\alpha \le \gamma^{-1}(\sigma(\|x\|_A))$. By Corollary 4.3.3 we obtain $\Phi(t, x, u, p[u]) \in B_{\alpha'(t)}$ weakly for all $t \ge 0$, with $\alpha'(t) = \max\{\vartheta(\alpha, t), \gamma^{-1}(\tilde{\nu}(p, t))\}$, hence

$$\|\Phi(t, x, u, p[u])\|_A \le \max\{\gamma(\vartheta(\alpha, t)), \tilde{\nu}(p, t)\} \le \max\{\mu(\sigma(\|x\|_A), t), \tilde{\nu}(p, t)\}$$

weakly for all $t \ge 0$, which implies wISDS.

Conversely, assume wISDS. Then we define the sets

$$B_\alpha := \left\{ x \in \mathbb{R}^n \;\middle|\; \begin{array}{l} \text{for all } p \in \mathcal{P} \text{ the system satisfies} \\ \|\Phi(t, x, u, p[u])\|_A \le \max\{\mu(\gamma(\alpha), t), \tilde{\nu}(p, t)\} \\ \text{weakly for all } t \ge 0 \end{array} \right\}$$

for all $\alpha \ge 0$. Obviously $B_{\alpha'} \subseteq B_\alpha$ for $\alpha' < \alpha$. The assumptions on the distance are immediate implying in particular that the sets shrink down to A. The fact that we can choose $\vartheta(\alpha, t) = \gamma^{-1}(\mu(\gamma(\alpha), \tau))$ follows directly from the construction. Closedness in the case that the system satisfies (2.14) follows immediately from (2.14). $\Box$

Unfortunately, the technique used in the proof of Proposition 3.3.4 in the last chapter does not yield a strictly contractible family for weakly attracting sets. Hence the existence of a strictly contractible family still remains an open question, see also the discussion after Theorem 4.5.5, below.

4.4 Relation between Robustness Concepts

In this section we show the relation between the various robustness concepts for weak attractivity and asymptotic stability that we have introduced in this chapter. We start by investigating the inherent robustness properties of weakly asymptotically stable sets.

Proposition 4.4.1 Consider a system Φ of type (2.6) or (2.22) satisfying (2.9) or (2.25), respectively, for some class $\mathcal{K}_\infty$ function ρ. Then any weakly asymptotically stable set A with rate β and some open attracted neighborhood B satisfying $\mathrm{dist}(B, A) < \infty$ is nested γ-robust for suitable functions γ and χ of class $\mathcal{K}_\infty$ and some perturbation range W with nonvoid interior. In particular, γ, χ and W depend only on β, ρ, $\mathrm{dist}(B, A) < \infty$ and on the Lipschitz constant L of the system on the (compact) set $\mathrm{cl}\,(\mathcal{B}(\beta(2r_0, 0), A) \setminus A)$.

Proof: Assume asymptotic stability of A with some B with $\mathrm{dist}(B, A) < \infty$ and some β of class $\mathcal{KL}$. Observe that by the assumption on an asymptotically stable set either A or A^c is bounded, hence $\mathrm{cl}\,(\mathcal{B}(\beta(2r_0, 0), A) \setminus A)$ is compact and the system is uniformly Lipschitz with some constant L on this set. Set $r_0 := \mathrm{dist}(B, A)$. Then for all $r \in (0, r_0]$ we can define

$$T(r) = \min\{t \in \mathbb{T}^+ \,|\, \beta(r, t) \leq \frac{r}{4}\}.$$

Note that T is finite for all $r > 0$ and w.l.o.g. monotone decreasing, furthermore we obtain $\beta(s, T(r) + t) \leq r/4$ for all $t \geq 0$ and all $s \in [0, r]$. Now for all $\alpha \leq \alpha_0 := \rho^{-1}(e^{-LT(r_0)} \min\{r_0, \beta(r_0, 0)\}/4)$ consider the sets

$$D_\alpha := \mathcal{B}(r(\alpha), A),$$

where $r(\alpha)$ is chosen such that $e^{LT(r(\alpha))}\rho(\alpha) \leq r(\alpha)/8$. Observe that both α_0 and $r(\cdot)$ only depend on β, r_0 and L, and $r(\alpha) \to 0$ as $\alpha \to 0$. We set $W = \mathcal{B}(\alpha_0, 0)$. For each point $x \in B$ we denote by $\mathcal{U}_x \subset \mathcal{U}$ the set of controls u such that

$$\|\varphi(t, x, u)\|_A \leq \beta(\|x\|_A, t).$$

Then by Gronwall's Lemma or by induction we obtain for $t \in [0, T(\|x\|_A)] \cap \mathbb{T}$

$$\|\Phi(t, x, u, p[u])\|_A \leq \beta(\|x\|_A, t) + e^{Lt}\alpha$$

weakly for all $u \in \mathcal{U}_x$ and all $p \in \mathcal{P}_\alpha$, which implies that for each point $x \in D_\alpha$ there exists $u_x \in \mathcal{U}_x$ (independent of p) such that

$$\Phi(T(r(\alpha)), x, u_x, p[u_x]) \in D_\alpha$$

and

$$\|\Phi(t, x, u_x, p[u_x])\|_A \leq \beta(r(\alpha), 0) + r(\alpha)/4 \text{ for all } t \in [0, T(r(\alpha))] \cap \mathbb{T}.$$

Furthermore, for any $p \in \mathcal{P}_\alpha$ and any $x \in B$ this inequality implies that the concatenated control $u = u_1 \&_{T_1} u_2 \ldots$ defined inductively by $x_0 = x$, $T_i = T(\|x_{i-1}\|_A)$, $x_i = \Phi(T_i, x, u, p[u])$ and $u_i = u_{x_{i-1}}$ satisfies

$$\|\varphi(T_i, x, u, p[u])\|_A \leq \max\{r_0/2^i, r(\alpha)\}$$

and hence steers x to D_α in some finite time. Now we set

$$A_\alpha := \bigcup_{x \in D_\alpha} \bigcup_{\substack{p \in \mathcal{P}_\alpha \\ t \in [0, T(r(\alpha))] \cap \mathbb{T}}} \Phi(t, x, u_x, p[u_x]).$$

This set is weakly α-forward invariant by construction, and because each point from B can be steered to A_α in some (uniform) finite time by Proposition 4.1.6 it is weakly α-attracting. Furthermore it satisfies $\mathcal{B}(r(\alpha), A) \subset A_\alpha$, $d_H(A_\alpha, A) \leq \gamma(\alpha) := \beta(r(\alpha), 0) + r(\alpha)/4$ and $d_{\min}(A_\alpha, A) \geq \chi(\alpha) := r(\alpha)$. Note that we can extend χ arbitrarily for $\alpha \geq \alpha_0$, in particular it can be chosen of class $\mathcal{K}_\infty$. It remains to show property (iii) of Definition 4.2.5. Hence let $\alpha'' > \alpha' > \alpha > 0$. We first restrict ourselves to the case $r(\alpha') \geq r(\alpha'')/2$. Consider $x \in A_{\alpha''}$ and $p \in \mathcal{P}_\alpha$. By construction of the A_α we can use u_x to steer x to some $y \in D_{\alpha''}$ without leaving $A_{\alpha''}$. Now we can apply u_y yielding

$$\|\Phi(T(r(\alpha'')), y, u_y, p[u_y])\|_A \leq r(\alpha'')/2 \leq r(\alpha'),$$

i.e., $\varphi(T(r(\alpha'')), y, u_y, p[u_y]) \in D_{\alpha'} \subset A_{\alpha'}$, and by construction of the A_α this trajectory stays inside $A_{\alpha''}$, which shows (iii) for $r(\alpha') \geq r(\alpha'')/2$. For arbitrary $\alpha'' > \alpha' > 0$, by continuity of $r(\cdot)$ we find a decreasing sequence α_i, $i = 0, \ldots, m$ such that $\alpha_0 = \alpha''$, $\alpha_m = \alpha'$ and $r(\alpha_{i+1}) \geq r(\alpha_i)/2$. Since furthermore the construction of the A_α implies $A_{\alpha_{i+1}} \subset A_{\alpha_i}$ we can extend the above construction by induction and obtain the desired property. $\qquad \square$

The relation between β and γ and χ is somewhat hidden since, in general, an explicit expression for $r(\alpha)$ cannot be derived. Under the assumption of exponential weak attraction we can, however, overcome this difficulty as the following example shows.

Example 4.4.2 Assume that A is a weakly exponentially attracting set, i.e., a weakly attracting set with

$$\beta(r, t) = c e^{-\lambda t}$$

for two constants $c, \lambda > 0$ and let $\rho(r) = r$. Then we obtain $T(r) = T := \ln(4c)/\lambda$ for all $r \geq 0$, and consequently $r(\alpha) = c_1 \alpha$ for $c_1 = 4e^{LT}$. Hence we obtain $\gamma(\alpha) = cr(\alpha) + r(\alpha)/4 = c_2 \alpha$, i.e. a linear robustness gain and $\chi(\alpha) = r(\alpha) = c_1 \alpha$, i.e. the distance $d_{\min}(A, A_\alpha)$ also grows linearly. $\qquad \square$

Next we show the relation between wISDS and nested γ-robustness. Observe that in Proposition 4.4.1 we obtained that the χ in Definition 4.2.5 can be chosen of class $\mathcal{K}_\infty$. We will also need this property in the following proposition, because it ensures that χ is invertible.

Proposition 4.4.3 If a weakly attracting set A is wISDS with attraction rate μ and gains σ and γ, then it is also nested γ-robust for the same robustness gain γ and $\chi = \sigma^{-1}(\gamma(\cdot))$.

Conversely, if A is nested γ-robust for some χ and γ of class $\mathcal{K}_\infty$ then for each class $\mathcal{K}_\infty$ function $\tilde\gamma$ with $\tilde\gamma(r) > \gamma(r)$ there exist $\tilde\mu$ of class $\mathcal{KLD}$ such that A is wISDS on $B = \mathrm{int}\,\bigcup_{\alpha \ge 0} A_\alpha$ with $\sigma = \tilde\gamma(\chi^{-1}(\cdot))$.

Proof: Assume wISDS. Then it is easily seen that the elements B_α of the α-contractible family of neighborhoods provided by Proposition 4.3.5 give the desired α-attracting sets A_α from Definitions 4.2.2 and 4.2.5.

Conversely, assume nested γ-robustness of A. We show the assertion by constructing an α-contractible family of neighborhoods meeting the assumptions of Proposition 4.3.5. For this we follow the proof of Proposition 3.4.4 which simplifies here since we have stronger assumptions on the given A_α. Consider again the two sided sequences α_i and δ_i, $i \in \mathbb{Z}$, as constructed in the proof of Proposition 3.4.4 and set $B_i = A_{\alpha_i}$.

Then by the assumption on the A_α for each $i \in \mathbb{Z}$ there exists a $\Delta t_i \in \mathbb{T}^+$ such that for each $x \in B_{i-1} \setminus B_i$ and each $p \in \mathcal{P}_{\alpha_{i+1}}$ there exists a $u_{x,p} \in \mathcal{U}$ with $\Phi(t, x, u_{x,p}, p[u_{x,p}]) \in B_{i-1}$ for all $t \in [0, \Delta t_i] \cap \mathbb{T}$ and $\Phi(\Delta t_i, x, u_{x,p}, p[u_{x,p}]) \in B_i$. Now for all $\alpha \in [\alpha_{i+2}, \alpha_{i+1}]$ we set

$$B_\alpha := \bigcup_{\substack{x \in B_{i-1} \setminus B_i \\ p \in \mathcal{P}_{\alpha_{i+1}}}} \Phi\left(\frac{\alpha_{i+1} - \alpha}{\alpha_{i+1} - \alpha_{i+2}}\Delta t_i, x, u_{x,p}, p[u_{x,p}]\right) \bigcup B_i$$

for continuous time systems and

$$B_\alpha := \bigcup_{\substack{x \in B_{i-1} \setminus B_i \\ p \in \mathcal{P}_{\alpha_i}}} \Phi\left(\left[\frac{\alpha_{i+1} - \alpha}{\alpha_{i+1} - \alpha_{i+2}}\Delta t_i\right]_h, x, u_{x,p}, p[u_{x,p}]\right) \bigcup B_i$$

for discrete time systems with time step h, where $[r]_h$ denotes the largest value $s \in h\mathbb{Z}$ with $s \le r$. As in the proof of the strong case this construction implies $B_{\alpha_i} = B_{i-2}$ and $B_\alpha \subseteq B_{\alpha'}$ for all $0 < \alpha \le \alpha'$.

Thus for all $\alpha \in [\alpha_{i+2}, \alpha_{i+1}]$ we obtain the inequality

$$d_H(B_\alpha, A) \le d_H(B_{i-1}, A) \le \gamma(\alpha_{i-1}) = \tilde\gamma(\alpha_{i+2}) \le \tilde\gamma(\alpha)$$

and the inclusion

$$\mathcal{B}(\tilde{\sigma}^{-1}(\gamma(\alpha)), A) = \mathcal{B}(\chi(\alpha), A) \subseteq \mathcal{B}(\chi(\alpha_{i+1}), A)$$
$$\subseteq A_{\alpha_{i+1}} = B_{i+1} = B_{\alpha_{i+3}} \subseteq B_\alpha.$$

Now we can define ϑ analogously to the proof of Proposition 3.4.4 to get the desired contractible family. $\square$

Let us continue Example 4.4.2 in order to see what happens in the case of exponential attraction.

Example 4.4.2 (continued) For ε sufficiently small we obtain $\alpha_{i+1} > \alpha_i/2$, for the α_i from the proof of Proposition 4.4.3, hence we also have $r(\alpha_{i+1}) > r(\alpha_i)/2$ and from the construction in the proof of Proposition 4.4.1 we obtain $\Delta t_i \leq T$. Hence we obtain $\vartheta(T, \alpha_i) \leq \alpha_{i+1} \leq c\alpha_i$ for $c = (1 + \varepsilon)^{-1/3}$ and consequently ϑ decays exponentially.

The following corollary follows immediately from Propositions 4.4.1 and 4.4.3.

Corollary 4.4.4 A weakly attracting set A is weakly asymptotically stable if and only if there exist functions χ and γ of class $\mathcal{K}_\infty$ such that A is nested γ-robust for some W with nonvoid interior.

We explicitly state another useful consequence of Propositions 4.4.1 and 4.4.3.

Theorem 4.4.5 Consider a system Φ of type (2.6) or (2.22) satisfying (2.9) or (2.25), respectively, for some ρ of class $\mathcal{K}_\infty$. Let A be a weakly asymptotically stable set for the corresponding unperturbed system (2.1) or (2.17), respectively, with open attracted neighborhood B satisfying $\text{dist}(B, A) < \infty$, and rate β of class $\mathcal{KL}$. Then A is wISDS for some suitable attraction rate μ of class $\mathcal{KLD}$ and suitable gains γ and σ of class $\mathcal{K}_\infty$ for some perturbation range W with nonvoid interior, where μ, γ, σ and W depend only on $\text{dist}(B, A)$, β, ρ and on the Lipschitz continuity of the unperturbed system.

Remark 4.4.6 Note that the assumption "$\text{dist}(B, A) < \infty$" is no real restriction since we can always restrict ourselves to arbitrary open subsets of B (observe that we do not assume weak forward invariance of B). Nevertheless, W may indeed depend on $\text{dist}(B, A)$ similarly to the strong case, cf. Example 3.4.3. $\square$

Let us again investigate the case of a weakly asymptotically stable set with exponential attraction.

Example 4.4.2 (continued) In the exponential case the considerations from above and the relation between ϑ and μ as well as χ and σ show that we obtain wISDS with exponential rate μ and linear gains γ and σ.

The following example illustrates the gap between γ-robust weak attraction and wISDS and also shows that if both properties hold then wISDS might only hold with a much larger robustness gain.

Example 4.4.7 Consider the system

$$\dot{x} = -\rho(y)x + h(x)$$
$$\dot{y} = -uy + (u-1)(y-3)$$

for $U = [0,1]$, ρ some monotone increasing Lipschitz continuous function with $\rho(y) = 0$, $y \in (-\infty, 1]$ and $\rho(y) = 1$, $y \in [2, \infty)$ and $h : \mathbb{R} \to \mathbb{R}$ is defined for $x \geq 0$ by

$$h(x) = \begin{cases} 3x - \frac{3}{2^n}, & x \in \left[\frac{3}{2^{n+2}}, \frac{1}{2^n}\right] \\[2mm] -3x + \frac{3}{2^{n+1}}, & x \in \left[\frac{1}{2^{n+1}}, \frac{3}{2^{n+1}}\right] \\[2mm] 0, & x = 0 \end{cases}$$

for all $n \in \mathbb{Z}$ and for $x < 0$ by $h(-x) := -h(x)$. Observe that $h(x) \geq 0$ for $x \leq 0$ and $h(x) \leq 0$ for $x \geq 0$.

We first show that for each $\varepsilon \in (0,1)$ the set $A_\varepsilon = [-\varepsilon, \varepsilon]^2$ is weakly asymptotically stable with attracted neighborhood $B = (-5,5)^2$ but that $A_0 = \{(0,0)^T\}$ is not even a weakly attracting set:

Clearly, for each $(x,y) \in A_\varepsilon$ the control $u \equiv 1$ yields a trajectory staying in A_ε for all future times. In order to see asymptotic stability, for $\varepsilon > 0$ consider the smallest $n_0 \in \mathbb{N}$ such that $\varepsilon \leq 1/2^{n_0}$. Then for all points $(x,y) \in B$ with $|x - \varepsilon| < 1/2^{n_0+1}$ the constant control $u \equiv 1$ yields a trajectory converging to A_ε exponentially fast. All other points in B can be controlled with $u \equiv 0$ into an ε-neighborhood of the point $(0,3)$ in bounded time, and from there using $u \equiv 1$ to A_ε.

In order to see that A_0 is not an attracting set, observe that the only way to reach A_0 asymptotically is to control the system to the y-axis. This, however, is impossible for points (x,y) with $x \neq 0$, hence no point not already lying on the y-axis can be controlled to A_0 asymptotically, hence A_0 is not weakly attracting and thus, in particular, not weakly asymptotically stable.

Now consider the perturbed system which we obtain by choosing $W = \mathbb{R}^2$ and adding w_1 and w_2 to the $\dot{x}$ and $\dot{y}$-equation, respectively. We show that we can also find α-asymptotically stable sets A_α converging to A_0: The same control strategy as in the unperturbed case shows that if we consider the inflated system with perturbations $p \in \mathcal{P}_\alpha$, $\alpha < 1/2$, then the sets $A_\alpha = [-3/2^{n(\alpha)+2}, 3/2^{n(\alpha)+2}] \times [-\alpha, \alpha]$ where $n(\alpha)$ is the maximal $n \in \mathbb{N}$ with $3/2^{n+2} > \alpha$, are α-asymptotically stable with the same B. Since $3/2^{n(\alpha)+2} \leq 2\alpha$ we obtain that $d_H(A_\alpha, A_0) \leq \gamma(\alpha) := \sqrt{5}\alpha$, hence A_0 is γ-robust although it is not even an attracting set.

Note that in this example the A_α do not provide nested γ-robustness since it is not possible to control the system from a bigger to a smaller A_α without leaving the bigger one. This is, of course, what is expected from Corollary 4.4.4, since otherwise A_0 would have to be weakly asymptotically stable.

Now consider some $\varepsilon > 0$ and some Lipschitz continuous function $g(x)$ with $g(x) = -x$, $x \in [-\varepsilon/2, \varepsilon/2]$, and $g(x) = 0$, $x \in (-\infty, -\varepsilon] \cup [\varepsilon, \infty)$. We modify the equation for $\dot{x}$ by setting

$$\dot{x} = -\rho(y)x + h(x) + g(x).$$

Now $A = \{0\}$ is weakly asymptotically stable, and hence γ-robust with $\gamma(\alpha) \leq \sqrt{5}\alpha$. Since it is weakly asymptotically stable it is also wISDS. However, for initial values (x, y) with $x \notin [-\varepsilon, \varepsilon]$ we still need to control the system to the half plane $y > 1$ in order to reach A asymptotically. Hence the gain γ for wISDS for $\alpha > \varepsilon$ has to be larger than 1, and thus is much larger than the gain for γ-robustness. Summarizing, we can say that—unlike the strong case—weak γ-robustness does not imply wISDS for some nearby $\tilde{\gamma}$.   □

4.5 Lyapunov Function Characterization

As for the strong ISDS case, we will now develop a characterization of the wISDS property by means of Lyapunov functions. The motivation for using these functions is exactly the same as for the strong case, cf. Section 3.5, i.e., we obtain a nonlinear distance function allowing us to remove the overshoot gain σ in the wISDS estimate.

In contrast to the strong case, here we do not obtain decay of the Lyapunov function along each trajectory but only for (at least) one control function u, which leads to what is usually called a control Lyapunov function. We will give a number of references to the literature related to control Lyapunov functions throughout this section, see in particular the discussion after Theorem 4.5.5.

We start this section by giving two sufficient conditions for wISDS in terms of Lyapunov functions.

Proposition 4.5.1 Let $A \subset \mathbb{R}^n$ be a closed c-bounded weakly forward invariant set. Assume there exist functions σ_1, σ_2 and $\tilde{\gamma}$ of class $\mathcal{K}_\infty$, $\tilde{\mu}$ of class $\mathcal{KLD}$, a set $O \subset \mathbb{R}^n$, an open subset $P \subset O$ with $A \subset P$ and a function $V : O \to R_0^+$ with

$$\sigma_1(\|x\|_A) \leq V(x) \text{ for all } x \in O,$$
$$V(x) \leq \sigma_2(\|x\|_A) \text{ for all } x \in P$$

such that for each $p \in \mathcal{P}$, each $x \in O$ and each $t \in \mathbb{T}^+$ there exists $u^* \in \mathcal{U}$ with $\Phi(t, x, u^*, p[u^*]) \in O$ and

$$V(\Phi(t, x, u^*, p[u^*])) \leq \max\{\mu(V(x), t), \tilde{\nu}(p, t)\} \tag{4.3}$$

for $\tilde{\nu}$ defined by (4.2) for $\tilde{\mu}$ and $\tilde{\gamma}$. Then the system is wISDS with attracted neighborhood $B = P$, attraction rate $\mu(r, t) = \sigma_1^{-1}(\tilde{\mu}(\sigma_1(r), t))$, overshoot gain $\sigma(r) = \sigma_1^{-1}(\sigma_2(r))$ and robustness gain $\gamma(r) = \sigma_1^{-1}(\tilde{\gamma}(r))$.

Proof: Consider the sublevel sets

$$B_\alpha := \{x \in \mathbb{R}^n \,|\, V(x) \le \tilde{\gamma}(\alpha)\}$$

Then we obtain

$$d_H(B_\alpha, A) \le \sup\{\|x\|_A \,|\, V(x) \le \tilde{\gamma}(\alpha)\} \le \sigma^{-1}(\tilde{\gamma}(\alpha)) = \gamma(\alpha)$$

and

$$B(\sigma^{-1}(\gamma(\alpha)), A) \cap P = B(\sigma_2^{-1}(\tilde{\gamma}(\alpha)), A) \cap P \subseteq \{x \,|\, V(x) \le \tilde{\gamma}(\alpha)\} \subseteq B_\alpha.$$

Now let $x \in B_\alpha$, $t \in \mathbb{T}^+$ and $p \in \mathcal{P}$. By the assumption we find $u^* \in \mathcal{U}$ such that

$$V(\Phi(t, x, u^*, p[u^*])) \le \max\{\tilde{\mu}(\tilde{\gamma}(\alpha), t), \tilde{\nu}(p, t)\}$$

which shows that $\Phi(t, x, u^*, p[u^*]) \in B_{\alpha'}$ with

$$\alpha' = \max\{\tilde{\gamma}^{-1}(\tilde{\mu}(\tilde{\gamma}(\alpha), t)), \tilde{\gamma}^{-1}(\tilde{\nu}(p, t))\}.$$

We set $\vartheta(\alpha, t) = \tilde{\gamma}^{-1}(\tilde{\mu}(\tilde{\gamma}(\alpha), t)) = \gamma^{-1}(\mu(\gamma(\alpha), t))$. Then if for all $u \in \mathcal{U}$ the strategy p satisfies

$$p[u](\tau) \le \vartheta(\alpha, \tau) \text{ for almost all } \tau \in [0, t]$$

we obtain

$$\gamma^{-1}(\tilde{\nu}(p, t)) \le \vartheta(\alpha, t)$$

which yields $\alpha' = \vartheta(\alpha, t)$. Thus the B_α form a contractible family meeting the assumption of Proposition 4.3.5 which shows the claim. □

The following proposition shows that we can slightly relax inequality (4.3) and still obtain a sufficient condition for wISDS. Here and in what follows we use the convention $\inf \emptyset = \infty$.

Proposition 4.5.2 Let $A \subset \mathbb{R}^n$ be a closed c-bounded weakly forward invariant set. Assume there exist functions σ_1, σ_2 and $\tilde{\gamma}$ of class $\mathcal{K}_\infty$, $\tilde{\mu}$ of class $\mathcal{KLD}$, a set $O \subset \mathbb{R}^n$, an open subset $P \subset O$ with $A \subset P$ and a function $V : O \to R_0^+$ with

$$\sigma_1(\|x\|_A) \le V(x) \text{ for all } x \in O,$$

$$V(x) \le \sigma_2(\|x\|_A) \text{ for all } x \in P$$

such that for each $p \in \mathcal{P}$, each $x \in O$ and each $t \in \mathbb{T}^+$ the inequality

$$\inf_{u \in \mathcal{U}_{p,x,t}} V(\Phi(t, x, u, p[u])) \le \max\{\mu(V(x), t), \tilde{\nu}(t, x)\} \tag{4.4}$$

holds for $\tilde{\nu}$ defined by (4.2) for $\tilde{\mu}$ and $\tilde{\gamma}$ and $\mathcal{U}_{p,x,t} := \{u \in \mathcal{U} \,|\, \Phi(t, x, u, p[u]) \in O\}$. Then for each $\varepsilon > 0$ the system is wISDS with attracted neighborhood $B = P$, attraction rate $\mu_\varepsilon(r, t) = \sigma_1^{-1}(\tilde{\mu}(\sigma_1(r), (1 - \varepsilon)t))$, overshoot gain $\sigma(r) = \sigma_1^{-1}(\sigma_2(r))$ and robustness gain $\gamma_\varepsilon(r) = \sigma_1^{-1}((1 + \varepsilon)\tilde{\gamma}(r))$.

If the system satisfies (2.14), then we also obtain wISDS for $\varepsilon = 0$.

Proof: We show that for each $x \in O$, each $t \in \mathbb{T}^+$ and each $p \in \mathcal{P}$ there exists a $u^* \in \mathcal{U}$ such that $\Phi(t, x, u^*, p[u^*]) \in O$ and the inequality

$$V(\Phi(t, x, u^*, p[u^*])) \leq \max\{\tilde{\mu}(V(x), (1 - \varepsilon)t),\, \tilde{\nu}_\varepsilon(p, t)\} \tag{4.5}$$

holds for $\tilde{\nu}_\varepsilon$ defined by (4.2) with $(1 + \varepsilon)\tilde{\gamma}$ and $\tilde{\mu}(\cdot, (1 - \varepsilon)\cdot)$. From this inequality Proposition 4.5.1 gives the assertion.

In order to show (4.5) pick $p \in \mathcal{P}$, $x \in O$ and $t \in \mathbb{T}^+$ and fix some $\delta > 0$. Then the assumption implies the existence of a $u^* \in \mathcal{U}$ such that $\Phi(t, x, u^*, p[u^*]) \in O$ and

$$V(\Phi(t, x, u^*, p[u^*])) \leq \max\{\tilde{\mu}(V(x), t),\, \tilde{\nu}(p, t)\} + \delta.$$

If the equality $\max\{\tilde{\mu}(V(x), t),\, \tilde{\nu}(p, t)\} = 0$ holds then we can conclude that $x \in A$ and $\sup_{u \in \mathcal{U}} \|p[u]\|_{[0,t]} = 0$. Since A is weakly forward invariant, this implies the existence of a control $u' \in \mathcal{U}$ such that $\Phi(t, x, u', p[u']) \in A$, i.e. $V(\Phi(t, x, u', p[u'])) = 0$. Otherwise, we can choose $\delta > 0$ so small that

$$\max\{\tilde{\mu}(V(x), t),\, \tilde{\nu}(p, t)\} + \delta \leq \max\{\tilde{\mu}(V(x), (1 - \varepsilon)t),\, \tilde{\nu}_\varepsilon(p, t)\}$$

holds. In both cases we find $u^* \in \mathcal{U}$ for which the inequality (4.5) holds.

The assertion for systems satisfying (2.14) follows easily from (2.14) applied to the wISDS estimate. $\qquad\qquad\Box$

For future reference we make the following definition.

Definition 4.5.3 (wISDS Lyapunov function)

A function V satisfying the assumptions of Proposition 4.5.1 or Proposition 4.5.2 is called a wISDS *Lyapunov function*. $\qquad\qquad\Box$

These two propositions show that the existence of a wISDS Lyapunov function is sufficient for the wISDS property of some set A. The following theorem gives a necessary and sufficient condition in terms of wISDS Lyapunov functions.

Theorem 4.5.4 Consider the system (2.6) or (2.22) and functions γ and σ of class $\mathcal{K}_\infty$ and μ of class $\mathcal{KLD}$. Let A be a closed c-bounded weakly forward invariant set for the corresponding unperturbed system (2.1) or (2.21), respectively, and let B be some open neighborhood of A. Then the following properties are equivalent:

(i) For each $\varepsilon > 0$ the set A is wISDS with attracted neighborhood B, robustness gain $(1+\varepsilon)\gamma$, overshoot gain σ and attraction rate $\mu(\cdot, (1-\varepsilon)\cdot)$.

(ii) For each $\varepsilon > 0$ there exists a set $O \subset \mathbb{R}^n$ with $B \subseteq O$ and a function $V_\varepsilon : O \to \mathbb{R}_0^+$ which satisfies

$$V_\varepsilon(x) \geq \|x\|_A \quad \text{for all } x \in O,$$

$$V_\epsilon(x) \leq \sigma(\|x\|_A) \text{ for all } x \in B$$

and for each $x \in B$ and each $p \in \mathcal{P}$ the inequality

$$\inf_{u \in \mathcal{U}_{p,x,t}} V_\epsilon(\Phi(t,x,u,p[u])) \leq \max\{\mu(V_\epsilon(x),(1-\epsilon)t), \tilde{\nu}_\epsilon(p,x)\}$$

holds for all $t \in \mathbb{T}^+$, where $\tilde{\nu}_\epsilon$ is defined by (4.2) with $(1+\epsilon)\gamma$ and $\mu(\cdot,(1-\epsilon)\cdot)$ and $\mathcal{U}_{p,x,t} := \{u \in \mathcal{U} \mid \Phi(t,x,u,p[u]) \in O\}$.

Proof: (i)$\Rightarrow$(ii): Fix $\epsilon > 0$ and set $\gamma_\epsilon := (1+\epsilon)\gamma$. Then by Proposition 4.3.5 there exists a contractible family of neighborhoods B_α with $\vartheta(\alpha,t) = \gamma_\epsilon^{-1}(\mu(\gamma_\epsilon(\alpha),(1-\epsilon)t))$. Now we set $O := \text{int} \bigcup_{\alpha > 0} B_\alpha$ and

$$V(x) := \gamma_\epsilon(\inf\{\alpha > 0 \mid x \in B_\alpha\}).$$

Note that since $d_H(B_\alpha, A) \leq \gamma_\epsilon(\alpha)$ and $\mathcal{B}(\sigma^{-1}(\gamma_\epsilon(\alpha)), A) \cap B \subset B_\alpha$, the function V satisfies the desired bounds.

It remains to show the estimate for $\inf_{u \in \mathcal{U}_{p,x,t}} V(\Phi(t,x,u,p[u]))$. For this consider $x \in O$ and choose α with $\gamma_\epsilon(\alpha) = V(x)$. Then the construction of V implies that $x \in B_{\alpha+\delta}$ for each $\delta > 0$. Hence by Lemma 4.3.2 for any $p \in \mathcal{P}$ and any $t \in \mathbb{T}^+$ we obtain a $u^* \in \mathcal{U}_{p,x,t}$ with $\Phi(t,x,u^*,p[u^*]) \in B_{\alpha'}$, where

$$\alpha' = \max\{\vartheta(\alpha+\delta,\tau), \gamma_\epsilon^{-1}(\tilde{\nu}(p,t))\},$$

which implies

$$\begin{aligned} V(\Phi(t,x,u^*,p[u^*])) &\leq \gamma_\epsilon(\alpha') \\ &\leq \max\{\gamma_\epsilon(\vartheta(\alpha+\delta,\tau)), \tilde{\nu}(p,t)\} \\ &\leq \max\{\mu(V(x)+\delta,\tau), \tilde{\nu}(p,t)\}. \end{aligned}$$

This shows the claim since this expression is continuous in δ and $\delta > 0$ was arbitrary.

(ii)$\Rightarrow$(i): Fix $\epsilon > 0$ and choose $\epsilon_1 > 0$ such that $(1+\epsilon_1)^2 \leq (1+\epsilon)$. Assuming (ii) with ϵ_1 and applying Proposition 4.5.2 with ϵ_1 immediately gives the assertion. $\qquad \square$

The slightly unsatisfactory ϵ-formulation is due to the fact that we cannot exactly represent the contractible family of neighborhoods as sublevel sets of V_ϵ. We can avoid this problem if we assume our system to satisfy the continuity assumption (2.14). In this case the B_α can be chosen to be closed, which leads to the following result.

Theorem 4.5.5 Consider the system (2.6) or (2.22) and functions γ and σ of class $\mathcal{K}_\infty$ and μ of class $\mathcal{KLD}$. Let A be a closed c-bounded weakly forward invariant set for the corresponding unperturbed system (2.1) or (2.21), respectively, let B be some open neighborhood of A and assume that the system satisfies (2.14). Then the following properties are equivalent:

(i) The set A is wISDS with attracted neighborhood B, robustness gain γ, overshoot gain σ and attraction rate μ.

(ii) There exists a set $O \subset \mathbb{R}^n$ with $B \subseteq A$ and a function $V : O \to \mathbb{R}_0^+$ which satisfies

$$V(x) \geq \|x\| \text{ for all } x \in O,$$

$$V(x) \leq \sigma(\|x\|) \text{ for all } x \in B$$

and for each $x \in \tilde{B}$ and each $p \in \mathcal{P}$ the inequality

$$\inf_{u \in \mathcal{U}_{p,x,t}} V(\Phi(t,x,u,p[u])) \leq \max\{\mu(V(x),t), \tilde{\nu}(p,x)\}$$

holds for all $t \in \mathbb{T}^+$, with $\tilde{\nu}$ defined by (4.2) and $\mathcal{U}_{p,x,t} := \{u \in \mathcal{U} \mid \Phi(t,x,u,p[u]) \in O\}$.

Proof: (i)$\Rightarrow$(ii): Assume wISDS and consider the α-contractible family of neighborhoods B_α with closed sets B_α given by Proposition 4.3.5. Let $O = \text{int} \bigcup_{\alpha > 0} B_\alpha$. Then for each point $x \in O \setminus A$ there exists a unique minimal $\alpha(x) > 0$ such that $x \in B_\alpha$ for all $\alpha > \alpha(x)$ and since the B_α shrink down to A we obtain $\alpha(x) \to 0$ as $x \to A$.

We set $V(x) = \gamma(\alpha(x))$ for $x \in \tilde{B} \setminus A$ and $V(x) = 0$ on A. Then the properties of the sets B_α imply the desired upper and lower bounds for V, and the last inequality for V follows from Lemma 4.3.4 similarly to the proof of Proposition 4.5.4.

(ii)$\Rightarrow$(i): Follows immediately from Proposition 4.5.2 with $\sigma_1(r) = r$ and $\varepsilon = 0$. □

At this point, a question arises naturally: Can we also construct a *continuous* wISDS Lyapunov function V still reflecting the wISDS gains, similar to Theorem 3.5.3(ii) in the strong case? This question is closely related to the strict contractability of the B_α, which could not be achieved since the trick of slowing down the rate of attraction used in the corresponding statement for strong ISDS (cf. Remark 3.5.4) does not work for the weak version.

It should be noted that for weak asymptotic stability of differential inclusions (i.e., without perturbations) a similar construction was used recently by Kellett and Teel [70] to construct a continuous (even Lipschitz) Lyapunov function. However, the regularization needed in this construction in order to obtain continuity make the bounds on V and on the decay of V uncontrollable.

Of course, for control systems the existence of continuous Lyapunov functions characterizing weak asymptotic stability (or asymptotic controllability) without perturbation is a now classical result by Sontag [101] (see also Artstein [4]) based on a suitable optimal control problem much in the spirit of Zubov's approach [129], which we will discuss in Chapter 7. The existence proof for Lipschitz continuous Lyapunov functions for weak asymptotic stability is much more recent: Apart from the result by Kellett and Teel mentioned above there exists another construction by Rifford [96] based on an optimal control technique combined with some regularization.

Note that the continuity problem does not mean that it is difficult to construct continuous wISDS Lyapunov functions at all. In fact, using the small gain estimate in Theorem 4.7.8, below, and constructing V via a suitable differential game analogous to the optimal control problem in Sontag [101] could be a promising approach (which, nevertheless, is beyond the scope of this book). The interesting problem to be solved lies in encoding the optimal rate and gains (at least up to some ε) in this function.

Let us proceed with the characterization of wISDS via Lyapunov functions. When we try to follow the idea to develop our "weak" results parallel to the their "strong" counterparts in Chapter 3, it is now tempting to apply the beautiful theory of viscosity solutions for Hamilton-Jacobi-Isaacs equations in order to get an analogous statement to Theorem 3.5.7 for the weak ISDS property. Unfortunately, two obstructions appear at this point:

First, as we are not able to prove the existence of a continuous Lyapunov function V, it is not possible to apply the corresponding theory at the current state. Of course, there are techniques to handle discontinuous viscosity solutions but we were not able to find results in the literature that cover those aspects we need, cf. Remark A.2.2 in Appendix A.

Secondly, and this is the more fundamental obstruction, any viscosity (sub-, super-) solution of some Hamilton-Jacobi-Isaacs equation is naturally linked to the fact that one player uses (0-)nonanticipating strategies from $\mathcal{P}^0$ as defined in Definition 2.1.1, and *not* the δ-nonanticipating strategies from $\mathcal{P}^\delta$. This fact is independent of the solution concept one uses because it is exclusively due to the infinitesimal formulation in any partial differential equation or inequality. Since we have already seen in Example 4.2.4 that the choice of either $\mathcal{P}^0$ or $\mathcal{P}^\delta$ for some arbitrary $\delta > 0$ makes a qualitative difference in the resulting robustness of weakly asymptotically stable set (and since we will see in Chapter 5 that we need $\delta > 0$ in order to cover certain numerical perturbations) we do not think that a comprehensive infinitesimal description of a wISDS Lyapunov function related to strategies $p \in \mathcal{P}^\delta$ is feasible.

Nevertheless, we can recover one direction of the argumentation of Chapter 3 and give a sufficient condition of wISDS in terms of viscosity solutions when we restrict ourselves to perturbation strategies from $\mathcal{P}^0$. As in Chapter 3 we start with a preliminary proposition.

Proposition 4.5.6 Consider system (2.6), some open set $O \subset \mathbb{R}^n$ and a class $\mathcal{KLD}$ function μ satisfying Assumption 3.5.5. Then a continuous function $V : \text{cl}\, O \to \mathbb{R}_0^+$ which is constant on ∂O satisfies the inequality

$$\inf_{u \in \mathcal{U}_{p,x,t}} V(\varphi(t,x,u,p[u])) \leq \max\{\mu(V(x),t), \tilde{\nu}(t,p)\}$$

for all $x \in O$, all $p \in \mathcal{P}^0$ and all $t \geq 0$ with $\tilde{\nu}$ from (4.2) and $\mathcal{U}_{p,x,t}$ given by $\mathcal{U}_{p,x,t} := \{u \in \mathcal{U} \mid \Phi(t,x,u,p[u]) \in O\}$, if and only if it is a viscosity supersolution of the Hamilton-Jacobi-Isaacs equation

$$\sup_{u \in U} \inf_{\|w\| < \gamma^{-1}(V(x))} \{-DV(x)f(x,u,w) - g(V(x))\} \geq 0.$$

Proof: Let V satisfy the inequality, fix $x \in O$ and for each $u \in U$ pick some $p_u^0 \in W$ such that the inequality $\sup_{u \in U} \|p_u^0\| < \gamma^{-1}(V(x))$ holds. Consider the perturbation strategy $p[u](t) = p_{u(t)}^0$ and choose for all $t \in [0,1]$ a control function $u_t \in \mathcal{U}_{p,x,t}$ such that

$$V(\varphi(t,x,u_t,p[u_t])) \leq \max\{\mu(V(x),t), \tilde{\nu}(t,p)\} + t^2$$

Note that by the choice of p_u^0 and by continuity there exists $t_0 > 0$ such that $\mu(V(x),t) \geq \sup_{u \in U} \gamma(\|p[u]\|) = \tilde{\nu}(t,p)$ for all $t \in [0,t_0]$. Observe that for all $u \in \mathcal{U}$ and all $t \in (0,1]$ we have that

$$\frac{1}{t} \int_0^t f(x,u(\tau),p[u])d\tau \in \text{co} \bigcup_{u \in U} f(x,u,p_u^0),$$

where "co" denotes the convex hull, and that for all row vectors $\xi \in \mathbb{R}^n$ we have the equality

$$\inf_{u \in U} \xi f(x,u,p_u^0) = \inf_{v \in \text{co} \bigcup_{u \in U} f(x,u,p_u^0)} \xi v.$$

Now let $\xi \in D^- V(x)$. Then for $t \in [0,t_0]$ by the considerations above we obtain

$$t \inf_{u^* \in U} \xi f(x,u^*,p_{u^*}^0) \leq \xi \left(\int_0^t f(x,u_t(\tau),p[u_t])d\tau \right)$$

$$\leq \xi \left(\int_0^t f(\varphi(\tau,x,u_t,p^0),u_t,p[u_t])d\tau \right) + O(t^2)$$

$$= \xi \left(\varphi(t,x,u_t,p[u_t]) - x \right) + O(t^2)$$

and hence by the definition of $D^- V$ and the choice of u_t

$$\inf_{u^* \in U} \xi f(x,u^*,p_{u^*}^0) \leq \limsup_{t \to 0} \frac{V(\varphi(t,x,u_t,p[u_t])) - V(x) + O(t^2)}{t}$$

$$\leq \limsup_{t \to 0} \frac{\mu(V(x),t) - V(x) + t^2 + O(t^2)}{t} = -g(V(x)),$$

implying

$$\inf_{u \in U} \sup_{\|w\| < \gamma^{-1}(V(x))} \xi f(x, u, w) \leq -g(V(x)),$$

which was the claim.

Let conversely V be a viscosity supersolution of the given equation and fix some $t > 0$. From Corollary A.2.5 applied with $b = V(x)$, $a = \mu(V(x), t)$ and $W = W_{\gamma^{-1}(\mu(V(x), t))}$ we obtain

$$\inf_{u \in \mathcal{U}_{p,x,t}} V(\varphi(t, x, u, p[u])) \leq \mu(V(x), t) \text{ for all } p \text{ with}$$
$$\gamma(\|p[u](\tau)\|) \leq \mu(V(x), t) \text{ for almost all } \tau \in [0, t] \text{ and all } u \in \mathcal{U}. \tag{4.6}$$

We claim that this implies

$$\inf_{u \in \mathcal{U}_{p,x,t}} V(\varphi(t, x, u, p[u])) \leq \mu(V(x), t) \text{ for all } p \text{ with}$$
$$\gamma(\|p[u](\tau)\|) \leq \mu(V(x), \tau) \text{ for almost all } \tau \in [0, t] \text{ and all } u \in \mathcal{U} \tag{4.7}$$

(note that the difference to (4.6) lies in the "τ" in the argument of the μ in the second line). In order to prove (4.7) fix some $t > 0$, let p satisfy this constraint, and assume the inequality $\inf_{u \in \mathcal{U}_{p,x,t}} V(\varphi(t, x, u, p[u])) > \mu(V(x), t)$. Then there exists $\rho > 0$ such that $\inf_{u \in \mathcal{U}_{p,x,t}} V(\varphi(t, x, u, p[u])) > \mu(V(x), t) + \rho$. Now pick an arbitrary $\varepsilon < \rho$ and choose $t^* > 0$ such that the equality $\inf_{u \in \mathcal{U}_{p,x,t}} V(\varphi(t^*, x, u, p[u])) = \mu(V(x), t^*) + \varepsilon$ holds and the inequality $\inf_{u \in \mathcal{U}_{p,x,t}} V(\varphi(\tau, x, u, p[u])) > \mu(V(x), \tau) + \varepsilon$ holds for all $\tau \in (t^*, t]$. From the assumption on p we obtain $\|p[u](\tau)\| \leq \inf_{u \in \mathcal{U}} V(\varphi(\tau, x, u, p[u])) - \varepsilon$ for almost all $\tau \in [t^*, t]$ and all $u \in \mathcal{U}$. Using the continuity of $\inf_{u \in \mathcal{U}} V(\varphi(\tau, x, u, p[u]))$ in τ and the Lipschitz property of g we can now conclude the existence of times t_i, $i = 0, \ldots, k$ such that $t_0 = t^*$, $t_k = t$ and $\inf_{u \in \mathcal{U}} \mu(V(\varphi(t_i, x, u, p[u]), t_{i+1} - t_i) \geq V(\varphi(t_i, x, u, p[u])) - \varepsilon$, which implies $\|p[u](\tau)\| \leq \mu(V(\varphi(t_i, x, u, p[u]))$ for almost all $\tau \in [t_i, t_{i+1}]$ and all $u \in \mathcal{U}$. Hence by (4.6) we can conclude

$$\inf_{u \in \mathcal{U}} V(\varphi(t_{i+1}, x, u, p[u])) \leq \inf_{u \in \mathcal{U}} \mu(V(\varphi(t_i, x, u, p[u])), t_{i+1} - t_i)$$

which, by induction, implies

$$\inf_{u \in \mathcal{U}} V(\varphi(t, x, u, p[u]) \leq \inf_{u \in \mathcal{U}} \mu(V(\varphi(t^*, x, u, p[u])), t - t^*).$$

From this inequality and (3.7) we obtain

$$\inf_{u \in \mathcal{U}} V(\varphi(t, x, u, p[u]) \leq \mu(\mu(V(x), t^*) + \varepsilon, t - t^*)$$
$$\leq \mu(V(x), t) + C_\mu(\mu(V(x), t), V(x) + \rho - \mu(V(x), t), t)\varepsilon$$

which contradicts the assumption as $\varepsilon \to 0$.

We finally use (4.7) to show the assertion. If $\tilde{\nu}(p, t) \leq \mu(V(x), t)$ then inequality (4.7) directly implies the assertion. Hence consider some $t_1 > 0$ such that

$$\inf_{u \in \mathcal{U}} V(\varphi(t_1, x, u, p[u])) > \tilde{\nu}(p, t_1) > \mu(V(x), t_1). \tag{4.8}$$

We set $r = \mu(\tilde{\nu}(w, t_1), -t_1)$ and choose $t_0 > 0$ minimal such that the inequality $\inf_{u \in \mathcal{U}} V(\varphi(t, x, u, p[u])) \geq \mu(r, t)$ holds for all $t \in [t_0, t_1]$. Since by the choice of r and by the second inequality in (4.8) we have $r > V(x)$, from (4.7) we obtain $\inf_{u \in \mathcal{U}} V(\varphi(t_0, x, u, p[u])) \leq \mu(r, t_0)$. We fix $\varepsilon > 0$ and choose some u_ε such that $V(\varphi(t_0, x, u_\varepsilon, p[u_\varepsilon])) \leq \mu(r, t_0) + \varepsilon$. By Lemma 4.2.7 (i) and (ii) we obtain

$$\tilde{\nu}(w(t_0 + \cdot), t) \leq \mu(r, t_0 + t) \leq \mu(V(x_0), t)$$

for $x_0 = \varphi(t_0, x, u_\varepsilon, p[u_\varepsilon])$ which by (4.7) implies

$$V(\varphi(t_1, x, u_\varepsilon, p[u_\varepsilon])) = V(\varphi(t_1 - t_0, x_0, u_\varepsilon(t_0 + \cdot), p[u_\varepsilon](t_0 + \cdot)))$$
$$\leq \mu(V(x_0), t_1 - t_0) \leq \mu(\mu(r, t_0) + \varepsilon, t_1 - t_0).$$

Since this last expression is continuous in ε we obtain

$$\inf_{u \in \mathcal{U}} V(\varphi(t_1, x, u, p[u])) \leq \mu(\mu(r, t_0), t_1 - t_0) = \mu(r, t_1) \leq \tilde{\nu}(p, t_1)$$

which contradicts (4.8) and hence shows the claim. □

Now we can state the sufficient Hamilton-Jacobi-Isaacs equation.

Theorem 4.5.7 Consider system (2.6) with perturbation strategies $\mathcal{P} \subseteq \mathcal{P}^0$ and let A be a weakly attracting set with attracted neighborhood B for the unperturbed system (2.1). Consider some open set $O \subset \mathbb{R}^n$ with $B \subset O$ and assume there exists a continuous function $V : \mathrm{cl}\, O \to \mathbb{R}_0^+$ which is constant on ∂O and satisfies

$$V(x) \geq \|x\|_A \text{ for all } x \in O,$$

$$V(x) \leq \sigma(\|x\|_A) \text{ for all } x \in B$$

and is a viscosity supersolution of the equation

$$\sup_{u \in U} \inf_{\|w\| \leq \gamma^{-1}(V(x))} \{-DV(x)f(x, u, w) - g(V(x))\} \geq 0.$$

Then for each $\varepsilon > 0$ the set A is wISDS with $\mu_\varepsilon(r, t) = \mu(r, (1 - \varepsilon)t)$ and μ from Assumption 3.5.5.

If, in addition, the system satisfies (2.14) then A is also wISDS for $\varepsilon = 0$, i.e., $\mu_\varepsilon = \mu$.

Proof: Follows immediately from Propositions 4.5.2 and 4.5.6. □

4.6 Stability of Robustness Concepts

In this section we will investigate the effect of additional external perturbations on weakly attracting, γ-robust and wISDS sets. We will show what happens if a sequence of systems possessing one of these sets converges to some limiting system, and we will see how far we can make statements for the existence of those sets for systems nearby some given reference system.

For the following proposition recall Definition 3.6.1 of asymptotic boundedness.

Proposition 4.6.1 Consider a system (2.6) or (2.22) given by f or Φ_h, respectively, and a sequence of approximating systems f_n or $\Phi_{h,n}$ with

$$\|f(x, u, w) - f_n(x, u, w)\| \leq \varepsilon_n$$

or, respectively,

$$\|\Phi_h(x, u, w) - \Phi_{h,n}(x, u, w)\| \leq h\varepsilon_n$$

for all $u \in \mathcal{U}$, $w \in \mathcal{W}$ $x \in \mathbb{R}^n$ and some sequence ε_n, $n \in \mathbb{N}$ with $\varepsilon_n \to 0$ as $n \to \infty$. Denote the trajectories by Φ_n and Φ and consider closed c-bounded sets A and A_n such that $d_H(A, A_n) \to 0$ as $n \to \infty$ and an open set B with $A_n \subset B$, $A \subset B$. Then the following properties hold
(i) If the unperturbed system satisfies (2.13) and there exists a sequence of functions $\beta_n : (0, \infty) \times (0, \infty) \to [0, \infty)$ which is asymptotically bounded by some class $\mathcal{KL}$ function β, and for each $T > 0$ there exists $N \in \mathbb{N}$ such that

$$\|\Phi_n(t, x, u, 0)\|_{A_n} \leq \beta_n(\|x\|_{A_n}, t) \quad \text{weakly for all } t \in [0, T]$$

for all $x \in B$ and all $n \geq N$, then A is an asymptotically stable set for Φ with attraction rate β.
(ii) If the system satisfies (2.14) and there exist sequences of functions μ_n, γ_n and σ_n which are (ISDS–) asymptotically bounded by μ, ρ and σ, and for each $T > 0$ there exists $N \in \mathbb{N}$ such that

$$\|\Phi_n(t, x, u, p[u])\|_{A_n} \leq \max\{\mu_n(\sigma_n(\|x\|_{A_n}), t), \tilde{\nu}_n(p, t)\}$$

weakly for all $t \in [0, T]$, for all $x \in B$, all $p \in \mathcal{P}$ and all $n \geq N$, then A is wISDS for Φ with attraction rate μ and gains σ and γ.

Proof: We show assertion (ii), (i) follows by similar arguments. Consider $(t_i)_{i \in \mathbb{N}_0} \in \mathcal{T}$, $p \in \mathcal{P}$ and $x \in B$. Then we find control functions $u_n \in \mathcal{U}$ such that

$$\|\Phi_n(t_i, x, u, p[u_n])\|_{A_n} \leq \max\{\mu_n(\sigma_n(\|x\|_{A_n}), t_i), \tilde{\nu}_n(p, t_i)\}$$

holds for all $i \in \mathbb{N}$ with $t_i < T(n)$, where $T(n) \to \infty$ as $n \to \infty$. Consider the limiting trajectory $\Phi(t, x, u, p[u])$ for $\Phi(t, x, u_n, p[u_n])$ provided by (2.14).

Then it is immediate from (2.14) and from the asymptotic bounds on the rate and the gains that this trajectory satisfies

$$\|\Phi(t, x, u, p[u])\|_A \leq \max\{\mu(\sigma(\|x\|_A), t), \, \check{\nu}(p, t)\},$$

i.e., the desired estimate.   □

The weak analogue of Proposition 3.6.4 does not hold in general as we have already seen in Example 4.4.7. We will see later (cf. Theorem 4.7.7, below) that a weakened version of Proposition 3.6.4 also holds for weakly asymptotically stable sets in the context of inflated systems. Here we shall only give a sufficient condition for limits of attracting sets being an asymptotically stable set.

Definition 4.6.2 (uniformly bounded overshoot)

A sequence of weakly α-attracting sets $A_n \subset \mathbb{R}^n$ with rate of attraction determined by β_n and C_n is said to have *uniformly bounded overshoot* if there exists $r_0 > 0$, a continuous function $R : [0, r_0] \to \mathbb{R}$ with $R(0) = 0$ and a sequence $\varepsilon_n \to 0$ such that

$$\beta_n(r + C_n, 0) \leq R(r) + \varepsilon_n \text{ for all } r \in [0, r_0].$$

□

Remark 4.6.3 Note that this condition is not satisfied for the weakly asymptotically stable sets A_ϵ in Example 4.4.7.   □

Proposition 4.6.4 Consider a system of type (2.1) or (2.17) satisfying (2.13), a closed and c-bounded set A and attracting sets A_n, all with the same attracted neighborhood B. Assume that $d_H(A, A_n) \to 0$ as $n \to \infty$ and suppose that the A_n have uniformly bounded overshoot. Then A is α-asymptotically stable.

Proof: Using (2.13) one easily shows weak α-forward invariance. In order to prove asymptotic stability we show the assumptions of Proposition 4.1.7. Set $\varepsilon_0 = r_0$. Then we find $n_0 \in \mathbb{N}$ such that $A_{n_0} \subset \mathcal{B}(\varepsilon_0/2, A)$, hence we can set $T_1(r) := \min\{t \in \mathbb{T}_0^+ \,|\, \beta_{n_0}(r, t) \leq \varepsilon_0/4\}$. It remains to construct T_2 and ρ. For this for each $r \in (0, \varepsilon_0]$ we choose $n > 0$ such that $\varepsilon_n < r/8$ and $d_H(A_n, A) < r/8$. Then we set $\rho(r) = R(r + r/8) + r/4$ and $T_2(r) = \min\{t \in \mathbb{T}_0^+ \,|\, \beta_n(r + r/8 + C_n, t) \leq r/4\}$. Thus we obtain for $x \in B$ with $\|x\|_A \leq \varepsilon_0$ and for a proper choice of u^* the inequalities

$$\|\Phi(t, x, u^*, p[u^*])\|_A \leq \|\Phi(t, x, u^*, p[u^*])\|_{A_n} + r/8$$
$$\leq \beta_n(\|x\|_A + C_n, t) + r/8 \leq R(\|x\|_{A_n} + r/8) + r/4$$

and

$$\|\Phi(T_2(r), x, u^*, p[u^*])\|_A \leq \beta_n(\|x\|_{A_n} + r/8 + C_n, t) + r/8 \leq r/4 + r/8 < r/2$$

which by Lemma 4.1.7 implies the assertion.   □

Similar to the strong case we now introduce an embedding concept allowing results about nearby systems.

Definition 4.6.5 (weak (α, C)-embedded system)

Consider two perturbed systems, both either of type (2.6) or of type (2.22) with same control range U and perturbations ranges W and W^*, respectively. Denote the trajectories of the systems by Φ and Ψ, respectively, and let $\alpha \geq 0$ and $C \geq 1$. Then we say that the second system Ψ is (α, C)-embedded in the first system Φ on some set $B \subseteq \mathbb{R}^n$, if for each $x \in B$ and each $p^* \in \mathcal{P}^*$ there exist $p \in \mathcal{P}$ with $\|p[u](t)\| \leq \alpha + C\|p^*[u](t)\|$ for almost all $t > 0$ and all $u \in \mathcal{U}$ and

$$\Psi(t, x, u, p^*[u]) = \Phi(t, x, u, p[u])$$

for all $u \in \mathcal{U}$ and all $t > 0$ satisfying $\Psi(\tau, x, u, p^*[u]) \in B$ for all $\tau \in [0, t]$.

Here we call Ψ the *embedded system* and Φ the *embedding system*.   □

Using this definition we can make statements on the dynamical behavior of systems nearby some reference system.

Proposition 4.6.6 Consider a system of type (2.6) or (2.22) with trajectories Ψ, which is (α, C)-embedded on some open set B in some other system of the same type with trajectories denoted by Φ for some $\alpha \geq 0, C \geq 1$. Assume that the embedding system Φ has a weakly attracting set A which is wISDS on B with rate μ and gains σ and γ. Then for each $D > 1$ the embedded system Ψ has a weakly attracting set $\tilde{A}$ which satisfies $d_H(\tilde{A}, A) \leq \gamma(D\alpha)$ and the "wISDS-like" estimate

$$\|\Psi(t, x, u, p^*[u])\|_{\tilde{A}} \leq \max\{\mu(\sigma(\|x\|_{\tilde{A}} + \gamma(D\alpha)), t), \tilde{\nu}(CDp^*/(D-1), t)\}.$$

weakly for all $t \geq 0$, for each $p^* \in \mathcal{P}^*$. If $\alpha = 0$ then the set $\tilde{A} = A$ for all $p^* \in \mathcal{P}^*$ satisfies the wISDS estimate

$$\|\Psi(t, x, u, p^*[u])\|_{\tilde{A}} \leq \max\{\mu(\sigma(\|x\|_{\tilde{A}}, t), \tilde{\nu}(Cp^*, t)\} \text{ weakly for all } t \geq 0.$$

Proof: Fix $x \in B$ and $p^* \in \mathcal{P}^*$ and let $p \in \mathcal{P}$ be the perturbation strategy for which the embedding is obtained. Consider the α-contractible family of neighborhoods from Proposition 4.3.5 and set $\tilde{A} := B_{D\alpha}$. Then the properties of the B_α imply that $d_H(A_\alpha, A) \leq \gamma(D\alpha)$. Now let $r := \|x\|_{\tilde{A}} \leq \|x\|_A$. Then by the properties of the B_α we can conclude that $x \in B_\rho$ for $\rho =$

$\gamma^{-1}(\sigma(r))$. Hence choosing some sequence $(t_i)_{i \in \mathbb{N}_0} \in \mathcal{T}$ Corollary 4.3.3 implies the existence of a $u^* \in \mathcal{U}$ such that

$$\Phi(t_i, x, u^*, p[u^*]) \in B_{\alpha'(t_i)}$$

for $\alpha'(\tau) = \max\{\vartheta(\alpha, \tau), \gamma^{-1}(\tilde{\nu}(p, \tau))\}$, implying

$$\|\Phi(t_i, x, u^*, p[u^*])\|_{\tilde{A}} \leq \max\{\mu(\sigma(\|x\|_A, t_i), \tilde{\nu}(p, t_i)\}$$
$$\leq \max\{\mu(\sigma(\|x\|_{\tilde{A}} + \gamma(D\alpha)), t_i), \tilde{\nu}(p, t_i)\} \quad (4.9)$$

for $\max\{\mu(\sigma(\|x\|_A, t_i), \tilde{\nu}(p, t_i)\} > D\alpha$ and

$$\|\Phi(t_i, x, u^*, p[u^*])\|_{\tilde{A}} = 0 \quad (4.10)$$

for $\max\{\mu(\sigma(\|x\|_A, t_i), \tilde{\nu}(p, t_i)\} \leq D\alpha$.

We claim that if $\tilde{\nu}(p, t_i) > D\alpha$ holds then the inequality

$$\tilde{\nu}(p, t_i) \leq \tilde{\nu}(CDp^*/(D-1), t) \quad (4.11)$$

is valid. In order to see this, recall that from the embedding property we obtain $\|p[u](t)\| \leq \alpha + C\|p^*[u]\|$ for almost all $t > 0$, and all $u \in \mathcal{U}$. We define $\tilde{p}$ by $\tilde{p}[u](t) = 0$, if $\|p[u](t)\| \leq D\alpha$ and $\tilde{p}[u](t) = p[u](t)$ if $\|p[u](t)\| > D\alpha$. Then for almost all $t > 0$ and all $u \in \mathcal{U}$ we obtain

$$CD\|p^*[u](t)\|/(D-1) \geq \max\{\|\tilde{p}[u](t)\|, 0\}.$$

which implies

$$\tilde{\nu}(\tilde{p}, t) \leq \tilde{\nu}(CDp^*/(D-1), t).$$

Since the definition of $\tilde{\nu}$ yields the implication

$$\tilde{\nu}(p, t) > D\alpha \Rightarrow \tilde{\nu}(p, t) = \tilde{\nu}(\tilde{p}, t)$$

this shows (4.11).

Combining (4.9), (4.10) and (4.11) we obtain

$$\|\Phi(t_i, x, u, p[u])\|_{\tilde{A}} \leq \max\{\mu(\sigma(\|x\|_{\tilde{A}} + \gamma(D\alpha)), t_i), \tilde{\nu}(CDp^*/(D-1), t_i)\}$$

which implies the assertion since $\Psi(t_i, x, u, p^*[u]) = \Phi(t_i, x, u, p[u])$. $\qquad \square$

As in the strong case, we can state a similar proposition for γ-robustness instead of wISDS.

Proposition 4.6.7 Consider a system of type (2.6) or (2.22) with trajectories Ψ, which is (α, C)-embedded on some open set B in some other system of the same type with trajectories denoted by Φ for some $\alpha \geq 0$, $C \geq 1$. Assume that the embedding system Φ has a weakly attracting set A which is γ-robust on B for some gain γ of class $\mathcal{K}_\infty$. Then for each $D > 1$ the embedded system Ψ has a weakly attracting set $\tilde{A}$, which is $\gamma(CD \cdot /(D-1))$-robust and satisfies $d_H(\tilde{A}, A) \leq \gamma(D\alpha)$. If $\alpha = 0$ then the set $\tilde{A} = A$ itself is a $\gamma(C \cdot)$-robust weakly attracting set for Ψ.

Proof: We set $\tilde{A} = A_{D\alpha}$. Fix $x \in B$. Then the assumption on the (α, C)-embedding implies, that for each $p^* \in \mathcal{P}^*_{\alpha'}$ with $\alpha' \in [0, (D-1)\alpha/C]$ there exists $p \in \mathcal{P}_{D\alpha}$ such that

$$\Psi(t, x, u, p^*[u]) = \Phi(t, x, u, p[u]) \text{ for all } u \in \mathcal{U}$$

and for each $p^* \in \mathcal{P}^*_{\alpha'}$ with $\alpha' > (D-1)\alpha/C$ there exists $p \in \mathcal{P}_{CD\alpha'/(D-1)}$ such that
$$\Psi(t, x, u, p^*[u]) = \Phi(t, x, u, p[u]) \text{ for all } u \in \mathcal{U}.$$

Hence setting $\tilde{A}_{\alpha'} = A_{D\alpha}$ for $\alpha' \in [0, (D-1)\alpha/C]$ and $\tilde{A}_{\alpha'} = A_{CD\alpha'/(D-1)}$ for $\alpha' \geq D\alpha/C$ gives α'-attracting sets $\tilde{A}_{\alpha'}$ for Ψ satisfying

$$d_H(\tilde{A}_{\alpha'}, \tilde{A}) \leq d_H(\tilde{A}_{\alpha'}, A) \leq CD\alpha'/(D-1) \text{ for all } \alpha' \geq 0.$$

This shows the claim.   □

If we do not require the weakly attracting set $\tilde{A}$ to be robust then we can obtain the corresponding statement without the constant D.

Proposition 4.6.8 Consider a system of type (2.6) or (2.22) with trajectories Ψ, which is (α, C)-embedded on some open set B in some other system of the same type with trajectories denoted by Φ for some $\alpha \geq 0, C \geq 1$. Assume that the embedding system Φ has a weakly attracting set A which is γ-robust on B. Then the embedded systems Ψ has a weakly attracting set $\tilde{A}$ with $A \subseteq \tilde{A}$ and $d_H(\tilde{A}, A) \leq \gamma(\alpha)$.

Proof: Consider the weakly α-attracting sets A_α from the definition of γ-robustness. Then it is immediate from the definitions that $\tilde{A} = A_\alpha$ is the desired weakly attracting set.   □

4.7 Inflated Systems

In this section we will investigate the special case of inflated systems, i.e., of systems (2.6) and (2.22) where the right hand side is given by (2.10) and (2.11) or (2.26) and (2.27), respectively.

We have already seen that the choice of $\delta > 0$ for the class of strategies $\mathcal{P}^\delta$ plays an important role in the behavior of the inflated control system. This will also be apparent in the following lemma, which establishes a priori bounds on the effect of the perturbation under assumption (2.8) or (2.24).

Lemma 4.7.1 Consider a discrete or continuous time α_0-inflated control system with right hand side given by (2.10) or (2.26) satisfying (2.8) or (2.24), respectively, and with perturbation strategies $p \in \mathcal{P}^\delta$ for some $\delta \geq 0$. Let

$\varepsilon > 0$ and $\alpha \geq 0$ such that $\alpha + \varepsilon \leq \alpha_0$. Then the following assertions hold.

(i) For each $p \in \mathcal{P}_{\alpha+\varepsilon}$ there exists $\tilde{p} \in \mathcal{P}_\alpha$ such that estimate

$$\Phi(t, x, u, p[u]) \in B(\varepsilon(e^{Lt} - 1)/L, \Phi(t, x, u, \tilde{p}[u]))$$

holds for all $x \in \mathbb{R}^n$, all $u \in \mathcal{U}$ and all $t \in \mathbb{T}^+$.

(ii) Consider the continuous time system $\Phi = \varphi$. Let $x \in \mathbb{R}^n$, $T \in \mathbb{T}^+$, $\tilde{p} \in \mathcal{P}_\alpha$ and consider a function $x : \mathbb{T} \times \mathcal{U} \to \mathbb{R}^n$ with $x(0, u) = 0$ for all $u \in \mathcal{U}$ and

$$\|x(t, u) - \varphi(t, x, u, \tilde{p}[u])\| \leq t\varepsilon/(Lt + 1)$$

for all $u \in \mathcal{U}$ and all $t \in \mathbb{T} \cap [0, T]$. Assume furthermore that x solves $\dot{x}(t, u) = b(t, u)$ for some essentially bounded, measurable and δ-nonanticipating (with respect to the same sequence $(t_i)_{i \in \mathbb{N}_0} \in \mathcal{T}^\delta$ as $\tilde{p}$) function $b : [0, T] \times \mathcal{U} \to \mathbb{R}^n$ with

$$\|b(t, u) - f(\varphi(t, x, u, \tilde{p}[u]), u(t), \tilde{p}[u](t))\| \leq \varepsilon/(Lt+1) \text{ for almost all } t \in [0, T].$$

Then there exists $p \in \mathcal{P}_{\alpha+\varepsilon}$ such that

$$\varphi(t, x, u, p[u]) = x(t, u)$$

for all $t \in \mathbb{T} \cap [0, T]$ and all $u \in \mathcal{U}$. If, in addition, $x(t, u)$ satisfies (2.12), then p can be chosen such that $\varphi(t, x, u, p[u])$ satisfies (2.12).

(iii) Consider the continuous time system $\Phi = \varphi$. Then for each $T > 0$, each $\tilde{p} \in \mathcal{P}_\alpha$ and each two points x, $x^* \in \mathbb{R}^n$ satisfying $\|x - x^*\| \leq e^{-TL}T\varepsilon/(LT + 1)^2$ there exists $p \in \mathcal{P}_{\alpha+\varepsilon}$ such that

$$\varphi(T, x^*, u, p[u]) = \varphi(T, x, u, \tilde{p}[u])$$

for all $u \in \mathcal{U}$. If, in addition, $\varphi(t, x, u, \tilde{p}[u])$ satisfies (2.12), then p can be chosen such that $\varphi(t, x, u, p[u])$ satisfies (2.12).

(iv) If $\delta > 0$ then for each $T \in \mathbb{T}^+ \cap [0, \delta]$, each $\tilde{p} \in \mathcal{P}_\alpha$ and each family of points $x_T(u)$ with $\|x_T(u) - \Phi(T, x, u, \tilde{p}[u])\| \leq T\varepsilon/(LT + 1)$ for all $u \in \mathcal{U}$ there exists $p \in \mathcal{P}_{\alpha+\varepsilon}$ such that $\Phi(T, x, u, p[u]) = x_T(u)$ for all $u \in \mathcal{U}$. If, in addition, $\Phi(t, x, u, \tilde{p}[u])$ and $x_T(u)$ satisfy (2.12) (for $I = \mathbb{T}$ and $I = \{T\}$, respectively), then p can be chosen such that $\Phi(t, x, u, p[u])$ satisfies (2.12). If $\Phi = \Phi_h$ is a discrete time system with time step $h > 0$ and $T = h$ then the assertion also holds if $\|x_T - \Phi_h(T, x, u, \tilde{p}[u])\| \leq T\varepsilon$.

Proof: (i) This inclusion follows from Gronwall's Lemma for continuous time systems and by induction for discrete time systems setting $\tilde{p}[u](t) := G(p[u](t))$ with $G : \mathbb{R}^n \to \mathbb{R}^n$ defined by $G(w) = w$, $\|w\| \leq \alpha$ and $G(w) = \alpha w/\|w\|$, $\|w\| \geq \alpha$.

(ii) Fix T, x, $\tilde{p}$ and $x(t, u)$ as in the assumption. We claim that p defined by $p[u](t) := \tilde{p}[u](t) + b(t, u) - f(x(t, u), u(t), \tilde{p}[u](t))$ satisfies the assertion.

Clearly, this strategy is δ-nonanticipating if $\tilde{p}$ and $b(t, u)$ are so with respect to the same sequence $(t_i)_{i \in \mathbb{N}_0} \in \mathcal{T}^\delta$. Furthermore, we have that

$$\frac{d}{dt}x(t, u) = b(t, u) = f(x(t, u), u(t), \tilde{p}[u](t)) + p[u](t) - \tilde{p}[u](t)$$
$$= f(x(t, u), u(t), p[u](t))$$

and

$$\frac{d}{dt}\varphi(t, x, u, p[u]) = f(\varphi(t, x, u, p[u]), u(t), p[u](t)).$$

Hence, since $x(0, u) = \varphi(0, x, u, p[u])$, by uniqueness of the solution to this differential equation we can conclude

$$\varphi(t, x, u, p[u]) = x(t, u)$$

for all $t \in [0, T]$. Since for all $u \in \mathcal{U}$ we have the inequality

$$\|p[u](t)| \leq \|\tilde{p}[u](t)\| + \|b(t, u) - f(x(t, u), u(t), \tilde{p}[u](t))\|$$
$$\leq \alpha + \|b(t, u) - f(\varphi(t, x, u, \tilde{p}[u]), u(t), \tilde{p}[u](t))\|$$
$$+ L\|\varphi(t, x, u, \tilde{p}[u]) - x(t, u)\|$$
$$\leq \alpha + \varepsilon/(Lt + 1) + Lt\varepsilon/(Lt + 1) \ \leq \ \alpha + \varepsilon$$

for almost all $t \in [0, T]$ we obtain that $p \in \mathcal{P}_{\alpha + \epsilon}$. Property (2.12) of $\varphi(\cdot, x, u, \tilde{p}[u])$ is immediate from the equality $\varphi(\cdot, x, u, p[u]) = x(\cdot, u)$ whenever $x(\cdot, u)$ satisfies (2.12).

(iii) From Gronwall's Lemma we obtain

$$\|\varphi(t, x^*, u, \tilde{p}[u]) - \varphi(t, x, u, \tilde{p}[u])\| \leq T\varepsilon/(LT + 1)^2 \text{ for all } t \in \mathbb{T} \cap [0, T], u \in \mathcal{U}.$$

Now we set

$$x(t, u) = \varphi(t, x, u, \tilde{p}[u]) + \frac{t}{T}(\varphi(t, x^*, u, \tilde{p}[u]) - \varphi(t, x, u, \tilde{p}[u]))$$

for all $t \in \mathbb{T} \cap [0, T]$ and all $u \in \mathcal{U}$. This implies

$$\|x(t, u) - \varphi(t, x, u, \tilde{p}[u])\| \leq t\varepsilon/(LT + 1)^2 \leq t\varepsilon/(Lt + 1).$$

Clearly, if $\varphi(t, x, u, \tilde{p}[u])$ satisfies (2.12) then $x(t, u)$ also does. Now for $b(t, u) := \dot{x}(t, u)$ we obtain the equation

$$b(t, u) = f(\varphi(t, x, u, \tilde{p}[u]), u(t), \tilde{p}[u](t)) + \frac{1}{T}(\varphi(t, x^*, u, \tilde{p}[u]) - \varphi(t, x, u), \tilde{p}[u])$$

$$+ \frac{t}{T}(f(\varphi(t, x^*, u, \tilde{p}[u]), u(t), \tilde{p}[u](t)) - f(\varphi(t, x, u, \tilde{p}[u]), u(t), \tilde{p}[u](t)))$$

which implies

$$\|b(t,u) - f(\varphi(t,x,u,\tilde{p}[u]), u(t), \tilde{p}[u](t))\|$$

$$\leq \|\frac{1}{T}(\varphi(t,x^*,u,\tilde{p}[u]) - \varphi(t,x,u,\tilde{p}[u]))$$

$$+\frac{t}{T}(f(\varphi(t,x^*,u,\tilde{p}[u]), u(t), \tilde{p}[u](t)) - f(\varphi(t,x,u,\tilde{p}[u]), u(t), \tilde{p}[u](t)))\|$$

$$\leq \frac{1}{T}(\varepsilon T/(LT+1)^2 + Lt\varepsilon T/(LT+1)^2) \quad \leq \quad \varepsilon/(1+Lt).$$

Thus we can apply (ii) which shows the claim.

(iv) Note that we do not have any "nonanticipating" restrictions in the choice of $p[u](t)$ for $t \in [0,\delta]$. Hence this property follows with exactly the same proof as Lemma 3.7.1(iii). If $\Phi(t,x,u,\tilde{p}[u])$ and $x_T(u)$ satisfy property (2.12) then an inspection of the construction in this proof easily reveals (2.12) for $\Phi(t,x,u,p[u])$. □

Remark 4.7.2 Note that assertions (ii) and (iii) in Lemma 4.7.1 are only formulated for continuous time systems and arbitrary $\delta \geq 0$. Similar statements could be made for discrete time systems (by straightforward inductive arguments based on statement (iv) of this Lemma) if we restricted ourselves to mh-nonanticipating strategies p for some $m \in \mathbb{N}$ with $\mathbf{t}(p) = (0, mh, 2mh, \ldots)$. We have omitted these statements because what we will need in the rest of this section is already covered by statement (iv).   □

Let us show an immediate consequence of Lemma 4.7.1 (iii) and (iv).

Lemma 4.7.3 Let A be a (not necessarily closed) weakly α-forward invariant set for the inflated system and let $\delta \geq 0$ with $\delta \geq h$ in the case of discrete time systems with time step $h > 0$. Let $\mathcal{P} = \mathcal{P}^\delta$ or $\mathcal{P} = \mathcal{P}^{\delta,c}$. Then the following properties hold.
(i) For each $\alpha' < \alpha$ there exists $\varepsilon > 0$ and $C > 0$ such that for all $x \in \mathbb{R}^n$ with $0 < \text{dist}(x,A) \leq \varepsilon$ and all $p \in \mathcal{P}_\alpha$ there exists a $u^* \in \mathcal{U}$ such that

$$\Phi(t_x, x, u^*, p[u^*]) \in A$$

for $t_x = C\text{dist}(x,A)$ for continuous time and $t_x = \max\{C\text{dist}(x,A), h\}$ for discrete time systems.
(ii) For each $x \in \partial A$, each $T \in \mathbb{T}^+$, each $\alpha' < \alpha$ and each $p \in \mathcal{P}_{\alpha'}$ there exists $u^* \in \mathcal{U}$ such that $\Phi(T, x, u^*, p[u^*]) \in \text{cl} A$.

Proof: We show the assertion for continuous time systems; for discrete time systems with $\delta \geq h$ both assertions follow similarly using Lemma 4.7.1 (iv).

(i) Fix $\alpha' \in [0,\alpha)$ and let $\varepsilon_1 = \alpha - \alpha'$. We set $C = 2(L+1)^2 e^L/\varepsilon_1$ and choose $\varepsilon = 1/C$. Now fix $x \in B(\varepsilon, A)$ with $\text{dist}(x,A) > 0$ and assume the existence of $\tilde{p} \in \mathcal{P}_{\alpha'}$ such that $\Phi(t_x, x, u, \tilde{p}[u]) \notin A$ for all $u \in \mathcal{U}$ and t_x from the assertion. Pick $x^* \in A$ with $\|x - x^*\| \leq 2\text{dist}(x,A)$. Setting $T_x = C\text{dist}(x,A) \leq 1$ gives

$$\|x - x^*\| \le 2T_x/C = e^{-L}T_x\varepsilon_1/(L+1)^2 \le e^{-LT_x}T_x\varepsilon_1/(LT_x+1)^2.$$

Thus Lemma 4.7.1 (iii) can be applied with $\varepsilon = \varepsilon_1$ and $T = T_x$ and yields the existence of $p \in \mathcal{P}_\alpha$ such that $\varphi(T_x, x^*, u, p[u]) = \varphi(T_x, x, u, \tilde{p}[u]) \notin A$, for all $u \in \mathcal{U}$ which contradicts the weak α-forward invariance of A.

(ii) Fix $x \in \partial A \setminus A$ and $T > 0$. Assume that there exists $p \in \mathcal{P}_{\alpha'}$ such that $\varphi(T, x, u, p[u]) \notin \mathrm{cl}\, A$ for all $u \in \mathcal{U}$. Pick an arbitrary $u' \in \mathcal{U}$ and let $d(t) := \|\varphi(T, x, u', p[u'])\|_A$. Since $d(0) = 0$, $d(T) > 0$ and $d(t)$ is continuous in t we find $t^* \in [0, T]$ such that $Cd(t^*) = T - t^*$ for the constant C from (i). Now we apply (i) to $x^* = \varphi(t^*, x, u', p[u'])$ and $\tilde{p} \in \mathcal{P}_{\alpha'}$ defined by $\tilde{p}[\tilde{u}](t) = p[u'\&_t\cdot\tilde{u}](t^*+t)$. Then we we find some $\tilde{u} \in \mathcal{U}$ such that $\varphi(T-t^*, x^*, \tilde{u}, \tilde{p}[u]) \in A$ implying

$$\varphi(T, x, u\&_t\cdot\tilde{u}, p[u'\&_t\cdot\tilde{u}]) = \varphi(T - t^*, x^*, \tilde{u}, \tilde{p}[u]) \in A$$

which contradicts the assumption $\varphi(T, x, u, p[u]) \notin \mathrm{cl}\, A$ for all $u \in \mathcal{U}$. $\square$

Using this lemma we obtain the following very useful result on the existence of weakly asymptotically stable set.

Proposition 4.7.4 Consider system (2.6) or (2.22) with α_0-inflated right hand side (2.10) or (2.26), respectively. Let $\mathcal{P} = \mathcal{P}^\delta$ or $\mathcal{P} = \mathcal{P}^{\delta,c}$ with $\delta \ge 0$ and $\delta \ge h$ for discrete time systems (2.22) with time step h. Let $\alpha \in [0, \alpha_0]$ and let A_α be a weakly α-attracting set with attracted neighborhood B. Then $\mathrm{cl}\, A_\alpha$ is a weakly α'-asymptotically stable set for all $\alpha' \in [0, \alpha)$. In addition, the class $\mathcal{KL}$ function β' can be chosen such that $\beta'(r, t) \le \beta(r + C, t)$ for all $r \ge 0$, where β and C characterize the weak α-attractivity of A_α.

Proof: From Lemma 4.7.3 we obtain that the assumption of Proposition 4.1.8 is satisfied for $\mathrm{cl}\, A$. This directly gives the assertion. $\square$

Using this proposition we can state the following variant of Proposition 4.6.8.

Proposition 4.7.5 Consider a system of type (2.6) or (2.22) with trajectories Ψ, which is (α, C)-embedded on some open set B in some other system of the same type with trajectories denoted by Φ for some $\alpha \ge 0$, $C \ge 1$. Let $\mathcal{P} = \mathcal{P}^\delta$ or $\mathcal{P} = \mathcal{P}^{\delta,c}$ for some $\delta \ge 0$ with $\delta \ge h$ in the case of discrete time systems with time step $h > 0$ and assume that the embedding system Φ has a weakly attracting set A which is γ-robust on B. Then for each $D > 1$ the embedded system Ψ has a weakly asymptotically stable set $\tilde{A}$ with $A \subseteq \tilde{A}$ and $d_H(\tilde{A}, A) \le \gamma(D\alpha)$.

Proof: Consider the weakly α-attracting sets A_α from the definition of γ-robustness. Then by Proposition 4.7.4 the set $A_{D\alpha}$ is a weakly α-asymptotically stable set. Then it is immediate from the definition of (α, C)-embedding that $\tilde{A} = A_{D\alpha}$ is the desired weakly asymptotically stable set for ψ. $\square$

We now turn to a weak version of Proposition 3.7.2. Since we know that a limit of a sequence of asymptotically stable sets might not be asymptotically stable, cf. Example 4.4.7, we introduce the following definition of weak practical attraction.

Definition 4.7.6 (weak practical attraction)

A set $A \subset \mathbb{R}^n$ is called weakly practically attracting with attracted neighborhood B if for each $\varepsilon > 0$ there exists a weakly attracting set A_ε with attracted neighborhood B satisfying $d_H(A_\varepsilon, A) \leq \varepsilon$. It is called γ-robust for some γ of class $\mathcal{K}_\infty$ if there exist weakly α-attracting sets $A_\alpha \supseteq A$ with attracted neighborhood B for all $\alpha \geq 0$ satisfying $d_H(A_\alpha, A) \leq \gamma(\alpha)$ for each $\alpha > 0$ and $A_{\alpha'} \subseteq A_\alpha$ for all $\alpha \geq \alpha' \geq 0$. $\square$

Using this definition we can obtain a statement of stability of weak γ-robustness under limits.

Proposition 4.7.7 Consider system (2.6) or (2.22) with inflated right hand side (2.10) or (2.26), respectively, with $\mathcal{P} = \mathcal{P}^\delta$ or $\mathcal{P} = \mathcal{P}^{\delta,c}$ for some $\delta \geq 0$ and $\delta \geq h$ for discrete time systems (2.22) with time step h. Consider a sequence of approximating systems f_n or $\Phi_{h,n}$ with

$$\|f(x, u, w) - f_n(x, u, w)\| \leq \varepsilon_n$$

or, respectively,

$$\|\Phi_h(x, u, w) - \Phi_{h,n}(x, u, w)\| \leq h\varepsilon_n$$

for all $u \in \mathcal{U}$, $w \in \mathcal{W}$, $x \in \mathbb{R}^n$ and some sequence ε_n, $n \in \mathbb{N}$ with $\varepsilon_n \to 0$ as $n \to \infty$. Assume that all these systems satisfy (2.14). Consider closed and c-bounded sets A and A_n such that $d_H(A, A_n) \to 0$ as $n \to \infty$ and an open set B with $A_n \subset B$, $A \subset B$, and assume that there exists a sequence of class $\mathcal{KL}$ functions γ_n which is asymptotically bounded by some class $\mathcal{KL}$ function γ such that for each n the set A_n is a γ_n-robust attracting set for the approximating systems f_n or $\Phi_{h,n}$. Then for each $D > 1$ the set A is a $\gamma(D\cdot)$-robust weakly practically attracting set. Furthermore, the A_α realizing the γ-robustness can be chosen to be weakly α-asymptotically stable.

Proof: Consider the α-weakly attracting sets $A_{\alpha,n}$ for A_n. We set

$$A_\alpha := \operatorname{Lim\,sup}_{n \to \infty} A_{n,\alpha}.$$

By (2.14) it is easily seen that these sets are weakly-α forward invariant, furthermore similar to the proof of Proposition 3.6.4 one sees that for each $\varepsilon > 0$ there exists $N \in \mathbb{N}$ such that $A_{\alpha,n} \subset B(\varepsilon, A_\alpha)$. Hence we can steer each point $x \in B$ into any neighborhood of A_α under perturbations $p \in \mathcal{P}_\alpha$, and consequently by Lemma 4.7.3 and Lemma 4.1.9 the set A_α is a weakly α'-asymptotically stable set for all $\alpha' < \alpha$. Clearly, these sets satisfy

$d_H(A_\alpha, A) \leq \gamma(\alpha) \leq \gamma(D\alpha')$ for $\alpha' < \alpha$, $\alpha - \alpha'$ sufficiently small and $A_{\tilde\alpha} \subseteq A_\alpha$ for all $\alpha > \tilde\alpha > 0$. This shows the claim. $\qquad\square$

Let us finally investigate the state dependent inflation (2.11) or (2.27). We first show how this perturbation affects an wISDS set. This is the weak version of the small gain theorem 3.7.4 for (strong) ISDS, whose proof, however, is slightly more complicated because we have to handle the $\sup_{u \in \mathcal{U}}$ in the definition of the $\tilde\nu$ term in the wISDS inequality and have to construct a control function $u \in \mathcal{U}$. For general systems the statement is slightly weaker than in the strong case, under the continuity assumption (2.14) we can, however, obtain the analogous statement as in the strong case.

Theorem 4.7.8 Consider an α_0-inflated system of type (2.6) or (2.22) with right hand side given by (2.10) or (2.26), and let A be an wISDS set with rate μ, gains γ and σ and attracted neighborhood B with respect to some sort of perturbations strategies $\mathcal{P}$. Consider the corresponding state dependent inflated system (2.11) or (2.27) with trajectories denoted by Φ and with b satisfying $b(x) \leq \alpha_0$ and $b(x) \leq \max\{\gamma^{-1}(\|x\|_A/(1+\varepsilon_1)), \rho\}$ for continuous time systems or $b(x) \leq \alpha_0$ and $b(x) \leq \max\{\gamma^{-1}(\mu(\|x\|_A/(1+\varepsilon_1), h)), \rho\}$ for discrete time systems, respectively, for all $x \in B$ and some $\rho > 0$ and $\varepsilon_1 > 0$. Then for each $\varepsilon > 0$, each $x \in B$ and each $p \in \mathcal{P}_1$ the inequality

$$\|\Phi(t, x, u, p[u])\|_A \leq \max\{\mu(\sigma(\|x\|_A, (1-\varepsilon)t), (1+\varepsilon)\gamma(\rho)\}$$

is weakly satisfied. If the system satisfies (2.14) then we also obtain the inequality for $\varepsilon = 0$ and $\varepsilon_1 = 0$.

Proof: We show the assertion for continuous time systems, for discrete time systems it follows similarly. First consider the case $\rho > 0$. Fix $x \in B$ and $p \in \mathcal{P}_1$. We set $\tilde p[u](t) = b(\varphi(t, x, u, p[u]))p[u](t)$. Then the global Lipschitz continuity of b implies the existence of $T > 0$ (independent of x) such that

$$\sup_{u \in \mathcal{U}} \|\tilde p[u]|_{[0,T]}\|_\infty \leq \max\{\gamma^{-1}(\|x\|_A/(1+\varepsilon_1/2)), \rho\}$$

Taking the wISDS Lyapunov function V_ε provided by Theorem 4.5.4 for $\varepsilon > 0$ so small that $(1+\varepsilon)/(1+\varepsilon_1/2) \leq 1/(1+\varepsilon_1/4)$ and proceeding as in the proof of Proposition 4.5.2 we obtain

$$V_\varepsilon(\varphi(t, x, u, p[u])) \leq \max\{\mu(V_\varepsilon(x), (1-\varepsilon)t), \tilde\nu_\varepsilon(\tilde p, t)\} \text{ weakly for all } t \in [0, T].$$

Now observe that the estimate above and the choice of ε implies

$$\tilde\nu_\varepsilon(\tilde p, t) \leq \max\{\sigma(\|x\|_A)/(1+\varepsilon_1/4), (1+\varepsilon)\gamma(\rho)\}$$

for all $t \in (0, T]$. Consider the interval $[\gamma(\rho), \sigma(\|x\|_A)]$. Then we find $T_1 > 0$ independent of r such that $\mu(r, t) \geq r/(1+\varepsilon_1/4)$ for all $t \in (0, T_1]$ and all $r \in [\gamma(\rho), \sigma(\|x\|_A)]$. Hence for $T_2 = \min\{T, T_1\}$ we obtain the inequality

$$V_\epsilon(\varphi(t, x, u, p[u])) \leq \max\{\mu(V_\epsilon(x), (1-\epsilon)t), \max\{\mu(V_\epsilon(x), t), (1+\epsilon)\gamma(\rho)\}\}$$

weakly for all $t \in [0, T_2]$ and hence

$$V_\epsilon(\varphi(t, x, u, p[u])) \leq \max\{\mu(V_\epsilon(x), (1-\epsilon)t), (1+\epsilon)\gamma(\rho)\} \qquad (4.12)$$

is weakly satisfied for all $t \in [0, T_2]$. Since T_2 does not depend on x (as long as $V_\epsilon(x) \geq \gamma(\rho)$) we can proceed inductively to obtain that (4.12) is weakly satisfied for all $t > 0$. Now the bounds on V give the desired inequality. For $\rho = 0$ we consider some monotone decreasing sequence $\rho_n \to 0$. Observe that T_2 now depends on ρ_n but is positive for each n. Hence we can still perform the induction using (4.12) and obtain the assertion.

For systems satisfying (2.14) consider a sequence $\epsilon^n \to 0$ and let $b_n(x) = b(x)/(1+\epsilon^n)$. For each of these systems from the general case we obtain a control satisfying the desired inequality for $\epsilon = \epsilon_1 = \epsilon^n$, hence the limiting control from (2.14) satisfies the inequality for $\epsilon = \epsilon_1 = 0$. $\qquad \square$

Observe that Remark 3.7.5 holds accordingly for this Theorem.

Analogously to the strong case, a similar small gain property holds true for γ-robust forward invariant sets.

Proposition 4.7.9 Consider an α_0-inflated system of type (2.6) or (2.22) with right hand side given by (2.10) or (2.26), and let C be a direct γ-robust weakly forward invariant set. Then the corresponding state dependent inflated system (2.11) or (2.27) with b satisfying $b(x) \leq \rho$ on $\mathcal{B}(\gamma(\rho), C)$ for some $\rho \in [0, \alpha_0]$ has a weakly forward invariant set $\tilde{C}$ which is weakly forward invariant for all $p \in \mathcal{P}_1$ and satisfies $d_H(\tilde{C}, C) \leq \gamma(\rho)$.

Proof: Completely analogous to Proposition 3.7.6. $\qquad \square$

4.8 Discrete and Continuous Time Systems

In this final section we will now investigate the interplay between weak robustness concepts for discrete and continuous time systems.

The motivation for this investigation is similar to the strong case: Since numerical approximations give "approximate information" only for the behavior of the time-h map φ^h defined by (2.19), we would like to be able to deduce information about the dynamical behavior of the continuous time system φ from φ^h.

Proposition 4.8.1 Consider a continuous time system (2.6) with inflated right hand side (2.10) satisfying (2.8). Let $h > 0$ and $\alpha > 0$ and let $A_{\alpha,h}$ be a weakly α-attracting set for the time-h map φ^h of the inflated system φ. Then there exists a set A_α with $d_H(A_\alpha, A_{\alpha,h}) \leq (M + \rho(\alpha))h$ which is weakly α'-asymptotically stable for (2.6) for each $\alpha' < \alpha$.

Proof: Since $A_{\alpha,h}$ is weakly α-forward invariant for Φ_h for each $x \in A_h$ and each $p \in \mathcal{P}_\alpha$ there exists a $u_{x,p} \in \mathcal{U}$ such that $\varphi(h, x, u_{x,p}, p[u_{x,p}]) \in A_{\alpha,h}$. We set

$$A_\alpha := \bigcup_{\substack{t \in [0,h] \\ x \in A_{\alpha,h}, p \in \mathcal{P}_\alpha}} \varphi(t, x, u_{x,p}, p[u_{x,p}]).$$

Then the distance estimate $d_H(A_\alpha, A_{\alpha,h}) \le (M + \rho(\alpha))h$ is immediate. The weak α'-asymptotic stability now follows from Proposition 4.1.9 whose assumptions are easily verified using Lemma 4.7.3, observing that by construction A_α is weakly α-forward invariant.   □

We can prove a similar result for γ-robust weakly attracting sets.

Proposition 4.8.2 Consider a continuous time system (2.6) with inflated right hand side (2.10) satisfying (2.8). Let $h > 0$ and $\alpha > 0$ and let A_h be a γ-robust weakly attracting set for the time-h map φ^h of the inflated continuous time system φ. Then for each $c > 0$ there exists a γ_c-robust weakly asymptotically stable set A_c for the inflated continuous time system φ with

$$d_H(A_c, A_h) \le \gamma(\rho^{-1}(2M/c)) + (2 + c)Mh/c$$

and

$$\gamma_c(r) = \gamma((1 + ch)r) + (1 + c)h\rho((1 + ch)r).$$

Proof: Consider the weakly α-attracting sets A_α^h for the time-h map φ^h which satisfy $d_H(A_\alpha^h, A_h) \le \gamma(\alpha)$. Since $A_{\alpha'}^h \subseteq A_\alpha^h$ for all $\alpha' \le \alpha$ we find control functions $u_{p,x} \in \mathcal{U}$ depending on x and p such that for each $\alpha > 0$, each $p \in \mathcal{P}_\alpha$ and each $x \in A_\alpha$ we have $\varphi(h, x, u_{x,p}, p[u_{x,p}]) \in A_\alpha^h$. We set

$$\tilde{A}_\alpha = \bigcup_{\substack{t \in [0,h] \\ x \in A_{\alpha,h}, p \in \mathcal{P}_\alpha}} \varphi(t, x, u_{x,p}, p[u_{x,p}]).$$

Then the choice of $u_{p,x}$ implies $\tilde{A}_{\alpha'} \subseteq \tilde{A}_\alpha$ for all $\alpha' \le \alpha$. Now fix some $c > 0$. We set $\alpha_0 = \rho^{-1}(2M/c)$ and $\varepsilon = ch\alpha_0/(1 + ch)$. We claim that $A = \tilde{A}_{\alpha_0}$ is a γ_c-robust asymptotically stable set for φ. By Proposition 4.8.1 we obtain that A is α'-asymptotically stable for all $\alpha' \in [0, \alpha_0)$ (hence in particular for $\alpha' = 0$, i.e., for the unperturbed system) and satisfies

$$d_H(A, A_h) \le d_H(A_{\alpha_0}^h, A) + d_H(A_{\alpha_0}^h, A_h)$$
$$\le \gamma(\alpha_0) + (M + \rho(\alpha_0))h = \gamma(\rho^{-1}(2M/c)) + (2 + c)Mh/c,$$

i.e., the desired distance. It remains to show that A is γ_c-robust. For this purpose let $\alpha > 0$. If $\alpha \le \alpha_0 - \varepsilon$ we obtain that $A_\alpha = A$ is weakly α-attracting with distance $d_H(A_\alpha, A) = 0 \le \gamma_c(\alpha)$. For $\alpha > \alpha_0 - \varepsilon$ we set

$A_\alpha = \tilde{A}_{\alpha+\epsilon}$. Again by Proposition 4.8.1 this set is weakly α-asymptotically stable (hence weakly α-attracting), and for the distance we obtain

$$d_H(A_\alpha, A) \le d_H(A_\alpha, A_h)$$
$$\le d_H(\tilde{A}_{\alpha+\epsilon}, A^h_{\alpha+\epsilon}) + d_H(A^h_{\alpha+\epsilon}, A_h)$$
$$\le (M + \rho(\alpha + \epsilon))h + \gamma(\alpha + \epsilon),$$

where for the first inequality we have used that $A_h \subseteq A \subseteq A_\alpha$. By the choice of α, α_0 and ϵ we obtain

$$\alpha + \epsilon \le (1 + ch)\alpha$$

and $\rho(\alpha + \epsilon) \ge \rho(\alpha_0) \ge 2M/c$, which implies

$$M + \rho(\alpha + \epsilon) \le (1 + c)\rho(\alpha + \epsilon) \le (1 + c)\rho((1 + ch)\alpha).$$

Inserting these inequalities into the estimate for $d_H(A_\alpha, A)$ proves the desired distance estimate. $\quad\square$

The next proposition summarizes the behavior of (robust) weakly attractive sets under limits of systems.

Proposition 4.8.3 Consider a continuous time system (2.6) with inflated right hand side (2.10) satisfying (2.8). Consider, furthermore, a sequence of time steps $h_n \to 0$, the time h_n-maps φ^{h_n} of φ, closed and c-bounded sets A and A_n such that $d_H(A, A_n) \to 0$ as $n \to \infty$ and an open set B with $A_n \subset B$, $A \subset B$. Then the following properties hold
(i) If the unperturbed system satisfies (2.13), $\beta_n : (0, \infty) \times (0, \infty) \to [0, \infty)$ is a sequence of functions which is asymptotically bounded by some class $\mathcal{KL}$ function β and for each $T > 0$ there exists $N \in \mathbb{N}$ such that the inequality

$$\|\varphi^{h_n}(t, x, u)\|_{A_n} \le \beta_n(\|x\|_{A_n}, t)$$

is weakly satisfied for all $x \in B$, all $n \ge N$ and all $t \in h_n\mathbb{Z} \cap [0, T]$ then A is a weakly attracting set for the continuous time system φ with attraction rate β.
(ii) If γ_n are functions which are asymptotically bounded by some class $\mathcal{KL}$ function γ and each A_n is a γ-robust attracting set for φ^{h_n} then for each $c > 1$ the set A is a $\gamma(c\,\cdot)$-robust weakly practically attracting set in the sense of Definition 4.7.6. Furthermore, the A_α realizing the γ-robustness can be chosen to be weakly α-asymptotically stable.
(iii) If the system satisfies (2.14), μ_n, γ_n and σ_n are functions which are (ISDS–) asymptotically bounded by functions μ, ρ and σ of class $\mathcal{KLD}$ and class $\mathcal{K}_\infty$ and for each $T > 0$ there exists $N \in \mathbb{N}$ such that the inequality

$$\|\varphi^{h_n}(t, x, u, p[u])\|_{A_n} \le \max\{\mu_n(\sigma_n(\|x\|_{A_n}), t), \tilde{\nu}_{h_n, n}(p, t)\}$$

is weakly satisfied for all $x \in B$, $p \in \mathcal{P}$, $n \ge N$ and all $t \in h_n\mathbb{Z} \cap [0, T]$ then A is a wISDS set for the continuous time system with rate μ and gains σ and γ.

Proof: (i) Observe that we have the inequality

$$\|\varphi(t,x,u) - \varphi^{h_n}(ih_n,x,u)\| \le Mh_n \text{ for all } t \in [ih_n,(i+1)h_n]$$

for all $u \in \mathcal{U}$. Now pick $x \in B$. Then for each $n \in \mathbb{N}$ we can pick $u_n \in \mathcal{U}$ such that the inequality from the assumption is satisfied and consequently the limiting control from (2.14) satisfies the needed inequality.

(ii) Completely analogous to the proof of Proposition 4.7.7.

(iii) Fix $t \ge 0$, $x \in B$, and $p \in \mathcal{P}$ and define the value

$$C := \sup_{u \in \mathcal{U}} \text{ess sup}_{\tau \in [0,t]} \|f(\varphi(t,x,u,p[u]),u(t),p[u](t))\| < \infty.$$

Then for all $u \in \mathcal{U}$ we obtain

$$\|\varphi(\tau,x,u,p[u]) - \varphi_{h_n}(ih_n,x,u,p[u])\| \le Ch_n$$

for all $\tau \in [ih_n,(i+1)h_n]$ with $\tau \le t$. Now for the fixed $t > 0$ we find a sequence $t_n \to t$ such that $t_n = i_n h_n$ for some $i_n \in \mathbb{N}$ and $t_n \le t \le t_n + h_n$. Hence we obtain the estimate

$$\|\varphi(t,x,u_n,p[u_n])\|_A \le \|\varphi_{h_n}(t_n,x,u_n,p[u_n])\|_{A_n} + \varepsilon_n$$
$$\le \max\{\mu(\sigma(\|x\|_{A_n}),t_n), \tilde{\nu}_{h_n}(p,t_n)\} + 2\varepsilon_n$$

for some suitable sequence $\varepsilon_n \to 0$. Considering a limiting control function u for the sequence u_n as provided by (2.14) implies for $n \to \infty$

$$\|\varphi(t,x,u,p[u])\|_A \le \limsup_{h \searrow 0} \max\{\mu(\sigma(\|x\|_A),t-h), \tilde{\nu}_h(p,t-h)\}.$$

Since $\mu(\tilde{\nu}_0(p,t),-h) \ge \tilde{\nu}_h(p,t)$ (which is easily seen from the definition of $\tilde{\nu}_0$ and $\tilde{\nu}_h$), we can use the continuity of μ and Lemma 4.2.7 (iii) to conclude the desired estimate. $\qquad\Box$

As in the strong case, we end this section by investigating the two different types of inflated systems. We consider the time-h map $\varphi^h(x,u,p[u])$ corresponding to the solution $\varphi(t,x,u,p[u])$ of the inflated system

$$\dot{x} = f^0(x,u) + p[u] \tag{4.13}$$

and the discrete time inflated system

$$\Phi_h(x,u,w) := \varphi^h(x,u,0) + \int_0^h p[u](t)dt \tag{4.14}$$

based on the time-h map $\varphi^h(x,u,0)$ corresponding to the solution $\varphi(t,x,u,0)$ of the unperturbed system (4.13) with $p[u] = 0$.

The following Lemma shows the relation between these systems. Again we set $W = \mathbb{R}^n$ here, the restriction to the case $W = B(\alpha_0,0) \subset \mathbb{R}^n$ is straightforward.

Lemma 4.8.4 Let $h > 0$ and consider the discrete time systems φ^h and Φ_h from (4.13) and (4.14). Assume that the continuous time system (4.13) satisfies (2.8). Let $\mathcal{P} = \mathcal{P}^\delta$ for some $\delta \geq 0$. Then for each $p \in \mathcal{P}$ and each $x \in \mathbb{R}^n$ there exists $\tilde{p} \in \mathcal{P}$ with $\|\tilde{p}[u](t)\| \leq \|p[u](t)\| + hL\|p[u]\|_{[ih,t]}$ for almost all $t \in [ih, (i+1)h]$ and all $u \in \mathcal{U}$ such that

$$\varphi^h(t, x, u, \tilde{w}) = \Phi_h(t, x, u, w) \text{ for all } t \in \mathbb{T}^+ \text{ and all } u \in \mathcal{U}. \qquad (4.15)$$

Conversely, for each $\tilde{p} \in \mathcal{P}$ and each $x \in \mathbb{R}^n$ there exists $p \in \mathcal{P}$ with $\|p[u](t)\| \leq \|\tilde{p}[u](t)\| + (e^{Lh} - 1)\|\tilde{p}[u]\|_{[ih,t]}$ for almost all $t \in [ih, (i+1)h]$ and all $u \in \mathcal{U}$ such that (4.15) holds.

Proof: It is sufficient to show (4.15) for $t = h$, since from that we obtain the assertion for arbitrary $t \in \mathbb{T}^+$ by a simple induction.

Let $x \in \mathbb{R}^n$ and $p \in \mathcal{P}$. For $t \in [0, h]$ we set

$$\tilde{p}[u](t)$$
$$= f^0(\varphi(t, x, u, 0), u(t)) + p[u](t) - f^0\left(\varphi(t, x, u, 0) + \int_0^t p[u](\tau)d\tau, u(t)\right).$$

Then we obtain

$$\frac{d}{dt}\left(\varphi(t, x, u, 0) + \int_0^t p[u](\tau)d\tau\right) = f^0(\varphi(t, x, u, 0), u(t)) + p[u](t)$$
$$= f^0\left(\varphi(t, x, u, 0) + \int_0^t p[u](\tau)d\tau, u(t)\right) + \tilde{p}[u](t)$$

and

$$\frac{d}{dt}\varphi(t, x, u, \tilde{p}[u]) = f^0(\varphi(t, x, u, \tilde{p}[u]), u(t)) + \tilde{p}[u](t),$$

which by the uniqueness of the solution to this differential equation implies

$$\varphi^h(x, u, \tilde{p}[u]) = \varphi(h, x, u, \tilde{p}[u]) = \varphi(h, x, u, 0) + \int_0^h p[u](\tau)d\tau = \Phi_h(x, u, p[u])$$

for all $u \in \mathcal{U}$. The estimate on the bound follows easily from the Lipschitz estimate on f^0.

Conversely, let $x \in \mathbb{R}^n$ and $\tilde{p} \in \mathcal{P}$. Setting

$$p[u](t) = f^0(\varphi(t, x, u, \tilde{p}[u]), u(t)) + \tilde{p}[u](t) - f^0(\varphi(t, x, u, 0), u(t))$$

similar arguments as above yield the assertion using Lemma 3.7.1(i) to obtain the estimate $\|\varphi(t, x, u, \tilde{p}[u]) - \varphi(t, x, u, 0)\| \leq \|\tilde{p}[u]\|_{[0,t]}(e^{Lt} - 1)/L$. $\square$

The following corollary is immediate from this lemma.

Corollary 4.8.5 Let $h > 0$ and consider the discrete time systems φ^h and Φ_h from (4.13) and (4.14). Assume that the continuous time system (4.13) satisfies (2.8) and let $\delta \geq 0$ and $\mathcal{P} = \mathcal{P}^\delta$ or $\mathcal{P} = \mathcal{P}^{\delta,c}$. Then, if A is a weakly attracting set for φ^h which is γ-robust or wISDS for some robustness gain γ of class $\mathcal{K}_\infty$ then A is also a γ-robust or wISDS weakly attracting set for Φ_h with robustness gain $\gamma(e^{Lt} \cdot)$.

Conversely, if A is a weakly attracting set for Φ_h which is γ-robust or wISDS for some robustness gain γ of class $\mathcal{K}_\infty$ then A is also a γ-robust or wISDS weakly attracting set for φ^h with robustness gain $\gamma((1 + Lh) \cdot)$.

Proof: Immediate from Lemma 4.8.4 and the definitions of γ-robustness and wISDS for discrete time systems. $\quad\square$

5 Relation between Discretization and Perturbation

In this chapter we investigate how numerical approximations can be embedded into the perturbed systems which have been considered in the last sections. "Embedding" here is to be understood in the sense of Definitions 3.6.5 and 4.6.5 and we will use this embedding in both directions, i.e., the numerical system is considered as a perturbation of the original one, and the original system in turn is interpreted as a perturbation of the numerical system. The combination of these two interpretations will then be used in the analysis of numerical dynamical behavior in the subsequent chapters.

In order to avoid too much technical overhead throughout this chapter we will assume that the system to be approximated satisfies the global bounds (2.8) or (2.24) for $w = 0$. Note that standard cutoff techniques allow to assume these bounds without loss of generality, provided the local estimates (2.7) or (2.23) hold and we are interested only in an approximation on some compact subset of the state space.

In the first section we will provide an abstract framework for one-step time discretization of system (2.1) and show how this can be embedded into the perturbed systems (2.21) and (2.22) based on the time-h map of (2.1) and vice versa. In the second section we will discuss several numerical schemes and show how they fit into the abstract framework of the first section.

The last section is devoted to the discussion of space discretization where we introduce abstract frameworks and (rather briefly) sketch how these can be realized by implementable schemes.

5.1 Time Discretization: Theoretical Framework

In this section we will investigate the time discretization of ordinary differential equations of type (2.1). We will mainly restrict ourselves to schemes with fixed time step, and only give some references on how to treat adaptive timestepping at the end of this section. Let us start by giving an abstract definition of a numerical one step approximation of (2.1).

Definition 5.1.1 (numerical one step approximation)

Consider a system (2.1) satisfying (2.8) with trajectories denoted by $\varphi(t, x, u)$. A family of discrete time systems $\widetilde{\Phi}_h$, $h \in (0, h_0]$ of type (2.17) satisfying (2.24) is called a *numerical one-step approximation* of order $q \in \mathbb{N}$ if there exists a constant $c > 0$ such that the inequality

$$\|\varphi(h, x, u) - \widetilde{\Phi}_h(x, u)\| \leq ch^{q+1}$$

is satisfied for all $x \in \mathbb{R}^n$ and all $u \in \mathcal{U}$. Each map $\widetilde{\Phi}_h$ in the approximating family is called a *numerical one-step system*.   □

A simple inductive proof shows the following error estimate.

Lemma 5.1.2 Consider system (2.1) satisfying (2.8) and a numerical one-step system $\widetilde{\Phi}_h$. Then the estimate

$$\|\varphi(t, x, u) - \widetilde{\Phi}_h(t, x, u)\| \leq \frac{c(e^{Lt} - 1)}{L} h^q$$

holds for each $x \in \mathbb{R}^n$, $u \in \mathcal{U}$ and each $t = ih$, $i \in \mathbb{N}$.

This estimate gives a useful a priori estimate for the discretization error on compact time intervals $[0, T]$, $T > 0$. For $t \to \infty$, however, it is useless since the constant in front of h^q blows up exponentially. Instead of using Lemma 5.1.2 we will use the embedding concept from Definitions 3.6.5 and 4.6.5. The following lemmata show the applicability of these definitions. We start with the case when the input u models a perturbation (which, in particular, applies to systems without input).

Lemma 5.1.3 Consider system (2.1) satisfying (2.8) and a numerical one step system $\widetilde{\Phi}_h$ of order $q \in \mathbb{N}$. Let $W = B(\alpha, 0) \subset \mathbb{R}^n$ and $W^* = B(\alpha - ch^q, 0) \subset \mathbb{R}^n$ for some $\alpha > ch^q$. Then the inflated system

$$x(t + h) = \widetilde{\Phi}_h(x(t), u(t + \cdot)) + \int_t^{t+h} w(\tau)d\tau \tag{5.1}$$

of type (2.21) with $w \in W^*$ is $(ch^q, 1)$-embedded in discrete time system induced by the inflated time-h map

$$x(t + h) = \varphi^h(x(t), u(t + \cdot)) + \int_t^{t+h} w(\tau)d\tau \tag{5.2}$$

with $w \in W$. Conversely, the inflated time-h map (5.2) with $w \in W^*$ is $(ch^p, 1)$-embedded in (5.1) with $w \in W$.

Proof: Immediate from Lemma 3.7.1(iii).   □

Similarly, we obtain the result for controlled systems.

Lemma 5.1.4 Consider system (2.1) satisfying (2.8) and a numerical one step system $\widetilde{\Phi}_h$ of order $q \in \mathbb{N}$. Let $W = \mathcal{B}(\alpha, 0) \subset \mathbb{R}^n$ and $W^* = \mathcal{B}(\alpha - ch^q, 0) \subset \mathbb{R}^n$ for some $\alpha > ch^q$. Then the inflated system

$$x(t+h) = \widetilde{\Phi}_h(x(t), u(t+\cdot)) + \int_t^{t+h} p[u](\tau)d\tau \qquad (5.3)$$

of type (2.22) with $p \in \mathcal{P}^{*,\delta}$ and $\delta = h$ is $(ch^q, 1)$-embedded in discrete time system induced by the inflated time-h

$$x(t+h) = \varphi^h(x(t), u(t+\cdot)) + \int_t^{t+h} p[u](\tau)d\tau \qquad (5.4)$$

with $p \in \mathcal{P}^\delta$ and $\delta = h$. Conversely, the inflated time-h map (5.4) with $p \in \mathcal{P}^{*,\delta}$ and $\delta = h$ is $(ch^p, 1)$-embedded in (5.3) with $p \in \mathcal{P}^\delta$ and $\delta = h$.

Proof: Immediate from Lemma 4.7.1(iv).   □

Note that in this Lemma we need δ-nonanticipating strategies for $\delta = h$ to ensure the embedding of the inflated numerical scheme. Since we have already seen that these are in general more powerful than 0-nonanticipating strategies, it is worth looking at approximations which satisfy the following property.

Definition 5.1.5 (nonanticipating approximations)

Consider system (2.1) satisfying (2.8) and a numerical one step system $\widetilde{\Phi}_h$ of order $q \in \mathbb{N}$. Let $W = \mathcal{B}(\alpha, 0) \subset \mathbb{R}^n$ and $W^* = \mathcal{B}(\alpha - ch^q, 0) \subset \mathbb{R}^n$ for some $\alpha > ch^q$. Then we say that $\widetilde{\Phi}_h$ is a *nonanticipating approximation* if the inflated system

$$x(t+h) = \widetilde{\Phi}_h(x(t), u(t+\cdot)) + \int_t^{t+h} p[u](\tau)d\tau \qquad (5.5)$$

of type (2.22) with $p \in \mathcal{P}^{*,0}$ is $(ch^q, 1)$-embedded in discrete time system induced by the inflated time-h

$$x(t+h) = \varphi^h(x(t), u(t+\cdot)) + \int_t^{t+h} p[u](\tau)d\tau \qquad (5.6)$$

with $p \in \mathcal{P}^0$ and, conversely, the inflated time-h map (5.6) with $p \in \mathcal{P}^{*,0}$ is $(ch^p, 1)$-embedded in (5.5) with $p \in \mathcal{P}^0$.

If the same properties hold with $\mathcal{P}^{*,0,c}$ and $\mathcal{P}^{0,c}$ from (2.16) instead of $\mathcal{P}^{*,0}$ and $\mathcal{P}^0$, respectively, then we call $\widetilde{\Phi}_h$ a *continuous nonanticipating approximation*.   □

The Lemmata 5.1.3 and 5.1.4 as well as Definition 5.1.5 provide all we need in order to apply the results from the Chapters 3 and 4 to our numerical schemes, which we will do in Chapter 6 and 7.

So far we have only provided a framework for numerical schemes with *fixed* equidistant time steps $h > 0$. For ordinary differential equations without inputs, however, nowadays most scientific and commercial software uses adaptive timestepping strategies which control the step size in order to ensure a user defined error tolerance. The problem with applying our techniques to schemes of this kind is that these schemes usually do not allow a representation as a discrete time dynamical system on $\mathbb{R}^n$. A remedy for this problem is provided by representing these schemes by suitable dynamical systems on an enlarged state space.

In Kloeden and Schmalfuss [80, 81] this is done by using the space $\mathcal{H}$ of all two sided sequences of time steps $\mathbf{h} = (h_i)_{i \in \mathbb{Z}}$, satisfying that $h_i \to -\infty$ as $i \to -\infty$, $h_i \to \infty$ as $i \to \infty$, and $|h_{i+1} - h_i|$ is bounded from above and below by some constant independent of $\mathbf{h}$. Using the usual shift dynamics θ on $\mathcal{H}$, i.e., $\theta((h_i)_{i \in \mathbb{Z}}) = (h_{i+1})_{i \in \mathbb{Z}}$, the numerical scheme is then interpreted as a discrete skew product dynamical system or discrete cocycle on $\mathbb{R}^n \times \mathcal{H}$. Consequently, the attracting sets for these discrete time systems have to be understood as cocycle or pullback attracting sets or attractors. These nonautonomous generalization of autonomous attracting sets and attractors are defined using the notion of pullback convergence which goes back at least to Krasnosel'skii [82] (an idea in that direction was used even earlier by Perron), and have recently been used by a number of authors both in random and deterministic settings, see, e.g., [3, 26, 75].

A different approach has recently been presented by Lamba [85]. Here the adaptive scheme is represented by a discrete dynamical system on $\mathbb{R}^n \times \mathbb{R}$ where the additional real state variable represents the time step chosen for the next time step by the step size controller. While with this concept we remain within the framework of autonomous discrete dynamical systems, any reasonable implementation of a step size control will lead to a discontinuous dynamical system. (Besides providing this framework, in [85] also the important question of reliability of adaptive timestepping is considered, i.e., it is investigated how far a small error on compact time intervals can be guaranteed when these kind of schemes are used, see also [86, 112].)

The results by Kloeden and Schmalfuss [80] and Lamba [85] show that under suitable assumptions both frameworks allow results on the approximation of the dynamical behavior of the original systems, more precisely, a generalization of the approximation result for attracting sets by Kloeden and Lorenz [77] for numerical one step approximations with fixed time steps. Both frameworks (i.e., the cocycle formalism and the discontinuous dynamical systems approach), however, lead to deviations from our setup and assumptions. Since our aim is to provide general results about the behavior of numerical approximations avoiding too much technical details, we have decided to restrict ourselves to the simpler case of fixed time steps and hence will not pursue the investigation of adaptive timestepping in what follows. It seems, however, reasonable to expect that most of the results in the subsequent chapters can be generalized to this kind of schemes.

5.2 Time Discretization: Numerical Schemes

In this section we will review several numerical schemes from the literature and show how they fit into the abstract framework from the last section.

For systems without inputs (i.e., $U = \{0\}$), there exists a vast amount of one-step approximations, which are by now standard and included in any textbook on numerical methods for ordinary differential equations, namely Runge-Kutta and Taylor methods. The simplest member of both of these families is the Euler scheme given by

$$\widetilde{\Phi}_h(x) = x + hf(x).$$

Since these schemes are so common we will not describe them in more detail here; the interested reader may consult, e.g., the textbooks [31, 64, 110]. It is immediate from their definition that all these schemes satisfy Definition 5.1.1 provided that the right hand side f of (2.1) satisfies the needed regularity properties (e.g., $q+1$-times continuous differentiability with globally bounded derivatives for a scheme of order q). It should be noted that there exists a detailed analysis about how schemes of this kind relate to nonautonomous perturbations of ordinary differential equations, see, for instance, [37, 128]. All we need for our analysis, however, is the bound on the local error as stated in Definition 5.1.1.

We will now turn to the more non-standard case of numerical schemes for systems with input, i.e., for systems with control or internal perturbation. The following scheme is a straightforward generalization of the Euler scheme for systems with input.

Definition 5.2.1 (Euler scheme for systems with input)

Consider system (2.1) and let $h > 0$. Let $G : \mathbb{R}^n \times \mathcal{U} \to U$ be a map satisfying the implication $u_1(t) = u_2(t)$ for almost all $t \in [0, h] \Rightarrow G(x, u_1) = G(x, u_2)$. Given $x \in \mathbb{R}^n$ and an input function $u \in \mathcal{U}$ we define

$$\widetilde{\Phi}_h(x, u) := x + hf(x, G(x, u)).$$

□

For suitable G this defines an approximation of order $q = 1$ if $f(x, U)$ is convex for each $x \in \mathbb{R}^n$.

Proposition 5.2.2 Consider system (2.1) and assume (2.8). If $f(x, U)$ is convex for each $x \in \mathbb{R}^n$ then there exists $c > 0$ and $h_0 > 0$ such that for each $h \in (0, h_0]$ there exists a map $G : \mathbb{R}^n \times \mathcal{U} \to U$ meeting the assumptions of Definition 5.2.1 and is such that $\tilde{\Phi}_h$ from that definition satisfies (2.8) and

$$\|\varphi(h, x, u) - \tilde{\Phi}_h(x, u)\| \le ch^2,$$

i.e., Definition 5.2.1 defines a one-step approximation of order $q = 1$.

Proof: Note that for each $u \in \mathcal{U}$ and each $x \in \mathbb{R}^n$ the integral

$$\frac{1}{h} \int_0^h f(x, u(t)) dt$$

lies in the convex hull of $f(x, U)$, hence by convexity of $f(x, U)$ we find a map $G : \mathbb{R}^n \times \mathcal{U} \to U$ such that

$$\int_0^h f(x, u(t)) dt = h f(x, G(x, u)).$$

The estimates (2.8) for $\tilde{\Phi}_h$ follow from the respective properties of f, hence of the integral. It remains to show the difference between $\varphi(h, x, u)$ and $\tilde{\Phi}_h(x, u)$. This estimate follows from

$$\|\varphi(h, x, u) - \tilde{\Phi}_h(x, u)\|$$

$$= \left\| x + \int_0^h f(\varphi(t, x, u), u(t)) dt - \tilde{\Phi}_h(x, u) \right\|$$

$$\le \left\| x + \int_0^h f(x, u(t)) dt + \int_0^h f(\varphi(t, x, u), u(t)) - f(x, u(t)) dt - \tilde{\Phi}_h(x, u) \right\|$$

$$= \left\| \int_0^h f(\varphi(t, x, u), u(t)) - f(x, u(t)) dt \right\|$$

$$\le \int_0^h L\|\varphi(t, x, u) - x\| dt \le M L h^2$$

which shows the assertion for $c = ML$. □

Remark 5.2.3 (i) The idea for this scheme was taken from Gonzalez and Tidball [44, Section 4], where a similar but technically more involved construction was used even without convexity assumption on f to yield an approximation of order $1/2$. We do not consider the construction from [44] in its full generality since we were not able to verify the Lipschitz continuity in x for the numerical one-step systems resulting from this construction.

(ii) Since every system of type (2.1) can be approximated by a system satisfying the convexity condition (see, e.g., [22, Proposition 3.2.29]), this scheme is in fact applicable to arbitrary systems of type (2.1).

(iii) The fact that G is not given explicitly may cause difficulties in the implementation when one wants to compute the trajectory for some given $u \in \mathcal{U}$. Nevertheless, in order to make statements about the relation of the dynamical behavior of φ and $\tilde{\Phi}_h$ an explicit knowledge is not necessary.   □

Note that if f is affine in u, i.e., of the form

$$f(x, u, p) = g_0(x) + \sum_{i=1}^{m} u_i g_i(x), \tag{5.7}$$

for globally Lipschitz and bounded vectorfields $g_i : \mathbb{R}^n \to \mathbb{R}^n$, and $U \subset \mathbb{R}^m$ is convex then it is easily seen that G can be chosen as

$$G(x, u) = \frac{1}{h} \int_0^h u(t)\, dt$$

i.e., we get an explicit expression in $u \in \mathcal{U}$ which in addition is independent of x.

We will now discuss a systematic way to obtain one–step approximations of arbitrary order for systems of type (5.7). This approach was developed by Kloeden and the author in [55], similar ideas were already used before for a more restricted class of systems by Ferretti [36].

In order to formulate these schemes it will be convenient to write the solutions of (5.7) in the form

$$\varphi(t, x, u) = x + \sum_{j=0}^{m} \int_{t_0}^{t} g_j(s, \varphi(s, x, u))\, u_j(s)\, ds \tag{5.8}$$

where we have introduced a fictitious input function $u_0(t) \equiv 1$ so that the first integral term can be included in the summation. Furthermore, we will use the notation $g_j = (g_{j,1}, \ldots, g_{j,n})^T$ for each vector field g_j, $j = 0, \ldots, m$.

The main principle for the construction of the schemes is a Taylor–like expansion which is adapted from stochastic calculus (or, more precisely, the numerical analysis of stochastic differential equations, see [79]). In mathematical control theory this is also known as the Fliess or Chen–Fliess expansion, see, e.g., [65, 104]. Here we use the notation from stochastic calculus, which will be more convenient for the actual derivation of the schemes. We need some preliminary definitions.

Definition 5.2.4 (multi–index)

We call a row vector $\mathbf{j} = (j_1, j_2, \ldots, j_l)$, where $j_i \in \{0, 1, \ldots, m\}$ for $i = 1$, $\ldots, l$, a *multi–index* of length $l = l(\mathbf{j}) \geq 1$ and for completeness we write $\odot$ for the multi–index of length zero, that is, with $l(\odot) = 0$. We denote the set of all such multi–indices by $\mathcal{M}_m$, so

$$\mathcal{M}_m = \Big\{ (j_1, j_2, \ldots, j_l) : j_i \in \{0, \ldots, m\}, i \in \{1, \ldots, l\}, l \in \{1, 2, \ldots\} \Big\} \cup \{\odot\}.$$

For any $\mathbf{j} = (j_1, j_2, \ldots, j_l) \in \mathcal{M}_m$ with $l(\mathbf{j}) \geq 1$, denote by $-\mathbf{j}$ and $\mathbf{j}-$ for the multi–index in $\mathcal{M}_m$ obtained by deleting the first and the last component, respectively, of $\mathbf{j}$, thus

$$-\mathbf{j} = (j_2, \ldots, j_l) \qquad \mathbf{j}- = (j_1, \ldots, j_{l-1}).$$

In addition, define the concatenation of any two multi–indices $\mathbf{j} = (j_1, j_2, \ldots, j_k)$ and $\bar{\mathbf{j}} = (\bar{j}_1, \bar{j}_2, \ldots, \bar{j}_l)$ in $\mathcal{M}_m$ by

$$\mathbf{j} * \bar{\mathbf{j}} = (j_1, j_2, \ldots, j_k, \bar{j}_1, \bar{j}_2, \ldots, \bar{j}_l), \tag{5.9}$$

that is, the multi–index formed by adjoining the two given multi–indices. Finally, define $n(\mathbf{j})$ to be the number of components of a multi–index $\mathbf{j} \in \mathcal{M}_m$ that are equal to 0. $\qquad\qquad\square$

In the following definition we decided to keep the term "multiple control integral" from [55], although in our context, of course, $u \in \mathcal{U}$ can be either control or internal perturbation.

Definition 5.2.5 (multiple control integrals)

For a multi–index $\mathbf{j} = (j_1, j_2, \ldots, j_l) \in \mathcal{M}_m$, some input function $u \in \mathcal{U}$ and an integrable function $f : [t_0, T] \to \mathbb{R}$ we define the *multiple control integral* $I_{\mathbf{j}}[f(\cdot)]_{t_0, t, u}$ recursively by

$$I_{\mathbf{j}}[f(\cdot)]_{t_0, t, u} := \begin{cases} f(t) & : l = 0 \\ \int_{t_0}^{t} I_{\mathbf{j}-}[f(\cdot)]_{t_0, s, u}\, u_{j_l}(s)\, ds & : l \geq 1 \end{cases}.$$

$\square$

Note that $I_{\mathbf{j}}[f(\cdot)]_{t_0, \cdot, u} : [t_0, T] \to \mathbb{R}$ is continuous, hence integrable, so the iterated integrals are well defined.

Example 5.2.6

$$I_\odot[f(\cdot)]_{t_0,t,u} = f(t), \quad I_{(0)}[f(\cdot)]_{t_0,t,u} = \int_{t_0}^{t} f(s)\, ds,$$

$$I_{(1)}[f(\cdot)]_{t_0,t,u} = \int_{t_0}^{t} f(s)\, u_1(s)\, ds,$$

$$I_{(0,1)}[f(\cdot)]_{0,t,u} = \int_{0}^{t} \int_{0}^{s_2} f(s_1) u_1(s_2)\, ds_1 ds_2,$$

$$I_{(0,2,1)}[f(\cdot)]_{0,t,u} = \int_{0}^{t} \int_{0}^{s_3} \int_{0}^{s_2} f(s_1)\, u_2(s_2)\, u_1(s_3)\, ds_1 ds_2 ds_3.$$

□

For simpler notation, we shall sometimes abbreviate $I_{\mathbf{j}}[f(\cdot)]_{t_0,t,u}$ to $I_{\mathbf{j},t_0,t,u}$ or just $I_{\mathbf{j},u}$ when $f(t) \equiv 1$.

The following Lemma gives an a priori estimate on $|I_{\mathbf{j}}[f(\cdot)]_{t_0,t,u}|$ which we will need in what follows.

Lemma 5.2.7 Assume that $|f|$ is globally bounded by some constant $M > 0$. Let $\bar{u} := \max_{u \in U, j=1,\ldots,m} |u_j|$ and consider a multi-index $\mathbf{j} \in \mathcal{M}_m$ with length $l = l(\mathbf{j})$. Then the inequality

$$|I_{\mathbf{j}}[f(\cdot)]_{t_0,t,u}| \leq \bar{u}^l (t - t_0)^l M$$

holds for all $u \in \mathcal{U}$ and all $t \geq t_0$ and, if $l \geq 1$, the inequality

$$\left| \frac{d}{dt} I_{\mathbf{j}}[f(\cdot)]_{t_0,t,u} \right| \leq \bar{u}^l (t - t_0)^{l-1} M$$

holds for all $u \in \mathcal{U}$ and almost all $t \geq t_0$.

Proof: We show the first assertion by induction over l. For $l = 0$ it is immediately clear. For $l \to l+1$ consider a multi-index $\mathbf{j}$ of length $l+1$. Then we obtain

$$|I_{\mathbf{j}}[f(\cdot)]_{t_0,t,u}| = \left| \int_{t_0}^{t} I_{\mathbf{j}-}[f(\cdot)]_{t_0,s,u}\, u_{j_l}(s)\, ds \right|$$

$$\leq \left| \int_{t_0}^{t} \bar{u}^l (t - t_0)^l M u_{j_l}(s)\, ds \right|$$

$$\leq \left| (t - t_0) \bar{u}^l (t - t_0)^l M \bar{u} \right| = \bar{u}^{l+1} (t - t_0)^{l+1} M$$

which shows the first assertion. Using the first inequality we obtain for any multi-index $\mathbf{j}$ of length $l = l(\mathbf{j}) \geq 1$

$$\left|\frac{d}{dt}I_{\mathbf{j}}[f(\cdot)]_{t_0,t,u}\right| = |I_{\mathbf{j}-}[f(\cdot)]_{t_0,t,u}\,u_{j_l}(t)|$$

$$\leq \bar{u}^{l-1}(t-t_0)^{l-1}M|u_{j_l}(t)| \leq \bar{u}^l(t-t_0)^{l-1}M$$

which shows the second assertion. ☐

Definition 5.2.8 (coefficient function)

For each $\mathbf{j} = (j_1, \ldots, j_l) \in \mathcal{M}_m$ and function $F : [t_0, T] \times \mathbb{R}^n \to \mathbb{R}$, the *coefficient function* $F_{\mathbf{j}}$ is defined recursively by

$$F_{\mathbf{j}} = \begin{cases} F & : \ l = 0 \\ L^{j_1} F_{-\mathbf{j}} & : \ l \geq 1. \end{cases} \qquad (5.10)$$

where the partial differential operators are defined by

$$L^0 = \frac{\partial}{\partial t} + \sum_{k=1}^n g_{0,k}\frac{\partial}{\partial x^k}, \qquad L^j = \sum_{k=1}^n g_{j,k}\frac{\partial}{\partial x^k}, \quad j = 1,\ldots,m. \qquad (5.11)$$

☐

Of course, this definition requires the functions $F, g_0, g_1, \ldots, g_m$ to be sufficiently smooth.

Example 5.2.9 In the one-dimensional case with $n = 1$ for the identity function $F(t, x) \equiv x$ we have

$$F_{(0)} = g_0, \quad F_{(j_1)} = g_{j_1}, \quad F_{(0,0)} = g_0 g_0',$$

$$F_{(0,j_1)} = g_0 g_{j_1}', \quad F_{(j_1,0)} = g_0' g_{j_1}, \quad F_{(j_1,j_2)} = g_{j_1} g_{j_2}',$$

where the dash $'$ denotes differentiation with respect to x. ☐

Definition 5.2.10 (hierarchical and remainder set)

A subset $\mathcal{H} \subset \mathcal{M}_m$ is called a *hierarchical set* if $\mathcal{H}$ is nonempty, if the multi–indices in $\mathcal{H}$ are uniformly bounded in length, that is $\sup_{\mathbf{j} \in \mathcal{H}} l(\mathbf{j}) < \infty$, and if

$$-\mathbf{j} \in \mathcal{H} \quad \text{for each} \quad \mathbf{j} \in \mathcal{H} \setminus \{\odot\},$$

where $\odot$ is the multi–index of length zero.

For a given hierarchical set $\mathcal{H}$ the corresponding *remainder set* $\mathcal{RS}(\mathcal{H})$ is defined by

$$\mathcal{RS}(\mathcal{H}) = \{\mathbf{j} \in \mathcal{M}_m \setminus \mathcal{H} : -\mathbf{j} \in \mathcal{H}\}.$$

☐

In other words, if a multi-index j belongs to an hierarchical set, then so does the multi-index $-j$ obtained by deleting the first component of j.

Accordingly, the remainder set consists of all of the next following multi-indices with respect to the given hierarchical set that do not already belong to the hierarchical set and is formed simply by adding a further component taking all possible values at the beginning of the "maximal" multi-indices in the hierarchical set.

Using this notation we can now state the following Taylor expansion for control systems.

Theorem 5.2.11 Let $F : \mathbb{R}_0^+ \times \mathbb{R}^n \to \mathbb{R}$ and let $\mathcal{H} \subset \mathcal{M}_m$ be an hierarchical set with remainder set $\mathcal{RS}(\mathcal{H})$. Then the following Taylor expansion corresponding to the hierarchical set $\mathcal{H}$

$$F(t, \varphi(t, x, u)) = \sum_{j \in \mathcal{H}} I_j [F_j(t_0, \varphi(t_0, x, u))]_{t_0, t} + \sum_{j \in \mathcal{RS}(\mathcal{H})} I_j [F_j(\cdot, \varphi(\cdot, x, u))]_{t_0, t}$$

(5.12)

holds for all $x \in \mathbb{R}^n$, all $t \geq t_0 \geq 0$ and all $u \in \mathcal{U}$, provided all of the derivatives of F, $g_0, g_1, \ldots, g_m$ and all of the multiple control integrals appearing here exist.

Proof: The proof follows that of the stochastic Ito-Taylor expansion from [79, Theorem 5.5.1].

We abbreviate $x(t) = \varphi(t, x_0, u)$ and apply the integrated version of the chain rule for the types of functions under consideration [45], that is

$$F(t, x(t)) = F(t_0, x(t_0)) + \sum_{j=0}^{m} I_{(j)}[L^{(j)}F(\cdot, x(\cdot))]_{t_0, t},$$ (5.13)

to the function F_j for some multi-index $j \in \mathcal{H}$ to obtain

$$I_j[F_j(\cdot, x(\cdot))]_{t_0, t} = I_j[F_j(t_0, x(t_0))]_{t_0, t} + I_j \left[\sum_{j=0}^{m} I_{(j)}[L^{(j)}F_j(\cdot, x(\cdot))]_{t_0, \cdot} \right]_{t_0, t}$$

$$= I_j[F_j(t_0, x(t_0))]_{t_0, t} + \sum_{j=0}^{m} I_{(j) \ast j}[F_{(j) \ast j}(\cdot, x(\cdot))]_{t_0, t} \quad (5.14)$$

We shall verify the expression in the theorem by induction over $k := \max\{l(j) \mid j \in \mathcal{H}\}$. For $k = 0$, the hierarchical set is simply $\mathcal{H} = \{\odot\}$, so the assertion follows directly from (5.13). For $k \geq 1$ consider the hierarchical set $\mathcal{E} := \{j \in \mathcal{H} \mid l(j) \leq k - 1\}$. Then

$$F(t, x(t)) = \sum_{j \in \mathcal{E}} I_j[F_j(t_0, x(t_0))]_{t_0, t} + \sum_{j \in \mathcal{RS}(\mathcal{E})} I_j[F_j(\cdot, x(\cdot))]_{t_0, t}$$

holds by the induction assumption and, since by the definition of a remainder set we know that $\mathcal{H} \setminus \mathcal{E} \subseteq \mathcal{RS}(\mathcal{E})$, we can conclude

$$F(t, x(t)) = \sum_{j \in \mathcal{E}} I_j[F_j(t_0, x(t_0))]_{t_0, t}$$

$$+ \sum_{j \in \mathcal{H} \setminus \mathcal{E}} I_j[F_j(\cdot, x(\cdot))]_{t_0, t} + \sum_{j \in \mathcal{RS}(\mathcal{E}) \setminus (\mathcal{H} \setminus \mathcal{E})} I_j[F_j(\cdot, x(\cdot))]_{t_0, t}$$

$$= \sum_{j \in \mathcal{H}} I_j[F_j(t_0, x(t_0))]_{t_0, t} + \sum_{j \in \mathcal{B}} I_j[F_j(\cdot, x(\cdot))]_{t_0, t}$$

with the last equality following from (5.14) with $\mathcal{B} = (\mathcal{RS}(\mathcal{E}) \setminus (\mathcal{H} \setminus \mathcal{E})) \cup \{(j) * j \mid j = 0, \ldots, m, j \in \mathcal{H} \setminus \mathcal{E}\}$. Finally, since the definition of a remainder set implies that $\mathcal{B} = (\mathcal{RS}(\mathcal{E}) \setminus \mathcal{H}) \cup \mathcal{RS}(\mathcal{H} \setminus \mathcal{E}) = \mathcal{RS}(\mathcal{H})$, we obtain the desired expression. $\qquad\square$

Note the difference between the two summands in (5.12): In the first the integrands are constant, since F_j depends on the fixed time t_0 while in the second the integrands are functions.

Let us illustrate this expansion by two examples.

Example 5.2.12 (i) For hierarchical and remainder sets given by

$$\mathcal{H} = \{\circ\} \quad \text{and} \quad \mathcal{RS}(\{\circ\}) = \{(0), \cdots, (m)\}$$

the Taylor expansion is

$$F(t, \varphi(t, x, u))$$

$$= I_\circ [F_\circ(t_0, \varphi(t_0, x, u))]_{t_0, t} + \sum_{j \in \mathcal{RS}(\{\circ\})} I_j [F_j(\cdot, \varphi(\cdot, x, u))]_{t_0, t}$$

$$= F(t_0, \varphi(t_0, x, u)) + \int_{t_0}^t L^0 F(s, x(s)) \, ds + \sum_{j=1}^m \int_{t_0}^t L^j F(s, \varphi(s, x, u)) u_j(s) ds$$

(ii) In the scalar case $n = m = 1$ with $F(t, x) \equiv x$, $t_0 = 0$ and hierarchical and remainder sets given by

$$\mathcal{H} = \{j \in \mathcal{M}_1 : l(j) \leq 2\} \quad \text{and} \quad \mathcal{RS}(\mathcal{H}) = \{j \in \mathcal{M}_1 : l(j) = 3\}$$

the Taylor expansion reads

$$\varphi(t, x, u) = x + g_0(x) I_{(0)} + g_1(x) I_{(1)} + g_0(x) g_0{}'(x) I_{(0,0)}$$

$$+ g_0(x) g_1(x)' I_{(0,1)} + g_1(x) g_0{}'(x) I_{(1,0)}$$

$$+ g_1(x) g_1{}'(x) I_{(1,1)} + R_3(t, t_0),$$

where the integrals are over the interval $[t_0, t]$, the dash ' denotes differentiation with respect to x and $R_3(t, t_0)$ is the corresponding remainder term collecting all the summands for $\mathbf{j} \in \mathcal{RS}(\mathcal{H})$. $\qquad\square$

Theorem 5.2.11 forms the basis for the construction of the following Taylor schemes for systems with input $u \in \mathcal{U}$.

Definition 5.2.13 (Taylor scheme)

For $q \in \mathbb{N}$, $h > 0$ and $x = (x_1, \ldots, x_n)^T \in \mathbb{R}^n$ we define the *Taylor scheme* $\widetilde{\Phi}_h = (\widetilde{\Phi}_{h,1}, \ldots, \widetilde{\Phi}_{h,n})^T$ of order q for the system (5.8) componentwise by

$$\widetilde{\Phi}_{h,k}(x, u) = x_k + \sum_{\mathbf{j} \in \mathcal{H}_q \setminus \{\mathbf{0}\}} F_{\mathbf{j},k}(0, x)\, I_{\mathbf{j},h,u} \qquad (5.15)$$

for $k = 1, \ldots, n$, where the coefficient functions $F_{\mathbf{j}} = (F_{\mathbf{j},1}, \ldots, F_{\mathbf{j},n})^T$ are constructed from $F(t, x) = x$, the hierarchical set $\mathcal{H}_q$ is given by

$$\mathcal{H}_q = \{\mathbf{j} \in \mathcal{M}_m : l(\mathbf{j}) \le q\}$$

and the multiple control integrals are

$$I_{\mathbf{j},h,u} = \int_0^h \int_0^{s_l} \cdots \int_0^{s_2} u_{j_1}(s_1) \cdots u_{j_l}(s_l)\, ds_1 \cdots ds_l.$$

$\qquad\square$

Note that for $q = 1$ we obtain

$$\widetilde{\Phi}_h(x, u) = hg_0(x) + \sum_{j=1}^m g_j(x) \int_0^h u_j(t) dt,$$

which is the generalized Euler scheme from Definition 5.2.1 with $G(x, u) = \frac{1}{h} \int_0^h u(t) dt$.

The remainder set $\mathcal{RS}(\mathcal{H}_q)$ is given by

$$\mathcal{RS}(\mathcal{H}_q) = \{\mathbf{j} \in \mathcal{M}_m : l(\mathbf{j}) = q + 1\}.$$

Hence, if the coefficient functions $g_0, g_1, \ldots, g_m$ of the system (5.8) are q times continuously differentiable with globally bounded derivatives then Lemma 5.2.7 yields the existence of $c > 0$ and $h_0 > 0$ such that for all $h \in (0, h_0]$ we obtain

$$\|R_q(h, x, u)\| \le ch^{q+1}$$

for all $u \in \mathcal{U}$, all $x \in \mathbb{R}^n$ and the remainder term $R_q = (R_{q,1}, \ldots, R_{q,n})^T$ given componentwise by

$$R_{q,k}(h, x, u) := \sum_{\mathbf{j} \in \mathcal{RS}(\mathcal{H}_q)} I_{\mathbf{j},h,u} \left[F_{k,\mathbf{j}}(\cdot, \varphi(\cdot, x, u)) \right] \tag{5.16}$$

with coefficient functions $F_{k,\mathbf{j}}$ corresponding to $F_k(t, x) = x_k$ for $k = 1, \ldots, n$. Hence Theorem 5.2.11 and Lemma 5.2.7 imply that Definition 5.2.13 indeed gives a scheme of order q. Similarly to what we discussed for assumption (2.8) in the introduction to this chapter, if the derivatives exist but are only locally bounded then we can apply a suitable cutoff technique to obtain the order of the scheme for x in arbitrary compact subsets of $\mathbb{R}^n$ with $c > 0$ depending on this set.

Clearly, under these global assumptions on the g_j these schemes are numerical one step approximations of order q in the sense of Definition 5.1.1. If we interpret the underlying system (2.1) as a control system we can prove the following proposition.

Proposition 5.2.14 Consider system (2.15) and the Taylor approximation $\widetilde{\Phi}_h$ from Definition 5.2.13 for some $q \in \mathbb{N}$. Assume that the coefficient functions g_j, $j = 0, \ldots, m$ are q times continuously differentiable with globally bounded derivatives. Then $\widetilde{\Phi}_h$ is a continuous nonanticipating approximation in the sense of Definition 5.1.5.

Proof: Let $x \in \mathbb{R}^n$, $u \in U$ and $p^* \in \mathcal{P}^{*,0,c}$ and consider the function

$$x(s) = \widetilde{\Phi}_s(x, u) + \int_0^s p^*[u](\tau) d\tau.$$

for $s \in [0, h]$. In order to prove the first $(ch^q, 1)$-embedding property from Definition 4.6.5 we have to show the existence of a $p \in \mathcal{P}^{0,c}$ such that $\|p[u](t)\| \leq ch^q + \|p^*[u](t)\|$ for all $u \in \mathcal{U}$ and almost all $t \in [0, h]$ and $\varphi(h, x, u) + \int_0^h p[u](\tau) d\tau = x(h)$.

By Theorem 5.2.11 we obtain that

$$x(s) = \varphi^s(x, u) + \int_t^{t+s} p^*[u](\tau) d\tau - R_q(s, x, u).$$

From Lemma 5.2.7 and the definition of R_q in (5.16) we obtain the inequality $\|\frac{d}{ds} R_q(s, x, u)\| \leq cs^q$ for almost all $s \geq 0$ and the constant c of the order estimate of the scheme. Hence setting $p[u](t) = -\frac{d}{dt} R_q(t, x, u) + p^*[u](t)$ (which defines $p[u]$ for almost all $t \in [0, h]$) we obtain

$$\varphi^h(x, u) + \int_0^h p[u](\tau) d\tau = x(h),$$

i.e., the desired equality. The fact that $p \in \mathcal{P}^{0,c}$ is immediate from its definition which only incorporates the values $p^*[u](t)$, $u(t)$ and multiple control integrals from 0 to t over u along the solution φ.

The second embedding property follows analogously.   □

Although the use of computer algebra systems like, e.g., MAPLE has greatly simplified the calculation of the derivatives occurring in these schemes (see for instance the MAPLE routines for stochastic differential equations described in [27] which can also be applied to deterministic systems with input), it might be favorable to have schemes which avoid the explicit use of derivatives of the g_j in the right hand side of our equations.

Here we will briefly sketch how a second order derivative free scheme can be derived.

Consider the second order Taylor scheme with order $q = 2$. In the scalar case with a one dimensional input u, that is with $n = m = 1$, this scheme reads

$$\widetilde{\Phi}_h(x, u) = x + g_0(x)\, h + g_1(x)\, I_{(1),h,u}$$

$$+ \frac{1}{2}\, g_0(x) g_0{}'(x)\, h^2 + g_0(x) g_1{}'(x)\, I_{(0,1),h,u}$$

$$+ g_1(x) g_0{}'(x)\, I_{(1,0),h,u} + g_1(x) g_1{}'(x)\, I_{(1,1),h,u},$$

where the dash $'$ denotes differentiation with respect to x.

By the ordinary Taylor expansion we have

$$g_j(x) g_i{}'(x) = \frac{1}{h}\left(g_i\left(x + g_j(x)\, h \right) - g_i(x) \right) + O(h),$$

so the (i, j) term in the above Taylor scheme reads

$$g_j(x) g_i{}'(x)\, I_{(i,j),h,u}$$

$$= \left(\frac{1}{h}\left(g_i\left(x + g_j(x)\, h \right) - g_i(x) \right) + O(h) \right) I_{(i,j),h,u}$$

$$= \frac{1}{h}\left(g_i\left(x + g_j(x)\, h \right) - g_i(x) \right) I_{(i,j),h,u} + O(h^3)$$

since $O(h)\, I_{(i,j),h,u} = O(h^3)$. The remainder here is of the same order as the local discretization error, so we can replace the term on the left by that on the right without reducing the order of the resulting scheme. In this way we obtain the second order derivative–free scheme

$$\widetilde{\Phi}_h(x, u) = x + g_0(x)\, h + g_1(x)\, I_{(1),h,u} + \frac{1}{2}\left(g_0\left(x + g_0(x)\, h \right) - g_0(x) \right) h$$

$$+ \frac{1}{h} \sum_{\substack{i,j=0 \\ i+j \neq 0}}^{1} \left(g_i\left(x + g_j(x)\, h \right) - g_i(x) \right) I_{(i,j),h,u}$$

$$= x + \frac{1}{2}g_0(x)\,h + g_1(x)\,I_{(1),h,u} + \frac{1}{2}g_0\left(x + g_0(x)\,h\right)h$$

$$+ \frac{1}{h}\sum_{\substack{i,j=0\\i+j\neq0}}^{1}\left(g_i\left(x + g_j(x)\,h\right) - g_i(x)\right)I_{(i,j),h,u}$$

in the case where $n = m = 1$. This was also obtained by Ferretti [36] when the vector field g_1 is equal to a constant.

Similarly one constructs the second order derivative–free scheme for an affine system with input for arbitrary $n, m \in \mathbb{N}$. This scheme has the components given by

$$\widetilde{\Phi}_{h,k}(x, u) = x_k + \frac{1}{2}g_{0,k}(x)\,h + \sum_{j=1}^{m}g_{j,k}(x)\,I_{(j),h,u} \tag{5.17}$$

$$+\frac{1}{2}g_{0,k}\left(x + g_0(x)\,h\right)h$$

$$+\frac{1}{h}\sum_{\substack{i,j=0\\i+j\neq0}}^{m}\left(g_{i,k}\left(x + g_j(x)\,h\right) - g_{i,k}(x)\right)I_{(i,j),h,u}$$

for $k = 1, \ldots, n$. In the case of an ordinary differential equation without input, that is with $g_j(x) \equiv 0$ for $j = 1, \ldots, m$, this is just the second order Runge–Kutta scheme known as the Heun scheme.

Clearly, (5.17) defines a numerical one-step approximation in the sense of Definition 5.1.1. Due to the structure of the error term also this scheme is a continuous nonanticipating approximation in the sense of Definition 5.1.5, which can be verified by the same proof as for Proposition 5.2.14.

We end this section by a brief discussion of some implementational issues. In a practical implementation of these schemes one also needs to evaluate the multiple control integrals $I_{j,h,u}$, or—what is sufficient if one is only interested in an approximation of the reachable set—calculate or approximate the set of input values

$$\mathcal{I}_{\mathcal{H},h,\mathcal{U}} := \bigcup\{((I_{j_1,h,u}, \ldots, I_{j_p,h,u}) \mid u \in \mathcal{U}\} \subset \mathbb{R}^{pm}$$

related to some hierarchical set $\mathcal{H} = \{j_1, \ldots, j_p\}$, which are needed as an input for the numerical schemes. It seems that a fully satisfactory solution to this problem has not been found until now, partial answers, however, can be found in [36] (explicit calculation of $\mathcal{I}$ for special cases of $\mathcal{H}$) and [55] (numerical approximation of I and $\mathcal{I}$ and use of MAPLE for the evaluation of I).

Note that under additional structural assumptions on the g_j the schemes above simplify considerably, see [55, Section 8]. For instance, if the vector fields g_j commute, i.e., $L^i g_{j,k} = L^j g_{i,k}$, then one can use integration by parts to obtain

$$L^i g_{j,k}(x) I_{(i,j),h} + L^j g_{i,k}(x) I_{(j,i),h} = L^i g_{j,k}(x) I_{(i),h} I_{(j),h}$$

which involves only multiple control integrals of lower multiplicity, which are easier to compute. This condition was also used by Veliov [124] to obtain a second order scheme for the approximation of the reachable set.

5.3 Space Discretization

In this section we will introduce two models for a space discretization of discrete time systems (2.17). For both models we will use the following very general underlying state space discretization.

Definition 5.3.1 (cells and cell covering)

Consider a compact set $\Omega \subset \mathbb{R}^n$. A *cell covering* $\mathbf{Q} = (Q_i)_{i=1,\ldots,P}$ of Ω is a finite family of closed sets Q_i, $i = 1, \ldots, P$, $P \in \mathbb{N}$, with nonvoid interior such that $\operatorname{int} Q_i \cap \operatorname{int} Q_j = \emptyset$ for all $i \neq j$ and $\bigcup_{i=1,\ldots,P} Q_i = \Omega$. The sets Q_i are called the *cells* of the discretization. With $\mathcal{C}_\mathbf{Q}$ we denote the family of all possible unions of cells in $\mathbf{Q}$, i.e.,

$$\mathcal{C}_\mathbf{Q} := \{ C \subseteq \Omega \mid C = \bigcup_{i \in I} Q_i \text{ for some index set } I \subseteq \{1, \ldots, P\}, I \neq \emptyset \}.$$

The value $\operatorname{diam}(Q_i) := \max_{x,y \in Q_i} \|x - y\|$ is called the *diameter* of Q_i, and the value $\operatorname{diam}(\mathbf{Q}) := \max_{i=1,\ldots,P} \operatorname{diam}(Q_i)$ is called the *maximal diameter* of $\mathbf{Q}$. □

We will later consider more specific examples of such a space discretization, namely we will choose the Q_i as simplices or cuboids. In this section, however, it will be convenient to consider this general model since in this way we can avoid unnecessary technicalities. The following simple lemma will be useful later.

Lemma 5.3.2 Consider a cell covering $\mathbf{Q}$ of some compact set $\Omega \subset \mathbb{R}^n$.

(i) Let $\varepsilon < \operatorname{dist}(Q_i, Q_j)$ for all $i, j = 1, \ldots, P$ with $i \neq j$. Then for any two sets C_1, $C_2 \in \mathcal{C}_\mathbf{Q}$ the implication

$$d_H(C_1, C_2) \leq \varepsilon \Rightarrow C_1 = C_2$$

holds.

(ii) Let C_i, $i \in \mathbb{N}$ be a sequence of sets in $\mathcal{C}_\mathbf{Q}$. Then there exists a subsequence $i_j \to \infty$ such that $C_{i_j} = C_{i_k}$ for all $j, k \in \mathbb{N}$.

Proof: (i) Follows immediately from the fact that any two sets $C_1, C_2 \in \mathcal{C}_\mathbf{Q}$ with $C_1 \neq C_2$ have a distance greater or equal $\mathrm{dist}(Q_i, Q_j)$ for some $i, j \in \{1, \ldots, P\}$ with $i \neq j$.

(ii) This holds since there are only finitely many different sets in $\mathcal{C}_\mathbf{Q}$.   □

Let us now define approximating maps for Φ_h on $\mathcal{C}_\mathbf{Q}$. For this we denote by $P(\mathbb{R}^n)$ the set of all subsets of $\mathbb{R}^n$.

Definition 5.3.3 (point–cell space discretization)

Consider a cell covering $\mathbf{Q}$ of some compact set $\Omega \subset \mathbb{R}^n$ and a discrete time system Φ_h^0 of type (2.17). Consider real numbers $\Delta_i \in \mathbb{R}^+$ for $i = 1, \ldots, P$. Then a set valued map $\overline{\Phi}_h : \Omega \times \mathcal{U} \to P(\mathbb{R}^n)$ is called a *point–cell space discretization* of Φ_h^0 with accuracy Δ_i if $\mathrm{diam}(Q_i) \leq \Delta_i$ and for all $x \in \Omega$ with $x \in Q_i$ and all $u \in \mathcal{U}$ the map $\overline{\Phi}_h$ satisfies

$$\overline{\Phi}_h(x, u) \cap \Omega \in \mathcal{C}_\mathbf{Q} \cup \{\emptyset\}$$

and

$$d_H(\{\Phi_h^0(x, u)\}, \overline{\Phi}_h(x, u)) \leq \Delta_i.$$

The point–cell space discretization $\overline{\Phi}_h(x, u)$ is called *rigorous*, if

$$\Phi_h^0(x, u) \in \overline{\Phi}_h(x, u)$$

for all $x \in \Omega$ and all $u \in \mathcal{U}$.   □

For the following definition recall the definition of $\mathcal{U}(B)$ from (2.3).

Definition 5.3.4 (cell-cell space discretization)

Consider a cell covering $\mathbf{Q}$ of some compact set $\Omega \subset \mathbb{R}^n$ and a discrete time system Φ_h^0 of type (2.17). Consider real numbers $\Delta_i \in \mathbb{R}^+$ for $i = 1, \ldots, P$. Then a set valued map $\widehat{\Phi}_h : \mathcal{C}_\mathbf{Q} \times \mathcal{U}(\Omega) \to P(\mathbb{R}^n)$ is called a *cell-cell space discretization* of Φ_h^0 with accuracy Δ_i if $\mathrm{diam}(Q_i) \leq \Delta_i$, for all $i = 1, \ldots, P$ and all $\bar{u} \in \mathcal{U}(\Omega)$ the map $\widehat{\Phi}_h$ satisfies

$$\widehat{\Phi}_h(Q_i, \bar{u}) \cap \Omega \in \mathcal{C}_\mathbf{Q} \cup \{\emptyset\}.$$

Furthermore, $\widehat{\Phi}_h(Q_i, \bar{u})$ is supposed to satisfy the following two conditions for $i = 1, \ldots, P$:

(i) For each $\bar{u} \in \mathcal{U}(\Omega)$ the inequality

$$\mathrm{dist}(\widehat{\Phi}_h(Q_i, \bar{u}), \Phi_h^0(Q_i, \bar{u})) \leq \Delta_i \tag{5.18}$$

holds, where for any $\bar{u} \in \mathcal{U}(\Omega)$ we denote

$$\Phi_h^0(Q_i, \bar{u}) := \bigcup_{x \in Q_i} \{\Phi_h^0(x, \bar{u}(x, \cdot))\}.$$

(ii) For each $u \in \mathcal{U}$ and each $x \in Q_i$ there exists $\bar{u} \in \mathcal{U}(\Omega)$ such that

$$\text{dist}(\{\Phi_h^0(x, u)\}, \widehat{\Phi}_h(Q_i, \bar{u})) \leq \Delta_i \tag{5.19}$$

holds and, conversely, for each $\bar{u} \in \mathcal{U}(\Omega)$ and each $x \in Q_i$ there exists $u \in \mathcal{U}$ such that (5.19) holds.

The cell-cell space discretization $\widehat{\Phi}_h(x, u)$ is called *rigorous*, if (ii) holds with 0 instead of Δ_i.

For $C \in \mathcal{C}_{\mathbf{Q}}$ and some sequence $\bar{\mathbf{u}} = (\bar{u}_j)_{j \in \mathbb{N}_0}$ with $\bar{u}_j \in \mathcal{U}(\Omega)$, $j = 0, 1, \ldots$ we define the *iterate* $\widehat{\Phi}_h(ih, C, \bar{\mathbf{u}})$, $i \in \mathbb{N}_0$, of $\widehat{\Phi}_h$ (relative to Ω) inductively by $\widehat{\Phi}_h(0, C, \bar{\mathbf{u}}) := C$ and

$$\widehat{\Phi}_h((i + 1)h, C, \bar{\mathbf{u}}) := \widehat{\Phi}_h(\widehat{\Phi}_h(ih, C, \bar{\mathbf{u}}) \cap \Omega, \bar{u}_i) \tag{5.20}$$

for $i \in \mathbb{N}_0$. □

Remark 5.3.5 (i) Note that $\widehat{\Phi}_h(ih, C, \bar{\mathbf{u}}) = \emptyset$ is possible.

(ii) The conditions (i)–(ii) are weaker than the single condition

$$d_H(\widehat{\Phi}_h(Q_i, \bar{u}), \Phi_h^0(Q_i, \bar{u})) \leq \Delta_i.$$

For systems without input, however, they are equivalent.

(iii) If Φ_h denotes the inflated system (2.26) corresponding to Φ_h^0 then it is immediate from Lemma 3.7.1(iii) and 4.7.1(iv) that for all $x \in \Omega$ with $x \in Q_i$, all $u \in \mathcal{U}$ and all $\bar{u} \in \mathcal{U}(\Omega)$ the inclusions

$$\overline{\Phi}_h(x, u) \subseteq \bigcup_{w \in W_{\Delta_i/h}} \{\Phi_h(x, u, w)\}$$

and

$$\widehat{\Phi}_h(Q_i, \bar{u}) \subseteq \bigcup_{w \in W_{\Delta_i/h}} \bigcup_{x \in Q_i} \{\Phi_h(x, \bar{u}(x, \cdot), w)\}$$

hold if u is considered as internal perturbation, and the inclusions

$$\overline{\Phi}_h(x, u) \subseteq \bigcup_{p \in \mathcal{P}_{\Delta_i/h}^h} \{\Phi_h(x, u, p[u])\}$$

and

$$\widehat{\Phi}_h(Q_i, \bar{u}) \subseteq \bigcup_{p \in \mathcal{P}_{\Delta_i/h}^h} \bigcup_{x \in Q_i} \{\Phi_h(x, \bar{u}(x, \cdot), p[\bar{u}(x, \cdot)])\}$$

hold if u is considered as a control function. □

Remark 5.3.6 Observe that any point-cell space discretization $\overline{\Phi}_h$ induces a cell-cell space discretization $\widehat{\Phi}_h$ by defining

$$\widehat{\Phi}_h(Q_i, \bar{u}) := \bigcup_{x \in Q_i} \overline{\Phi}_h(x, \bar{u}(x, \cdot)).$$

□

Let us briefly discuss how space discretizations meeting the Definitions 5.3.3 and 5.3.4 can be constructed for a given map Φ_h^0. For a given cell covering $\mathbf{Q}$ of some set Ω the natural way to construct a rigorous point–cell discretization is given by

$$\overline{\Phi}_h(x, u) := \bigcup_{i \in I(x)} Q_i \cup \Phi_h^0(x, u) \tag{5.21}$$

where $I(x) := \{i \in \{1, \ldots, P\} \mid \Phi_h^0(x, u) \in Q_i\}$. Here we can guarantee the accuracy $\Delta_i \leq \mathrm{diam}(\mathbf{Q})$.

In order to define an implementable cell–cell space discretization, to simplify the presentation we first consider systems without input u. We want to base our construction on $\overline{\Phi}_h$ from (5.21). Clearly, the construction from Remark 5.3.6 does not lead to an implementable map, since it involves the evaluation of $\overline{\Phi}_h$ at infinitely many points. The idea lies in the evaluation of $\overline{\Phi}_h$ in a finite number of *test points* in each Q_i. If, for example, we pick just one arbitrary point x_i in each Q_i and define

$$\widehat{\Phi}_h(Q_i) := \overline{\Phi}_h(x_i)$$

then we can estimate

$$\mathrm{dist}(\Phi_h^0(Q_i), \widehat{\Phi}_h(Q_i)) \leq \max_{x \in Q_i} \|\Phi_h^0(x) - \Phi_h^0(x_i)\| \leq (1 + Lh)\mathrm{diam}(Q_i), \tag{5.22}$$

where L is the Lipschitz constant of Φ_h^0 on Ω and, conversely

$$\mathrm{dist}(\widehat{\Phi}_h(Q_i), \Phi_h^0(Q_i)) \leq \mathrm{diam}(\mathbf{Q})$$

which implies the accuracy $\Delta_i \leq \max\{(1+Lh)\mathrm{diam}(Q_i), \mathrm{diam}(\mathbf{Q})\}$. Choosing a larger number of (suitably distributed) points for the evaluation of $\overline{\Phi}_h$ in each cell Q_i increases the accuracy since the constant in front of $\mathrm{diam}(Q_i)$ in (5.22) becomes smaller.

The more difficult part is to obtain a rigorous cell–cell discretization. A construction for such a discretization has been developed by Junge in [68, Section 2.2] and [69]; here we sketch the basic idea. For this we assume that the set Ω is rectangular and the cells Q_i are rectangular boxes of identical size with diameter $\mathrm{diam}(Q_i) = r$ for all i. The idea lies in the selection of test point x_i^k in each box such that the estimate

$$\text{dist}\left(\Phi_h^0(Q_i), \bigcup_k \overline{\Phi}_h(x_i^k)\right) \leq r \tag{5.23}$$

is guaranteed. For example, the simplest (and probably least efficient) choice would be an equidistant distribution of the test points such that the desired distance follows from the Lipschitz estimate on Φ_h^0 similarly to (5.22). Setting

$$\widehat{\Psi}_h(Q_i) = \bigcup_k \overline{\Phi}_h(x_i^k)$$

we define $\widehat{\Phi}_h$ by

$$\widehat{\Phi}_h(Q_i) := \bigcup_{j \in I(Q_i, r)} Q_j \tag{5.24}$$

where

$$I(Q_i, r) := \left\{ j \in \{1, \ldots, P\} \, \Big| \, Q_j \cap B\left(r, \widehat{\Psi}_h(Q_i)\right) \neq \emptyset \right\}.$$

This construction ensures that $\widehat{\Phi}_h$ is a rigorous discretization; furthermore from the immediate estimate

$$\text{dist}\left(\widehat{\Psi}_h(Q_i), \Phi_h^0(Q_i)\right) \leq \text{diam}(\mathbf{Q}) = r$$

and from the construction we obtain that

$$d_H(\widehat{\Phi}_h(Q_i), \Phi_h^0(Q_i)) \leq 3r$$

which gives a bound for the accuracy. An efficient selection of the test points x_i^k is an interesting and nontrivial problem in its own right; we will not go into details here but refer to [68, Section 2.2] for techniques using, e.g., matrix valued Lipschitz estimates and singular value decompositions.

For systems with input u the situation becomes slightly more complicated. Here we have to make an additional discretization, i.e., for each selection of input functions $\bar{u} \in \mathcal{U}(Q_i)$ we will have to select a finite subset of inputs together with an appropriate choice of the test points x_i^k as above. A comprehensive treatment of this topic is beyond the scope of this discussion, in particular since the solution to this problem strongly depends on the structure of the u–dependence on Φ_h^0. We only sketch a possible construction for a simple model problem: Assume that U is one–dimensional and that Φ_h^0 has the form

$$\Phi_h^0(x, u) = F(x) + G(x) \int_0^h u(t)dt$$

(which is for example the case when Φ_h^0 is the Euler discretization of an affine system with one–dimensional input). We pick a cell Q_i, parameters ε_1 and $\varepsilon_2 > 0$, divide Q_i into subcells R_k of radius less or equal ε_1, pick a (arbitrary) test points $x_k \in R_k$ and chose a finite set $\widetilde{U} \subset U$ with $d_H(\widetilde{U}, U) \leq \varepsilon_2$. Then for each $u \in U$ we find $\tilde{u}(u) \in \widetilde{U}$ with $|u - \tilde{u}| \leq \varepsilon_2$. Now for $\bar{u} \in \mathcal{U}(Q_i)$ we set

$$u_k := \tilde{u}\left(\frac{1}{h}\int_0^h \bar{u}(x_k,t)dt\right)$$

and

$$\widehat{\Psi}_h(Q_i,\bar{u}) := \bigcup_k \overline{\Phi}_h(x_k,u_k),$$

interpreting the value u_k as a constant control function. A straightforward computation yields

$$\text{dist}(\widehat{\Psi}_h(Q_i,\bar{u}),\Phi_h^0(Q_i,\bar{u})) \leq \max\{r,(1+Lh)\varepsilon_1 + M_G^i h\varepsilon_2\},$$

where M_G^i is a bound on $\|G\|$ on Q_i. This implies (5.18). Now, given some $u \in \mathcal{U}$ we define $\bar{u}(x,t) := u(t)$ for all $x \in Q_i$ and, conversely, given some $\bar{u}(x,t) \in \mathcal{U}(\Omega)$ and some $x \in R_k \subset Q_i$ we define $u(t) = \bar{u}(x_k,t)$. Then the construction of $\widehat{\Psi}_h$ yields

$$\text{dist}(\{\Phi_h^0(x,u))\},\widehat{\Psi}_h(Q_i,\bar{u})) \leq \max\{r,(1+Lh)\varepsilon_1 + M_G^i h\varepsilon_2\}$$

for all $x \in Q_i$, i.e., (5.19), implying that $\widehat{\Psi}_h$ is a cell–cell discretization with $\Delta_i = \max\{r,(1+Lh)\varepsilon_1 + M_G^i h\varepsilon_2\}$. For ε_1 and ε_2 sufficiently small we obtain

$$\max\{r,(1+Lh)\varepsilon_1 + M_G^i h\varepsilon_2\} \leq r$$

and thus we are in the situation of estimate (5.23) and can construct a rigorous discretization $\widehat{\Phi}_h$ from $\widehat{\Psi}_h$ similar to (5.24). Clearly, this procedure is far from optimal (we conjecture that ideas from [68] can be applied to obtain much more efficient constructions for $\widetilde{\Phi}_h$), but it shows that in principle rigorous cell–cell space discretization can be constructed also for systems with input u.

Although the spatial discretizations from Definitions 5.3.3 and 5.3.4 do not exactly meet our definition of discrete time systems it is easy to see what a suitable definition of an attracting set must look like. In view of Remark 5.3.6 we define it for cell–cell space discretizations.

Definition 5.3.7 (attracting set for space discretization)

Consider a cell covering $\mathbf{Q}$ of some compact set $\Omega \subset \mathbb{R}^n$ and a cell–cell space discretization $\widehat{\Phi}_h : \mathcal{C}_\mathbf{Q} \times \mathcal{U}(\Omega) \to P(\mathbb{R}^n)$.

Then a set $A \in \mathcal{C}_\mathbf{Q}$ is called strongly attracting if $\widehat{\Phi}_h(A,\bar{u}) \subseteq A$ for all $\bar{u} \in \mathcal{U}(\Omega)$ and there exists a set $B \in \mathcal{C}_\mathbf{Q}$ which contains a neighborhood of A such that the inclusion

$$\widehat{\Phi}_h(t,B,\bar{u}) \subseteq \Omega$$

holds for all $t \in \mathbb{T}^+$ and there exists a $T \in \mathbb{T}^+$ such that the inclusion

$$\widehat{\Phi}_h(t,B,\bar{u}) \subseteq A$$

holds for all $t \geq T$ and all sequences $\bar{u} = (\bar{u}_j)_{j \in \mathbb{N}_0}$ with $\bar{u}_j \in \mathcal{U}(\Omega)$.

A set $A \in \mathcal{C}_\mathbf{Q}$ is called weakly attracting if there exists a $\bar{u} \in \mathcal{U}(\Omega)$ with $\widehat{\Phi}_h(A, \bar{u}) \subseteq A$ and there exists a set $B \in \mathcal{C}_\mathbf{Q}$ which contains a neighborhood of A and a sequence $\bar{u} = (\bar{u}_j)_{j \in \mathbb{N}_0}$ with $\bar{u}_j \in \mathcal{U}(\Omega)$ such that the inclusion

$$\widehat{\Phi}_h(t, B, \bar{u}) \subseteq \Omega$$

holds for all $t \in \mathbb{T}^+$ and there exists $T \in \mathbb{T}^+$ such that the inclusion

$$\widehat{\Phi}_h(t, B, \bar{u}) \subseteq A$$

holds for all $t \geq T$.   □

Remark 5.3.8 Note that this definition demands that A lies in the interior of Ω, since $\operatorname{int} B \subseteq \operatorname{int} \Omega$ is supposed to be a neighborhood of A, and that no part of B is mapped outside Ω. One could relax this definition by requiring only that $\operatorname{int} B$ is a neighborhood of A relative to $\operatorname{int} \Omega$ and allowing that parts of B are mapped outside Ω. In the strong case this leads to the concept of attractors or attracting sets *relative to* Ω, see [29], while in the weak case we end up with *viability kernels*, see [5, 99, 116, 117]. Since these objects in general do not correspond to the attracting sets we have discussed so far we will not use this more general definition. We conjecture, however, that similar ideas to those we use here can be utilized in order to analyze these objects.

□

Having introduced these abstract settings we can now recall Example 4.2.4 to see why we need δ-nonanticipating strategies for $\delta = h > 0$ to model these state space discretizations.

Example 5.3.9 Recall Example 4.2.4 and consider the unperturbed system

$$\dot{x}(t) = u(t)$$

from this example along with its inflated time-h map

$$\Phi_h(x, u, w) = x + \int_0^h u(t)dt + \int_0^h w(t)dt, \tag{5.25}$$

which for this simple example coincides with the time-h map of the continuous inflated system $\dot{x}(t) = u(t) + w(t)$. Recall that $A = \{0\}$ is a weakly asymptotically stable set for the unperturbed system Φ_h^0.

Consider a cell covering $\mathbf{Q}$ with cells Q_i for $i = 1, \ldots, P$ and some cell-cell space discretization $\widehat{\Phi}_h$. Since $\widehat{\Phi}_h$ can only take values in $\mathcal{C}$ it is immediate (no matter how $\widehat{\Phi}_h$ is chosen) that $d_H(\widehat{\Phi}_h(x, u), \{0\}) \geq \varepsilon > 0$ for $\varepsilon := \min_{i=1,\ldots,P} \operatorname{diam}(Q_i)/2$, in particular each attracting set for the space discretization has a distance of at least ε from the origin. Hence any inflated discrete time system "containing" the dynamics of $\widehat{\Phi}_h$ (in the sense of Remark 5.3.5(iii)) cannot be controllable to $B(\varepsilon', 0)$ for all $\varepsilon' < \varepsilon$. Since we have seen in Example 4.2.4 that for $\mathcal{P} = \mathcal{P}^0$ the inflated system (5.25) is controllable to any neighborhood of 0, obviously the (0-)nonanticipating strategies are not sufficient to capture the discretization error caused by this space discretization.   □

Remark 5.3.10 A particular case of a space discretization is obtained when the discrete time system Φ_h^0 itself is defined by a numerical approximation of the time-h map φ^h of an ordinary differential equation (2.1). In this case one can consider $\overline{\Phi}_h$ or $\widehat{\Phi}_h$ either as space discretization of Φ_h^0 or of φ^h, i.e., the accuracy and rigorosity concepts from the Definitions 5.3.4 and 5.3.3, respectively, can be defined either for Φ_h^0 or for φ^h instead of Φ_h^0.

We will use the terminologies *total accuracy* and *totally rigorous* when want to emphasize that we refer to these properties with respect to φ^h.   □

The idea of a totally rigorous space discretization seems natural when combined space–time discretizations are considered. Essentially, it demands that the time and space discretizations are coupled such that the error of the time discretization is compensated by the blowing up of the space discretization in such a way that the exact solution is contained in the numerical one. In fact, this is not a new idea; it is used, for instance, in the numerical approximations of viability kernels for differential inclusions, see Saint–Pierre [99], where it is a crucial ingredient in order to obtain a convergence result.

Remark 5.3.11 The approximation Φ_h^0 in Remark 5.3.10 could simply be chosen as $\Phi_h^0 = \widetilde{\Phi}_h$ for some one–step approximation $\widetilde{\Phi}_h$ of φ^h with time step h, but could also be obtained from the numerical approximation $\widetilde{\Phi}_{h_1}(h, u)$ for some one–step scheme $\widetilde{\Phi}_{h_1}$ with smaller time step $h_1 < h$ and $h = jh_1$ for some $j \in \mathbb{N}$. In other words, when coupling time and space discretizations, the time step h used in discrete time system which is discretized in space does not need to coincide with the time step h_1 of the underlying time discretization.   □

6 Discretizations of Attracting Sets

In this chapter we will apply the results about the relation between discretization and perturbations from Chapter 5 in order to obtain convergence results for numerical approximations of attracting sets and attractors based on the theoretical results from the Chapters 3 and 4.

The main idea for the development of these results is to interpret the dynamical systems induced by the numerical schemes as perturbations of the original system *and* interpret the original system as a perturbation of the numerical systems. Proceeding this way we are able to give necessary and sufficient conditions for the convergence of attracting sets under discretization and give estimates for the rate of convergence and the discretization error.

We will consider both time and space discretizations of both strongly and weakly attracting sets. In particular, for space discretizations we will not only present a general approximation theorem, but will also analyze the convergence of a subdivision algorithm for the computation of strong attractors, which will be described in Section 6.3.

As in the previous chapter, in order to avoid technical cutoff techniques, throughout this chapter we will assume that the systems to be approximated satisfy the global bounds (2.8) or (2.24) for $w = 0$.

We start with considering strongly attracting sets.

6.1 Strongly Attracting Sets

In this section we will give several existence, convergence and approximation results for strongly attracting sets under one-step approximations. The "philosophy" behind all of these results is to embed the numerical system induced by the one-step approximation into the inflated time-h map via Lemma 5.1.3 and vice versa, and then use results from Chapter 3 to obtain the respective results.

We first give a criterion for the existence of strongly attracting sets for numerical one–step approximations.

Theorem 6.1.1 Consider a system φ of type (2.1) satisfying (2.8) and a numerical one step approximation $\widetilde{\Phi}_h$ of order $q \in \mathbb{N}$. Assume that φ has a strongly attracting set A which is γ-robust for the α_0-inflated system (2.10) for some γ of class $\mathcal{KL}$ and some $\alpha_0 > 0$. Then for all $h > 0$ with $e^{Lh} ch^q \leq \alpha_0$ the one step scheme $\widetilde{\Phi}_h$ has an attracting set A_h satisfying $A \subseteq A_h$ and $d_H(A_h, A) \leq \gamma(e^{Lh} ch^q)$.

Proof: By Lemma 5.1.3 for all h sufficiently small the inflated numerical system is $(ch^q, 1)$-embedded in the inflated time-h map (3.16). Since by Corollary 3.8.5 the set A is a $\gamma(e^{Lh} \cdot)$-robust strongly attracting set for (3.16) the assertion follows from Theorem 3.6.8. $\qquad\qquad\square$

Now we turn to the converse question: When does the existence of a strongly attracting set in the numerical system imply the existence of a strongly attracting set for the approximated system. We first investigate the time-h map of the original system.

Lemma 6.1.2 Consider a system φ of type (2.1) satisfying (2.8) and a numerical one step approximation $\widetilde{\Phi}_h$ of order $q \in \mathbb{N}$. Let $h > 0$ and assume that $\widetilde{\Phi}_h$ has a strongly attracting set A_h which is γ-robust for the α_0-inflated numerical system (2.26) for some γ of class $\mathcal{KL}$ and some $\alpha_0 > ch^q$. Then the time-h map φ^h of φ has a strongly attracting set $\tilde{A}_h$ satisfying $\tilde{A}_h \subseteq A_h$ and $d_H(\tilde{A}_h, A_h) \leq \gamma(ch^p)$.

Proof: Follows immediately from Lemma 5.1.3 and Proposition 3.6.8. $\qquad\square$

For the continuous time system we can deduce the following result.

Theorem 6.1.3 Consider a system φ of type (2.1) satisfying (2.8) and a numerical one step approximation $\widetilde{\Phi}_h$ of order $q \in \mathbb{N}$. Let $h > 0$ and assume that $\widetilde{\Phi}_h$ has a strongly attracting set A_h which is γ-robust for the α_0-inflated system for some γ of class $\mathcal{KL}$ and some $\alpha_0 > ch^q$. Assume furthermore that the attracted neighborhood B of A_h satisfies $d_{\min}(A_h, B) \geq \gamma(ch^q) + Mh$. Then φ has a strongly attracting set $\tilde{A}$ satisfying $A_h \subseteq \tilde{A}$ and $d_H(\tilde{A}, A_h) \leq \gamma(ch^q) + Mh$.

Proof: The assertion follows from Proposition 3.8.2 applied to the attracting set $\tilde{A}_h$ from Lemma 6.1.2. $\qquad\qquad\square$

The following Theorem gives several necessary and sufficient conditions under which the limit of a sequence of "numerical" strongly attracting sets is a strongly attracting set for the approximated system.

Theorem 6.1.4 Consider a system φ of type (2.1) satisfying (2.8) and a numerical one step approximation $\widetilde{\Phi}_h$ of order $q \in \mathbb{N}$. Consider furthermore a sequence of time steps $h_n \to 0$ and a sequence of strongly attracting sets

A_n for the numerical system $\widetilde{\Phi}_{h_n}$, all with the same attracted neighborhood B. Assume that there exists a closed and c-bounded set $A \subset B$ such that $d_H(A_n, A) \to 0$. Then the following statements are equivalent.

(i) A is a strongly attracting set for the continuous time system.

(ii) There exists $N \in \mathbb{N}$, β of class $\mathcal{KL}$, a sequence of real numbers $\rho_n \to 0$ as $n \to \infty$ and a sequence of attracting sets $\hat{A}_n$, $n \geq N$, for the one step discretizations $\widetilde{\Phi}_{h_n}$ with attracted neighborhood B, which satisfy $d_H(\hat{A}_n, A_n) \to 0$ and whose rates of attraction β_n satisfy $\beta_n(r, t) \leq \beta(r + \rho_n, t)$.

(iii) There exist $N \in \mathbb{N}$, $\alpha_0 > 0$, γ of class $\mathcal{K}_\infty$, a sequence of real numbers $\rho_n \to 0$ and a sequence of $\gamma((1+\rho_n) \cdot)$-robust strongly attracting sets $\hat{A}_n$, $n \geq N$, for the α_0-inflated one step discretizations $\widetilde{\Phi}_{h_n}$ with attracted neighborhood B, which satisfy $d_H(\hat{A}_n, A_n) \to 0$.

(iv) There exist $N \in \mathbb{N}$, $\alpha_0 > 0$, μ of class $\mathcal{KLD}$, γ and σ of class $\mathcal{K}_\infty$, a sequence of real numbers $\rho_n \to 0$ as $n \to \infty$ and a sequence of attracting sets $\hat{A}_n$, $n \geq N$, for the α_0-inflated one step discretizations $\widetilde{\Phi}_{h_n}$ with attracted neighborhood B, which satisfy $d_H(\hat{A}_n, A_n) \to 0$ and the ISDS-like estimate

$$\|\widetilde{\Phi}_{h_n}(t, x, u, w)\|_{\hat{A}_n} \leq \max\{\mu(\sigma(\|x\|_{\hat{A}_n} + \rho_n), t), \nu((1+\rho_n)w, t)\}.$$

In addition, if (ii) holds, then A is attracting with rate β, if (iii) holds, then A is γ-robust for the α_0-inflated continuous time system and if (iv) holds it is ISDS for the α_0-inflated continuous time system with rate μ and gains γ and σ.

Proof: (i) $\Rightarrow$ (iv): First recall that if A is an attracting set then by Theorem 3.4.6 it is ISDS for the inflated system for suitable gains μ, σ and γ and $W = \mathcal{B}(\varepsilon, 0)$ for some $\varepsilon > 0$. Then by Corollary 3.8.5 it is also an ISDS set for each inflated time-h_n map with gain $\gamma(e^{Lh} \cdot)$. Since by Lemma 5.1.3 the inflated numerical system is $(ch_n^q, 1)$-embedded in the inflated time-h_n map, by Proposition 3.6.6 applied with $\alpha = ch_n^q$, $C = 1$ and $D = 1/\sqrt{\alpha}$, for all $h_n > 0$ sufficiently small there exists an attracting set $\hat{A}_n$ for $\widetilde{\Phi}_{h_n}$ satisfying the estimate

$$\|\widetilde{\Phi}_{h_n}(t, x, u, w^*)\|_{\hat{A}_n} \leq \max\{\mu(\sigma(\|x\|_{\hat{A}_n} + \gamma(\sqrt{\alpha})), t), \nu(w^*/(1 - \sqrt{\alpha}), t)\}.$$

Setting $\rho_n := \max\{\sqrt{\alpha}/(1 - \sqrt{\alpha}), \gamma(\sqrt{\alpha})\}$ hence shows the claim.

(iv) $\Rightarrow$ (iii) and (iv) $\Rightarrow$ (ii) are obvious.

(ii) $\Rightarrow$ (i): Consider a sequence $T_n \to \infty$ such that $c(e^{LT_n} - 1)h_n^q/L \to 0$ as $n \to \infty$. We set $\tilde{\beta}_n(r, t) = \beta_n(r + \rho_n, t) + c(e^{LT_n} - 1)h_n^q/L$. Then Lemma 5.1.2 shows that

$$\|\tilde{\Phi}_{h_n}(t, x, u)\|_{\hat{A}_n} \leq \tilde{\beta}_n(\|x\|_{\hat{A}_n}, t) \text{ for all } t \in [0, T_n] \cap \mathbb{T}.$$

Since the $\tilde{\beta}_n$ are asymptotically bounded by β the assertion follows from Proposition 3.8.3 (i).

(iii) $\Rightarrow$ (i): By Lemma 5.1.3 the time-h_n map φ^{h_n} is ch_n^q-embedded in the inflated system $\tilde{\Phi}_{h_n}$. Hence by Proposition 3.6.7 applied to $A = \hat{A}_n$ with $\alpha = ch_n^q$, $C = 1$ and $D = 1/\sqrt{\alpha}$ we obtain the existence of $\tilde{\gamma}_n$-robust attracting sets $\tilde{A}_n$ for the inflated time-h_n map φ^{h_n} with $d_H(\tilde{A}_n, \hat{A}_n) \leq \gamma_n(\sqrt{ch_n^p})$ and $\tilde{\gamma}_n(r) = \gamma_n(1/(1 - \sqrt{ch_n^p})r)$. By Corollary 3.8.5 these sets are $\tilde{\gamma}_n((1 + Lh_n) \cdot)$-robust for the time-$h_n$ map of the continuous time inflated system. By the assumption on the $\hat{A}_n$ and γ_n we obtain that $d_H(\tilde{A}_n, A) \to 0$ as $n \to \infty$ and that the functions $\tilde{\gamma}_n((1 + Lh_n) \cdot)$ are asymptotically bounded by γ. Thus Proposition 3.8.3 (ii) yields the assertion.

(iv) $\Rightarrow$ (i): Similar to "(ii) $\Rightarrow$ (i)" but using Proposition 3.8.3 (iii). $\qquad\square$

Remark 6.1.5 One might argue about how useful these rather complicated conditions are for the actual interpretation of numerical results. It is clear that an assumption "for arbitrarily small time steps" will be hard to check rigorously in practice. Nevertheless, the results suggest justified heuristic checks for numerical simulations like, e.g., the following procedure: When an attracting is observed in a numerical system then repeat the computation with different time steps and observe the rates of attraction. If these rates vary for different time steps then the observed attractor is likely to be a numerical artifact and the numerical results should be interpreted with care. $\qquad\square$

Let us now show that robust attracting sets are also approximated by space discretizations. Here we start from some discrete time system, which could, of course, also be a time-h map in which case we need the respective total accuracy from Remark 5.3.10.

Theorem 6.1.6 Consider a discrete time system Φ_h^0 of type (2.17), a compact set $\Omega \subset \mathbb{R}^n$, a cell covering $\mathbf{Q}$ of Ω and a cell–cell space discretization $\hat{\Phi}_h$ on $\mathbf{Q}$.

Assume that $A \subset \text{int } \Omega$ is a strongly attracting set for Φ_h^0 which is ISDS with respect to α_0-inflation with rate μ of class $\mathcal{KLD}$ and gains σ and γ of class $\mathcal{K}_\infty$ and some $\alpha_0 > 0$. Assume furthermore that $\hat{\Phi}_h$ has accuracy Δ_i satisfying $\Delta_i/h \leq \max\{\Delta, \gamma^{-1}(\|x\|_A)\}$ for some $\Delta > 0$.

Then, if $\Delta > 0$ and $\text{diam}(\mathbf{Q})$ are sufficiently small there exists a strongly attracting set $\hat{A} \in \mathcal{C}_\mathbf{Q}$ for $\hat{\Phi}_h$ which satisfies

$$\text{dist}(\hat{A}, A) \leq \gamma(\Delta).$$

Proof: From Remark 5.3.5(iii) we obtain the inclusion

$$\bigcup_{\bar{u} \in \mathcal{U}(\Omega)} \widehat{\Phi}_h(Q_i, \bar{u}) \subseteq \Phi_{h,1}(Q_i),$$

where Φ_h denotes the state dependent inflated system (2.27) related to Φ_h^0 with $b(x) = \Delta_i/h$ for $x \in Q_i$. Now from Theorem 3.7.4 for $\Delta < \alpha_0$ we obtain the existence of a Δ-strongly attracting set $\tilde{A}$ with $d_H(\tilde{A}, A) \leq \gamma(\Delta)$. Let $\tilde{B}$ be an attracted neighborhood. Note that we can restrict $\tilde{B}$ such that $\Phi_{h,\Delta_i/h}(t, \tilde{B}) \subseteq \Omega$ for all $t \in \mathbb{T}^+$. Now we choose $\widehat{A} \in \mathcal{C}_\mathbf{Q}$ and $\widehat{B} \in \mathcal{C}_\mathbf{Q}$ to be the biggest sets (w.r.t. set inclusion) in $\mathcal{C}_\mathbf{Q}$ contained in $\tilde{A}$ and $\tilde{B}$, respectively. Since $\tilde{B}$ contains a neighborhood of $\tilde{A}$ also $\widehat{B}$ contains a neighborhood of $\widehat{A}$ provided $\mathrm{diam}(\mathbf{Q})$ is smaller than $d_{\min}(B, A)$. We claim that then $\widehat{A}$ is the desired strongly attracting set for $\widehat{\Phi}_h$. Since

$$\widehat{A}_1 := \bigcup_{\bar{u} \in \mathcal{U}(\Omega)} \widehat{\Phi}_h(\widehat{A}, \bar{u}) \subseteq \Phi_{h,\Delta/h}(\widehat{A}, \bar{u})$$

$$\subseteq \Phi_{h,\Delta/h}(A, \bar{u}) \subseteq \tilde{A}$$

and $\widehat{A}_1 \in \mathcal{C}_\mathbf{Q}$ we can conclude $\widehat{A}_1 \subseteq \widehat{A}$. Furthermore, again from the embedding we obtain that

$$\mathrm{dist}\left(\bigcup_{\bar{\mathbf{u}}} \widehat{\Phi}_h(ih, \widehat{B}, \bar{\mathbf{u}}), \tilde{A}\right) \to 0$$

as $i \to \infty$, where the union is taken over all sequences $\bar{\mathbf{u}} = (\bar{u}_i)_{i \in \mathbb{N}_0}$ with $\bar{u}_i \in \mathcal{U}(\Omega)$. Hence from Lemma 5.3.2(ii) we obtain

$$B_i := \bigcup_{\bar{\mathbf{u}}} \widehat{\Phi}_h(ih, \widehat{B}, \bar{\mathbf{u}}) \subseteq \tilde{A}$$

for some $i_0 > 0$ and all $i \geq i_0$ which implies $B_i \subseteq \widehat{A}$ since $B_i \in \mathcal{C}_\mathbf{Q}$. $\square$

Note that this result is weaker that the corresponding result for time discretizations in Theorem 6.1.1, in the sense that here we only obtain an estimate for "dist" instead of "d_H". This is due to the fact that we cannot take the Δ/h-strongly attracting set $\tilde{A}$ itself as the resulting attracting set. Instead, we have to use its approximation via cells, whose d_H-distance from A we cannot control. On the other hand, the use of the small gain estimate from Theorem 3.7.4 enables us to allow "large" errors away from A. Essentially, this means that a fine approximation of Φ_h^0 is only needed in a neighborhood of A.

We expect that an analogous result to Theorem 6.1.4 holds also for space discretizations, however, for this we would need suitable robustness concepts for attracting sets of space discretizations. Since this would require a lot of additional technical definitions we do not want to pursue this idea here.

6.2 Strong Attractors

In this section we will consider strongly attracting sets which in addition are compact and invariant, i.e., they are exactly mapped onto themselves under $\Phi_{\mathcal{U}}$. These objects play an important role in understanding the long time behavior of dynamical systems, as typically all trajectories not diverging to infinity end up inside or near an attractor for sufficiently large times, cf., e.g., [62]. It is a well known result in the numerical analysis of dynamical systems that attractors in general are not correctly reproduced by one-step approximations, see, e.g., Example 1.1.1 in Chapter 1, [42, Example (0.12)] or [113, Sections 7.5 ff.]. More precisely, if a sequence of attractors A_n for a numerical approximation converges to some compact set A in the Hausdorff sense, this set A might not be an attractor for the original continuous time system (in fact, it might not even be an attracting set). Using our concepts we will be able to give necessary and sufficient conditions under which this convergence holds. Unlike other approaches (see, e.g., [113, Chapter 7]) this condition does not involve assumptions on the attractor A (which is, of course, in general unknown) but only on the behavior of the approximating numerical system. It is therefore conceptually different from the structural stability conditions that are often used in the theory of dynamical systems, although the techniques presented here and certain structural stability conditions can be effectively combined, see Remark 6.2.10, below.

Usually, attractors are defined for dynamical systems, i.e., systems without input. It turns out that we can include the internal perturbations $\mathcal{U}$ in most of the results in what follows by extending the usual attractor definition in a natural way.

For the definition of what we call a strong attractor, recall the notation $\Phi_{\mathcal{U}}(t, x) := \bigcup_{u \in \mathcal{U}} \Phi(t, x, u)$.

Definition 6.2.1 (strong invariance and attractor)

(i) A closed set $A \subset \mathbb{R}^n$ is called *(strongly) invariant* if $\operatorname{cl} \Phi_{\mathcal{U}}(t, A) = A$ for all $t \in \mathbb{T}^+$.

(ii) A compact strongly attracting set A is called a *(strong) attractor* if it is strongly invariant. □

We will state a sequence of results on strong attractors that will be useful in what follows.

Lemma 6.2.2 Each compact strongly attracting set $\tilde{A}$ contains a strong attractor A with same attracted neighborhood.

Proof: Let B be the attracted neighborhood of $\tilde{A}$. Consider a monotone increasing sequence $T_i \to \infty$, a subset $\tilde{B} \subset B$ with $A \subset \operatorname{int} \tilde{B}$ and

$\operatorname{dist}(\tilde{B}, \tilde{A}) < \infty$ and define the sets $B_i := \bigcup_{t \geq T_i} \Phi_{\mathcal{U}}(t, \tilde{B})$. Now consider the set

$$A := \operatorname{Lim\,sup}_{i \to \infty} B_i.$$

Then by Lemma 2.3.6(iii) and the attractivity of $\tilde{A}$ we obtain $A \subseteq \tilde{A}$.

Now by compactness of A and Lemma 2.3.5 for each $\varepsilon > 0$ there exists $T \in \mathbb{T}^+$ such that $\operatorname{cl} \bigcup_{t \geq T} \Phi_{\mathcal{U}}(t, \tilde{B}) \subset B(\varepsilon, A)$, hence, since $\tilde{A}$ is attracting and contained in $\operatorname{int} \tilde{B}$, for any subset $\overline{B} \subset \operatorname{cl} B$ with $d_H(\overline{B}, A) < \infty$ and any $\varepsilon > 0$ there exists $T \in \mathbb{T}^+$ such that $\operatorname{cl} \bigcup_{t \geq T} \Phi_{\mathcal{U}}(t, \overline{B}) \subset B(\varepsilon, A)$, which shows that A is attracting with attracted neighborhood B.

In order to see the invariance of A, we show the inclusions $\operatorname{cl} \Phi_{\mathcal{U}}(t, A) \subseteq A$ and $\operatorname{cl} \Phi_{\mathcal{U}}(t, A) \supseteq A$. For the first inclusion, since A is closed it is sufficient to show that $\Phi_{\mathcal{U}}(t, x, u) \subseteq A$ for all $x \in A$, $u \in \mathcal{U}$ and all $t \in \mathbb{T}^+$. Hence let $x \in A, t \in \mathbb{T}^+$ and $u \in \mathcal{U}$. Then there exist sequences $t_n \to \infty$, $u_n \in \mathcal{U}$ and $x_n \in \tilde{B}$ such that $\Phi(t_n, x_n, u_n) \to x$. Hence $\Phi(t_n + t, x_n, u_n \&_{t_n} u) \to \Phi(t, x, u)$ as $n \to \infty$, and since $\Phi(t_n + t, x_n, u_n \&_{t_n} u) \in \operatorname{cl} \Phi_{\mathcal{U}}(t_n + t, \tilde{B})$ this implies $\Phi(t, x, u) \in A$.

For the second inclusion, let again $x \in A$. Again we find sequences $t_n \to \infty$, $u_n \in \mathcal{U}$ and $x_n \in \tilde{B}$ such that $\Phi(t_n, x_n, u_n) \to x$. Now fix $t \in \mathbb{T}^+$ and consider the sequence $y_n = \Phi(t_n - t, x_n, u_n)$ for $t_n > t$. W.l.o.g. this sequence converges to some $y \in A$. For this the sequence of perturbations $\tilde{u}_n$ given by $\tilde{u}_n(\tau) = u_n(t_n - t + \tau)$ yields $\Phi(t, y_n, \tilde{u}_n) \to x$, hence by continuous dependence on the initial value we obtain $\Phi(t, y, \tilde{u}_n) \to x$. This yields $x \in \operatorname{cl} \Phi_{\mathcal{U}}(t, y)$, i.e. the desired property. $\qquad\square$

Lemma 6.2.3 A compact strongly attracting set A for $\Phi_{\mathcal{U}}$ with attracted neighborhood B is a strong attractor with attracted neighborhood B if and only if it is the minimal compact strongly attracting set (w.r.t. set inclusion) with attracted neighborhood B. In particular for each open set $B \subset \mathbb{R}^n$ there exists at most one strong attractor with attracted neighborhood B.

Proof: Let A be a strong attractor with attracted neighborhood B. Then in particular A is strongly invariant. Now assume that $\tilde{A} \subset A$, $\tilde{A} \neq A$, is a strongly attracting set. Then there exists a neighborhood $\mathcal{N} \supset \tilde{A}$ with $\operatorname{int} A \not\subset \mathcal{N}$, such that $\Phi_{\mathcal{U}}(t, B) \subset \mathcal{N}$ for some $t \in \mathbb{T}^+$, i.e. in particular $\operatorname{cl} \Phi_{\mathcal{U}}(t, A) \neq A$ which contradicts the invariance of A.

Let conversely A be a minimal strongly attracting set. Then by Lemma 6.2.2 the set A contains a strong attractor which again is a strongly attracting set and by minimality it coincides with A. $\qquad\square$

Lemma 6.2.4 Let A be a strong attractor with attracted neighborhood B for $\Phi_{\mathcal{U}}$. Then each compact strongly invariant set $D \subset B$ is contained in A.

Proof: Let $D \subset B$ be a compact strongly invariant set. Then $D = \mathrm{cl}\,\Phi_{\mathcal{U}}(t, D) \subset \mathrm{cl}\,\Phi(t, \widetilde{B})$ for all $t \in \mathbb{T}^+$ and all sufficiently large compact sets $\widetilde{B} \subseteq \mathrm{cl}\,B$. On the other hand, for each neighborhood $\mathcal{N} \supset A$ we know that $\Phi_{\mathcal{U}}(t, \widetilde{B}) \subset \mathcal{N}$ for all $t \in \mathbb{T}^+$ sufficiently large. Hence $D \subset \mathrm{cl}\,\mathcal{N}$ for each neighborhood $\mathcal{N} \supset A$ which implies the assertion. $\square$

In the next two lemmata we investigate the relation between strongly attracting sets and strong attractors for the continuous time system and its time-h map.

Lemma 6.2.5 Consider the continuous time system (2.1). Then a strongly forward invariant set A is a strongly attracting set with attracted neighborhood B if and only if for each compact set $\widetilde{B} \subseteq \mathrm{cl}\,B$ there exists $T > 0$ such that

$$\lim_{i \to \infty, i \in \mathbb{N}} \mathrm{dist}(\varphi_{\mathcal{U}}(iT, \widetilde{B}), A) = 0. \tag{6.1}$$

Proof: Obviously, if A is strongly attracting then (6.1) holds for all $T > 0$. Conversely consider a compact set $\widetilde{B} \subseteq \mathrm{cl}\,B$ and let (6.1) hold for some $T > 0$. Then forward invariance of A and continuous dependence on the initial value imply that for each $\delta > 0$ there exists $\varepsilon > 0$ with

$$d_H(D, A) < \varepsilon \quad \Rightarrow \quad d_H(\varphi_{\mathcal{U}}(t, D), A) < \delta$$

for all $t \in [0, T]$. Thus (6.1) implies $\lim_{t \to \infty} \mathrm{dist}(\varphi(t, \widetilde{B}), A) = 0$, hence A is an attracting set. $\square$

Lemma 6.2.6 Let $h > 0$ and A_h be a strong attractor with attracted neighborhood B for the time-h map φ^h of the continuous time system (2.1). Then A_h is also a strong attractor with attracted neighborhood B for the continuous time system (2.1).

Proof: We first show strong forward invariance of A_h for $\varphi_{\mathcal{U}}$, i.e., $\varphi_{\mathcal{U}}(t, A_h) \subseteq A_h$ for each $t > 0$. By invariance of A_h for φ^h we know $\varphi_{\mathcal{U}}^h(\varphi_{\mathcal{U}}(t, A_h)) = \varphi_{\mathcal{U}}(t, \varphi_{\mathcal{U}}^h(A_h)) = \varphi_{\mathcal{U}}(t, A_h)$, hence $\varphi_{\mathcal{U}}(t, A_h)$ is strongly invariant for φ^h, and by Lemma 6.2.4 it is contained in A_h.

Since A_h is strongly forward invariant for $\varphi_{\mathcal{U}}$ and a strongly attracting set for its time-h map, by Lemma 6.2.5 it is also a strongly attracting set for $\varphi_{\mathcal{U}}$ with attracted neighborhood B.

Now by Lemma 6.2.2 there exists a strong attractor $A \subseteq A_h$. Clearly, this is also a strong attractor for the time-h map, hence by Lemma 6.2.3 we obtain $A = A_h$ which finishes the proof. $\square$

Using these lemmata and the results from the previous section we can now analyze the behavior of attractors under one-step discretization.

Theorem 6.2.7 Consider a system φ of type (2.1) satisfying (2.8) and a numerical one step approximation $\widetilde{\Phi}_h$ of order $q \in \mathbb{N}$. Assume that $\widetilde{\Phi}_h$ has a strongly attracting set A_h which is γ-robust for the α_0-inflated system for some γ of class $\mathcal{KL}$ and some $\alpha_0 > ch^q$. Then φ has a strong attractor A satisfying $\mathrm{dist}(A, A_h) \leq \gamma(ch^q)$.

Proof: Consider the strongly attracting set $\tilde{A}_h$ for φ^h from Lemma 6.1.2. By Lemma 6.2.2 this set contains a strong attractor A for φ^h which by Lemma 6.2.6 is also a strong attractor for φ. $\qquad\square$

Next we state attractor version of Theorem 6.1.4.

Theorem 6.2.8 Consider a system φ of type (2.1) satisfying (2.8) and a numerical one step approximation $\widetilde{\Phi}_h$ of order $q \in \mathbb{N}$. Consider furthermore a sequence of time steps $h_n \to 0$ and a sequence of strong attractors A_n for $\widetilde{\Phi}_{h_n}$, all with the same attracted neighborhood B. Assume that there exists a compact set $A \subset B$ such that $d_H(A_n, A) \to 0$. Then the following statements are equivalent.

(i) A is a strong attractor for the continuous time system φ.

(ii) There exists $N \in \mathbb{N}$, β of class $\mathcal{KL}$, a sequence of real numbers $\rho_n \to 0$ as $n \to \infty$ and a sequence of strongly attracting sets $\hat{A}_n$, $n \geq N$, for the one step discretizations $\widetilde{\Phi}_{h_n}$ with attracted neighborhood B, which satisfy $d_H(\hat{A}_n, A_n) \to 0$ and whose rates of attraction β_n satisfy $\beta_n(r, t) \leq \beta(r + \rho_n, t)$.

(iii) There exist $N \in \mathbb{N}$, $\alpha_0 > 0$, γ of class $\mathcal{K}_\infty$, a sequence of real numbers $\rho_n \to 0$ and a sequence of $\gamma((1+\rho_n)\,\cdot)$-robust strongly attracting sets $\hat{A}_n$, $n \geq N$, for the α_0-inflated one step discretizations $\widetilde{\Phi}_{h_n}$ with attracted neighborhood B, which satisfy $d_H(\hat{A}_n, A_n) \to 0$.

(iv) There exist $N \in \mathbb{N}$, $\alpha_0 > 0$, μ of class $\mathcal{KLD}$, γ and σ of class $\mathcal{K}_\infty$, a sequence of real numbers $\rho_n \to 0$ as $n \to \infty$ and a sequence of strongly attracting sets $\hat{A}_n$, $n \geq N$, for the α_0-inflated one step discretizations $\widetilde{\Phi}_{h_n}$, with attracted neighborhood B, which satisfy $d_H(\hat{A}_n, A_n) \to 0$ and the ISDS-like estimate

$$\|\widetilde{\Phi}_{h_n}(t, x, u, w)\|_{\hat{A}_n} \leq \max\{\mu(\sigma(\|x\|_{\hat{A}_n} + \rho_n), t), \nu((1 + \rho_n)w, t)\}.$$

In addition, if (ii) holds, then A is attracting with rate β, if (iii) holds, then A is γ-robust for the α_0-inflated continuous time system and if (iv) holds it is ISDS for the α_0-inflated continuous time system with rate μ and gains γ and σ.

Proof: From Theorem 6.1.4 we obtain the implications (i) $\Rightarrow$ (ii), (i) $\Rightarrow$ (iii) and (i) $\Rightarrow$ (iv). Furthermore, this theorem shows that if (ii), (iii) or (iv) holds then A is an attracting set with the respective additional property. Hence it remains to show that A is an attractor, i.e., the minimality of A. In order to accomplish this, assume that A is not an attractor. Then by Lemma 6.2.2 A contains an attractor $\tilde{A} \subset A$ with $\mathrm{dist}(A, \tilde{A}) =: \varepsilon > 0$. By Theorem 3.4.6 this attractor $\tilde{A}$ is an ISDS attracting set, hence by Corollary 3.8.5 for suitable $\alpha_0 > 0$ and γ of class $\mathcal{KL}$ it is a γ-robust strongly attracting set for each α_0-inflated time-h_n map φ^{h_n}. Thus by Proposition 3.6.7 for all n sufficiently large there exist strongly attracting sets $\tilde{A}_n$ for the one step schemes $\tilde{\Phi}_{h_n}$ with $d_H(\tilde{A}_n, \tilde{A}) \leq \varepsilon/2$. Since the A_n are attractors for the one step schemes $\tilde{\Phi}_{h_n}$ they must satisfy $A_n \subseteq \tilde{A}_n$. Thus we obtain

$$d_H(A, A_n) \geq \mathrm{dist}(A, A_n) \geq \mathrm{dist}(A, \tilde{A}_n)$$
$$\geq \mathrm{dist}(A, \tilde{A}) - d_H(\tilde{A}, \tilde{A}_n) \geq \varepsilon - \varepsilon/2 = \varepsilon/2 > 0$$

for all n sufficiently large. This contradicts the assumption $d_H(A, A_n) \to 0$ as $n \to \infty$ and hence shows the claim. $\qquad\qquad\qquad\qquad\qquad\square$

For the practical application of these criteria we refer to Remark 6.1.5, which holds accordingly here.

Note that it is not necessary that the numerical attractors A_h themselves have the stated properties on the attraction rate, the γ-robustness or the ISDS properties. If, however, we assume such a property of the A_h we can obtain further implication as stated in the following theorem. We formulate it for γ-robustness.

Theorem 6.2.9 Consider a system φ of type (2.1) satisfying (2.8) and a numerical one step approximation $\tilde{\Phi}_h$ of order $q \in \mathbb{N}$. Consider a positive sequence $h_n \to 0$ as $n \to \infty$, $\alpha_0 > 0$ and a function γ of class $\mathcal{K}$. Assume there exist γ-robust strong attractors A_n with attracted neighborhood B for the α_0-inflated numerical system $\tilde{\Phi}_{h_n}$ and let $A \subset B$ be a compact set. Then the following four statements are equivalent.

(i) A is a strong attractor with attracted neighborhood B for the continuous time system (2.1).

(ii) $d_H(A, A_{h_n}) \to 0$ as $n \to \infty$.

(iii) A is a γ-robust strong attractor with attracted neighborhood B for the continuous time inflated system (2.5) with right hand side (2.10).

(iv) For each $K > 1$ there exists $n_0 \in \mathbb{N}$ such that $d_H(A, A_{h_n}) \leq \gamma(K c h_n^q)$ for all $n \geq n_0$.

Proof: The implications "(iv) $\Rightarrow$ (ii)" and "(iii) $\Rightarrow$ (i)" are obvious and "(ii) $\Rightarrow$ (iii)" follows immediately from Theorem 6.2.8.

We now show "(iii) $\Rightarrow$ (iv)":

By the assumption the set A is a γ-robust attractor for the time h_n-map of the inflated system, hence by Corollary 3.8.5 it is a $\gamma(e^{Lh_n} \cdot)$-robust attractor for the inflated time-h_n map. By Lemma 5.1.3 the numerical scheme is $(ch_n^q, 1)$-embedded in the inflated time-h_n map. Hence by Proposition 3.6.8 with $h = h_n$ we obtain the existence of attracting sets $\tilde{A}_{h_n}$ for the numerical systems with

$$d_H(A, \tilde{A}_{h_n}) \leq \gamma(e^{Lh_n} ch_n^q),$$

and by Lemma 6.2.3 we know $A_{h_n} \subseteq \tilde{A}_{h_n}$, hence

$$\mathrm{dist}(A_{h_n}, A) \leq \gamma(e^{Lh_n} ch_n^q).$$

Conversely, Theorem 6.2.7 implies

$$\mathrm{dist}(A, A_{h_n}) \leq \gamma(e^{Lh_n} ch_n^q),$$

which together yields the assertion since $e^{Lh_n} \to 1$ as $h_n \to 0$.

Finally, we show "(i) $\Rightarrow$ (ii)", which finishes the proof:

Observe by Proposition 3.4.1 that there exists a class $\mathcal{K}_\infty$ function $\tilde{\gamma}$ such that A is a $\tilde{\gamma}$-robust attractor for the inflated continuous time system. Without loss of generality we may assume $\tilde{\gamma} \geq \gamma$. Hence also the A_{h_n} are $\tilde{\gamma}$-robust attractors, and by the same arguments as for "(ii) $\Rightarrow$ (iv)", above, we obtain

$$d_H(A, A_{h_n}) \leq \tilde{\gamma}(e^{Lh_n} ch_n^q),$$

which implies (ii). □

Remark 6.2.10 We have already mentioned that these convergence criteria are different from the structural stability assumptions on the attractor A which are often imposed in the theory of dynamical systems, because here we obtain conditions only on the numerical attractors A_h and not on A. We can, however, combine assumptions of this kind with our results:

A typical assumption for dynamical systems (without inputs) which is structurally stable (i.e., which persists under small perturbations) is uniform hyperbolicity, i.e., the property that along each trajectory the state space can be "decomposed" into an (exponentially) stable and an (exponentially) unstable manifold, where the exponential rates are independent of the trajectory.

If we consider a system of type (2.1) without inputs and assume uniform hyperbolicity in a neighborhood of A, then (due to the structural stability of the uniform hyperbolicity) also the numerical systems induced by the one-step approximations are uniformly hyperbolic for $h > 0$ sufficiently small. In this case the numerical attractors attract exponentially (see [63]), i.e. we obtain $\beta(r, t) = ce^{\lambda t} r$ for suitable $c, \lambda > 0$ independent of h and hence, as shown in Example 4.4.2, the ISDS property for the numerical attractors with linear robustness gain $\gamma(r) = Cr$ for some $C > 0$ independent of h. Thus from Theorem 6.2.9 we can conclude that hyperbolicity implies that for each $K > C$ we obtain the estimate

$$d_H(A_h, A) \leq K c h^q$$

for the numerical attractors A_h for all h sufficiently small.   □

6.3 Subdivision Algorithm

Instead of giving an "attractor version" of the space discretization Theorem 6.1.6 (whose statement and proof should be clear from the results for time discretizations and is left to the interested reader) we will now turn to a more specific algorithm for the computation of strong attractors. For systems without input this algorithm was first presented by Dellnitz and Hohmann [29] and developed further by Dellnitz and Junge [30] and Junge [68, 69]. For systems with input u a version of the algorithm has been developed by Szolnoki [115, 116], to which we will come back in Chapter 7, see Algorithm 7.5.1. Here we discuss a version for strong attractors based on the version presented in [68, 69]. Formulated in the abstract space discretization framework of Chapter 5 the algorithm can be described as follows.

Algorithm 6.3.1 Let Φ_h^0 be a discrete time system of type (2.17). Consider a compact set Ω, a cell covering $\mathbf{Q}^0$ of Ω with P^0 cells Q_i^0, $i = 1, \ldots, P^0$, and a cell-cell space discretization $\widehat{\Phi}_h^0$ on $\mathbf{Q}^0$ with accuracy $\Delta_i^0 \leq h\varepsilon^0$ for all $i = 1, \ldots, P^0$ and some $\varepsilon^0 > 0$. Let $A^0 = \Omega$, $j = 0$ and proceed iteratively:

(1) **(Selection Step)**
   Compute
   $$A^{j+1} := \bigcup_{\bar{u} \in \mathcal{U}(\Omega)} \widehat{\Phi}_h^j(A^j, \bar{u}) \bigcap A^j \in \mathcal{C}_{\mathbf{Q}^j}.$$

(2) **(Refinement Step)**
   Consider a new cell covering $\mathbf{Q}^{j+1}$ of Ω with P^{j+1} cells Q_i^{j+1}, $i = 1, \ldots, P^{j+1}$, satisfying $\mathcal{C}_{\mathbf{Q}^j} \subset \mathcal{C}_{\mathbf{Q}^{j+1}}$ and a new cell-cell space discretization $\widehat{\Phi}_h^{j+1}$ on $\mathbf{Q}^{j+1}$ with accuracy $\Delta_i^{j+1} \leq h\varepsilon^{j+1}$ for all $i = 1, \ldots, P^{j+1}$ for some $\varepsilon^{j+1} < \varepsilon^j$.

(3) Set $j := j + 1$ and continue with (1).

     □

Remark 6.3.2 (i) In the practical implementation in the references mentioned above the set Ω is a rectangular domain, and the cells Q_i are rectangular boxes. Then the new cell covering $\mathbf{Q}^{j+1}$ in step (2) is obtained by subdividing each cell Q_i^j in the coordinate direction x_{k+1} with $k = j \pmod{n}$, where n is the dimension of the system. This motivates the name *subdivision algorithm*.

(ii) Note that even if formally we have defined each $\mathbf{Q}^j$ to be a cell–covering .of the whole set Ω, it is in fact sufficient in step (2) to refine only those cells of $\mathbf{Q}^j$ which lie in A^j, since only these cells will be used in the subsequent iterations of the algorithm.

(iii) Instead of using $\widehat{\Phi}_h^j(A^0, u)$ in step (1) one could also use an iterate (5.20) of this map for some larger $t = hi$, $i \in \mathbb{N}$, see also Remark 6.3.7, below. □

It turns out that our concepts allow a simple and straightforward proof of the convergence of this algorithm as well as an estimate for its convergence rate as stated in the following theorem.

Theorem 6.3.3 Consider a discrete time system Φ_h^0 of type (2.17). Assume that Φ_h^0 has a strong attractor $A \subset \mathrm{int}\,\Omega$ with an attracted neighborhood containing Ω. Assume furthermore that for some $\alpha_0 > 0$ the set A is ISDS for the α_0-inflated system with rate μ and gains σ and γ, and that the cell–cell space discretizations $\widehat{\Phi}_h^j$ in Algorithm 6.3.1 are rigorous. Then for the sets A^j from Algorithm 6.3.1 with $\varepsilon^0 \leq \alpha_0$ we obtain the estimate

$$d_H(A^j, A) \leq \max\{\mu(\sigma(d_H(\Omega, A)), jh), \max_{k=0,\dots,j-1} \mu(\gamma(\varepsilon^{j-k-1}), kh)\}$$

for all $j \in \mathbb{N}_0$. In particular, we obtain convergence $d_H(A^j, A) \to 0$ as $j \to \infty$.

Proof: Since the space discretizations are assumed to be rigorous and A is strongly invariant we obtain $A \subseteq A^j$ for all $j \in \mathbb{N}_0$. Hence we only have to show the inequality for "dist" instead of "d_H". To this end consider the ISDS Lyapunov function V from Theorem 3.6(iii) and define $V^j := \sup_{x \in A^j} V(x)$. We show by induction that the inequality

$$V^j \leq \max\{\mu(V^0, jh), \max_{k=0,\dots,j-1} \mu(\gamma(\varepsilon^{j-k-1}), kh)\} \tag{6.2}$$

holds for all $j \in \mathbb{N}_0$ which implies the desired inequality by the bounds on V. For $j = 0$ the assertion is immediate. From the ISDS Lyapunov function property and the embedding from Remark 5.3.5(iii) we obtain

$$V^{j+1} \leq \max\{\mu(V^j, h), \gamma(\varepsilon^j)\}.$$

Hence, if (6.2) holds for $j \in \mathbb{N}_0$ we obtain

$$V^{j+1} \leq \max\left\{\mu\Big(\max\{\mu(V^0, jh), \max_{k=0,\dots,j-1} \mu(\gamma(\varepsilon^{j-k-1}), kh)\}, h\Big), \gamma(\varepsilon^j)\right\}$$

$$\leq \max\left\{\mu(V^0, (j+1)h), \max_{k=0,\dots,j-1} \mu(\gamma(\varepsilon^{j-k-1}), (k+1)h), \gamma(\varepsilon^j)\right\}$$

$$\leq \max\left\{\mu(V^0, (j+1)h), \max_{k=0,\dots,j} \mu(\gamma(\varepsilon^{j-k}), kh)\right\}$$

i.e., inequality (6.2).   □

The following corollary is an immediate consequence of this theorem.

Corollary 6.3.4 Consider a discrete time system Φ_h^0 of type (2.17). Assume that Φ_h^0 has a strong attractor $A \subset \mathrm{int}\,\Omega$ with some attracted neighborhood containing Ω. Assume that the cell–cell space discretizations $\widehat{\Phi}_h^j$ in Algorithm 6.3.1 are rigorous. Then for the sets A^j from Algorithm 6.3.1 we obtain the convergence $d_H(A^j, A) \to 0$ as $j \to \infty$.

Proof: Follow from the preceding theorem by observing that by Theorem 3.4.6 any attractor is an ISDS set for suitable rate and gains.   □

Remark 6.3.5 If we apply Algorithm 6.3.1 with non rigorous space discretizations then we still obtain

$$\mathrm{dist}(A^j, A) \leq \max\{\mu(\sigma(d_H(\Omega, A)), jh), \max_{k=0,\dots,j-1} \mu(\gamma(\varepsilon^{j-k-1}), kh)\}$$

and $\mathrm{dist}(A^j, A) \to 0$ as $j \to \infty$.

Note that in practice this algorithm shows good results also in this case, see, for instance, the numerical experiments in [29].   □

Remark 6.3.6 One can modify the algorithm by choosing $\widehat{\Phi}_h$ as the total discretization of a time-h map φ^h and by requiring the accuracy and rigorous discretization property in the total sense of Remark 5.3.10. In this case the A^j converge to a strong attractor of φ^h which by Lemma 6.2.6 is also an attractor for the underlying continuous time system. Hence the algorithm also allows the computation of (strong) attractors for ordinary differential equations. Note, however, that the condition on the total accuracy implies that the time discretization needs to become more and more accurate during the iterative process in order to ensure convergence.   □

An interesting modification of Algorithm 6.3.1 is obtained when the Selection Step (1) is performed iteratively until it becomes stationary. We obtain this version when we replace step (1) by

(1') **(Selection Step)**
    Let $A_0^{j+1} := A^j$, $l := 0$ and compute iteratively

$$A_{l+1}^{j+1} := \bigcup_{\bar{u} \in \mathcal{U}(\Omega)} \widehat{\Phi}_h^j(A_l^{j+1}, \bar{u}) \bigcap A^j \in \mathcal{C}_{\mathbf{Q}^j}$$

until $A_{l+1}^{j+1} = A_l^{j+1}$, and set $A^{j+1} = A_l^{j+1}$.

Since we work on a finite set of P^j cells the termination of this algorithm is guaranteed after at most P^j steps. In practice, the convergence of the iteration in step (1') can be expected to be much faster. Intuitively, it depends on the rate of attraction μ of the ISDS set A, although an exact description of this dependence seems to be difficult.

It is easily seen from the ISDS property that for the modified algorithm applied with a rigorous space discretization we obtain the estimate $d_H(A^j, A) \leq \gamma(\varepsilon^{j-1})$. Hence, compared with the original algorithm here the same accuracy of $\widehat{\Phi}_h$ gives a more accurate approximation at the expense of doing more iterations on each level. Since the accuracy of $\widehat{\Phi}_h$ essentially corresponds to the number of cells in the cell covering one can expect that the modified algorithm needs less memory for the same accuracy of approximation which can be a great advantage especially in higher dimensions.

Remark 6.3.7 As far as the number of evaluations of $\widehat{\Phi}_h$ is concerned, for general systems it is difficult to say which version of the algorithm is the most efficient. For a variant of this algorithm designed for the computation of viability kernels (see Algorithm 7.5.1, below) a comparative study has been carried out in [118]. There it turns out that neither step (1) nor step (1') is optimal, but instead a suitable choice of an iterate of $\widetilde{\Phi}_h$ as sketched in Remark 6.3.2 (iii) minimizes the evaluations of the system.   □

6.4 Weakly Attracting Sets

In this section we will develop the weak analogues to the results obtained for strongly attracting sets in Section 6.1, at least as far as this is possible. Again, the basic idea behind these results is to embed the numerical system induced by the one-step approximation into the inflated time-h map via Lemma 5.1.4 and vice versa, and then use results from Chapter 4 to obtain the respective results.

We will formulate the results in this section for numerical approximations satisfying Definition 5.1.1. If the approximations under consideration are nonanticipating or even continuous nonanticipating approximations in the sense of Definition 5.1.5, then we can relax the respective robustness assumptions, cf. Remark 6.4.6, below.

We start by giving an existence criterion for a weakly attracting set for the numerical system.

Theorem 6.4.1 Consider a system φ of type (2.1) satisfying (2.8) and a numerical one step approximation $\widetilde{\Phi}_h$ of order $q \in \mathbb{N}$. Assume that φ has a weakly attracting set A which is γ-robust for the α_0-inflated system with

$\mathcal{P} = \mathcal{P}^\delta$ for some $\delta > 0$, some γ of class $\mathcal{K}_\infty$ and some $\alpha_0 > 0$. Then for all $h \in (0, \delta]$ with $e^{Lh}ch^q \leq \alpha_0$ the one step scheme $\tilde{\Phi}_h$ has a weakly attracting set A_h satisfying $A \subseteq A_h$ and $d_H(A_h, A) \leq \gamma(e^{Lh}ch^q)$ and for each $K > 1$ it has a weakly asymptotically stable set A_h^K satisfying $A \subseteq A_h^K$ and $d_H(A_h^K, A) \leq \gamma(Ke^{Lh}ch^q)$.

Proof: By Lemma 5.1.4 for all h sufficiently small the inflated numerical system is $(ch^q, 1)$-embedded in the inflated time-h map (4.14). Since by Corollary 4.8.5 the set A is a $\gamma(e^{Lh} \cdot)$-robust weakly attracting set for (4.14) the assertion follows from Propositions 4.6.8 and 4.7.5.   □

Next we study the converse question, i.e., the existence of a weakly attracting set for the original system. Again, we start with the time-h map.

Lemma 6.4.2 Consider a system φ of type (2.1) satisfying (2.8) and a numerical one step approximation $\tilde{\Phi}_h$ of order $q \in \mathbb{N}$. Let $h > 0$ and assume that $\tilde{\Phi}_h$ has a weakly attracting set A_h which is γ-robust for the α_0-inflated numerical system with $\mathcal{P} = \mathcal{P}^\delta$ for some $\delta \geq h$, some γ of class $\mathcal{K}_\infty$ and some $\alpha_0 > ch^p$. Then for each $D > 1$ the time-h map φ^h of φ has a weakly asymptotically stable set $\tilde{A}_h^D$ which is $\gamma(D \cdot /(D-1))$-robust for the inflated time-h map φ^h and satisfies $\tilde{A}_h^D \subseteq A_h$ and $d_H(\tilde{A}_h^D, A_h) \leq \gamma(Dch^p)$.

Proof: Follows immediately from Lemma 5.1.4 and Proposition 4.6.7.   □

Now we carry over this result to the continuous time system.

Theorem 6.4.3 Consider a system φ of type (2.1) satisfying (2.8) and a numerical one step approximation $\tilde{\Phi}_h$ of order $q \in \mathbb{N}$. Let $h > 0$ and assume that $\tilde{\Phi}_h$ has a weakly attracting set A_h which is γ-robust for the α_0-inflated system with $\mathcal{P} = \mathcal{P}^\delta$ for some $\delta \geq h$, some γ of class $\mathcal{K}_\infty$ and some $\alpha_0 > ch^p$. Assume furthermore that the attracted neighborhood B of A_h satisfies $d_{\min}(A_h, B) \geq \gamma(ch^p) + Mh$. Then for each $K > 1$ the continuous time system φ has a weakly asymptotically stable set A^K satisfying $A_h \subseteq A^K$ and $d_H(A^K, A_h) \leq \gamma(Kch^p) + Mh$.

Proof: Choose $D \in (1, K)$ and pick the weakly asymptotically stable set $\tilde{A}_h^D$ from Lemma 6.4.2. Then for each $r > 0$ Proposition 4.8.2 (with $\rho = \mathrm{id}_\mathbb{R}$ and $c = r$) ensures the existence of a weakly asymptotically stable set A_r with

$$d_H(A_r, A_h) \leq d_H(A_r, \tilde{A}_h^D) + d_H(\tilde{A}_h^D, A_h)$$

$$\leq \gamma(Dch^p) + \gamma\left(\frac{D2M}{(D-1)r}\right) + M_h(2+r)/r.$$

For $r > 0$ sufficiently large this last expression is bounded by $\gamma(Kch^p) + Mh$ which shows the assertion be setting $A^K = A_r$.   □

Now we turn to the investigation of limits of numerical weakly attracting sets.

Theorem 6.4.4 Consider a system φ of type (2.1) satisfying (2.8) and (2.13) and a numerical one step approximation $\widetilde{\Phi}_h$ of order $q \in \mathbb{N}$. Let $\delta > 0$ and consider a sequence of time steps $h_n \to 0$, $h_n \leq \delta$ and a sequence of weakly attracting sets A_n for the one step discretization, all with the same attracted neighborhood B. Assume that there exists a set $A \subset B$ such that $d_H(A_n, A) \to 0$. Then the following statements are equivalent.

(i) A is a weakly asymptotically stable set for the continuous time system.

(ii) There exists $N \in \mathbb{N}$, β of class $\mathcal{KL}$, a sequence of real numbers $\rho_n \to 0$ as $n \to \infty$ and a sequence of weakly attracting sets $\hat{A}_n$, $n \geq N$, for the one step discretizations $\widetilde{\Phi}_{h_n}$, with attracted neighborhood B, which satisfy $d_H(\hat{A}_n, A_n) \to 0$ and whose rates of attraction β_n satisfy $\beta_n(r, t) \leq \beta(r + \rho_n, t)$.

(iii) There exist $N \in \mathbb{N}$, $\alpha_0 > 0$, μ of class $\mathcal{KLD}$, γ and σ of class $\mathcal{K}$, a sequence of real numbers $\rho_n \to 0$ as $n \to \infty$ and a sequence of weakly attracting sets $\hat{A}_n$, $n \geq N$, for the α_0-inflated one step discretizations $\widetilde{\Phi}_{h_n}$ with $\mathcal{P} = \mathcal{P}^\delta$, with attracted neighborhood B, which satisfy $d_H(\hat{A}_n, A_n) \to 0$ and which weakly satisfy the wISDS-like estimate

$$\|\widetilde{\Phi}_{h_n}(t, x, u, p[u])\|_{\hat{A}_n} \leq \max\{\mu(\sigma(\|x\|_{\hat{A}_n} + \rho_n), t), \tilde{\nu}((1 + \rho_n)p, t)\}$$

for each $p \in \mathcal{P}$.

In addition, if (ii) holds then A is attracting with rate β and if (iii) holds then A is wISDS for the α_0-inflated continuous time system with rate μ and gains γ and σ for all sets of perturbations strategies $\mathcal{P} \subseteq \mathcal{P}^\delta$ for which this inflated system satisfies (2.14).

Proof: (i) $\Rightarrow$ (iii): First recall that if A is a weakly asymptotically stable set then by Theorem 4.4.5 it is wISDS for the inflated system for suitable gains μ, σ and γ and $W = \mathcal{B}(\varepsilon, 0)$ for some $\varepsilon > 0$. Hence it is also a wISDS set for each time-h_n map of the inflated system, and by Corollary 4.8.5 it is also wISDS for the inflated time-h_n map with robustness gain $\gamma(e^{Lh} \cdot)$. Since by Lemma 5.1.4 the inflated numerical system is $(ch_n^q, 1)$-embedded in the inflated time-h_n map, by Proposition 4.6.6 applied with $\alpha = ch_n^q$, $C = 1$ and $D = 1/\sqrt{\alpha}$, for all $h_n > 0$ sufficiently small there exists an attracting set $\hat{A}_n$ weakly satisfying the estimates

$$\|\widetilde{\Phi}_{h_n}(t, x, u, p[u])\|_{\hat{A}_n} \leq \max\{\mu(\sigma(\|x\|_{\hat{A}_n} + \gamma(\sqrt{\alpha})), t), \tilde{\nu}(p[u]/(1 - \sqrt{\alpha}), t)\}$$

and $d_H(\hat{A}_n, A) \leq \gamma(\sqrt{\alpha})$.

Setting $\rho_n := \max\{\sqrt{\alpha}/(1-\sqrt{\alpha}), \gamma(\sqrt{\alpha})\}$ hence shows the claim.

(iii) $\Rightarrow$ (ii) is obvious.

(ii) $\Rightarrow$ (i): Consider a sequence $T_n \to \infty$ such that $c(e^{LT_n} - 1)h_n^q/L \to 0$ as $n \to \infty$. We set $\tilde{\beta}_n(r, t) = \beta_n(r+\rho_n, t) + c(e^{LT_n} - 1)h_n^q/L$. Then Lemma 5.1.2 yields

$$\|\tilde{\Phi}_{h_n}(t, x, u, p[u])\|_{\hat{A}_n} \leq \tilde{\beta}_n(\|x\|_{\hat{A}_n}, t) \text{ weakly for all } t \in [0, T_n] \cap \mathbb{T}.$$

Since the $\tilde{\beta}_n$ are asymptotically bounded by β the assertion follows from Theorem 4.8.3 (i).

(iii) $\Rightarrow$ (i): Similar to "(ii) $\Rightarrow$ (i)" but using Theorem 4.8.3 (iii). $\square$

Since we cannot expect the limit of a sequence of asymptotically stable sets to be asymptotically stable, cf. Example 4.4.7, we can only give a weak version of the equivalence (i) $\Leftrightarrow$ (iii) in Theorem 6.1.4 using the practical attraction from Definition 4.7.6.

Theorem 6.4.5 Consider a system φ of type (2.1) satisfying (2.8) and a numerical one step approximation $\tilde{\Phi}_h$ of order $q \in \mathbb{N}$. Let $\delta > 0$, $\mathcal{P} = \mathcal{P}^\delta$ and consider a sequence of time steps $h_n \to 0$, $h_n \leq \delta$ and a sequence of weakly attracting sets A_n for the one step discretization, all with the same attracted neighborhood B. Assume that there exists a set $A \subset B$ such that $d_H(A_n, A) \to 0$ and let furthermore γ be a class $\mathcal{K}_\infty$ function. Then the following statements are equivalent.

(i) For each $K > 1$ the set A is a $\gamma(K \cdot)$-robust weakly practically attracting set for the α_0-inflated continuous time system, where the A_α realizing the $\gamma(K \cdot)$-robustness can be chosen to be weakly α-asymptotically stable.

(ii) There exist $N \in \mathbb{N}$, $\alpha_0 > 0$, γ of class $\mathcal{K}$, a sequence of real numbers $\rho_n \to 0$ and a sequence of $\gamma((1+\rho_n) \cdot)$-robust weakly attracting sets $\hat{A}_n$, $n \geq N$, for the α_0-inflated one step discretizations $\tilde{\Phi}_{h_n}$ with attracted neighborhood B, which satisfy $d_H(\hat{A}_n, A_n) \to 0$.

Proof: (i) $\Rightarrow$ (ii): Let $C > 1$ and consider the weakly asymptotically stable sets A_α realizing the γ-robustness. Fix some $\alpha_0 > 0$ and consider the sets $\tilde{A} = A_{\alpha_0}$, $\tilde{A}_\alpha = A_{\alpha_0}$, $\alpha \leq \alpha_0$, and $\tilde{A}_\alpha = A_\alpha$ for $\alpha > \alpha_0$. Then it is immediate that $\tilde{A}$ is a $\gamma(C \cdot)$-robust attracting set for the inflated continuous time system, hence by Corollary 4.8.5 it is also $\gamma(e^{Lh_n}C \cdot)$-robust for the inflated time-h_n map. By Lemma 5.1.3 and Proposition 4.6.7 (applied with $\alpha = ch_n^q$, $C = 1$ and $D = 1/\sqrt{\alpha}$) we obtain that $\tilde{\Phi}_{h_n}$ has a $\gamma(e^{Lh_n}C \cdot /(1-\sqrt{\alpha}))$-robust attracting set $\tilde{A}_{h_n}$ with $d_H(\tilde{A}_{h_n}, A) \leq \gamma(C\sqrt{\alpha})$. Choosing $C = 1 + h_n$ the assertion follows by a proper choice of ρ_n.

(ii) $\Rightarrow$ (i): By Lemma 5.1.4 the inflated time-h_n map of (2.1) is $(ch_n^q, 1)$-embedded in the inflated numerical system $\widetilde{\Phi}_{h_n}$. Hence by Proposition 4.6.7 applied to $A = \hat{A}_n$ with $\alpha = ch_n^q$ and $D = 1/\sqrt{\alpha}$ we obtain the existence of $\tilde{\gamma}_n$-robust weakly attracting sets $\tilde{A}_n$ with $d_H(\tilde{A}_n, \hat{A}_n) \leq \tilde{\gamma}_n(\sqrt{ch_n^q})$ and $\tilde{\gamma}_n(r) = \gamma_n(1/(1 - \sqrt{ch_n^p})r)$. By Corollary 4.8.5 these sets are $\tilde{\gamma}_n((1 + Lh_n)\cdot)$-robust weakly attracting sets for the time-h_n map of the inflated continuous time system. By the assumption on the $\hat{A}_n$ and γ_n we obtain that $d_H(\tilde{A}_n, A) \to 0$ as $n \to \infty$ and that the $\tilde{\gamma}_n((1 + Lh_n)\cdot)$ are asymptotically bounded by γ. Thus Proposition 4.8.3 (ii) yields the assertion. $\square$

Remark 6.4.6 We have seen in Chapter 5 that several numerical one-step approximations for control systems are nonanticipating or even continuous nonanticipating approximations in the sense of Definition 5.1.5.

If we consider such approximations, it is easily seen that it is sufficient to require the respective robustness properties in the assumptions of all results in this section only for $\mathcal{P} = \mathcal{P}^0$ or, in the case of continuous nonanticipating approximations, for $\mathcal{P} = \mathcal{P}^{0,c}$ instead of $\mathcal{P} = \mathcal{P}^\delta$. In these cases the time steps $h > 0$ for which the results hold is, of course, independent of $\delta = 0$.

$\square$

Let us now investigate the space discretization of weakly attracting sets.

Theorem 6.4.7 Consider a discrete time system Φ_h^0 of type (2.17), a compact set $\Omega \subset \mathbb{R}^n$, a cell covering $\mathbf{Q}$ of Ω and a cell–cell space discretization $\widehat{\Phi}_h$ on $\mathbf{Q}$.

Assume that $A \subset \operatorname{int} \Omega$ is a weakly attracting set for Φ_h^0 which is wISDS with respect to α_0-inflation with rate μ of class $\mathcal{KLD}$ and gains σ and γ of class $\mathcal{K}_\infty$ and some $\alpha_0 > 0$ and perturbations from $\mathcal{P} = \mathcal{P}^h$. Assume furthermore that $\widehat{\Phi}_h$ has accuracy Δ_i satisfying $\Delta_i/h \leq \max\{\Delta, \gamma^{-1}(\|x\|_A)\}$ for some $\Delta > 0$.

Then, if $\Delta > 0$ and $\operatorname{diam}(\mathbf{Q})$ are sufficiently small there exists a strongly attracting set $\widehat{A} \in \mathcal{C}_\mathbf{Q}$ for $\widehat{\Phi}_h$ which satisfies

$$\operatorname{dist}(\widehat{A}, A) \leq \gamma(\Delta).$$

Proof: Analogous to the proof of Theorem 6.1.6 with the obvious modifications from "strong" to "weak". $\square$

Following the results for strongly attracting sets, it would now of course be nice to have an analogous result to Theorem 6.2.9 for weak attractors, but this would require to have a suitable definition of a "weak attractor".

The main properties of a strong attractor we have used in the proof of this theorem are its existence and uniqueness (provided we have an attracting set) which are basically obtained from its minimality.

In Example 4.4.7 each set $A_\varepsilon = [-\varepsilon, \varepsilon]^2$ for $\varepsilon > 0$ sufficiently small is weakly asymptotically stable, but $A_0 = \{(0,0)^T\}$ is not even weakly attracting, which shows that in general neither minimal weakly asymptotically stable sets nor minimal weakly attracting sets exist.

The following simple example illustrates that even if minimal weakly asymptotically stable sets exist, they need not be unique. In particular, we will see that a sequence of γ-robust numerical minimal weakly asymptotically stable sets does not need to converge, and if it converges, then it can do so arbitrary slow.

Example 6.4.8 Consider the control system

$$\dot{x} = -x + u$$

with $U = [-1, 1]$, along with its Euler discretization

$$\widetilde{\Phi}_h(x, u) = (1 - h)x + \int_0^h u(t)dt.$$

For each constant control $u \equiv u_0$ with $u_0 \in [-1, 1]$ we find that for $h \in (0, 1)$ the numerical system has the exponentially stable equilibrium $x_{u_0} = u_0$. Hence each set $A_h = \{x^*\}$, $x^* \in [-1, 1]$ is a weakly asymptotically stable set for $\widetilde{\Phi}_h$, and a short computation reveals that it is γ-robust for $\gamma(r) = r$ for the inflated system. Thus, for any sequence of time steps $h_n \to 0$ we can build arbitrary sequences $A_n = \{x_n^*\}$, $x_n^* \in [-1, 1]$ of γ-robust attracting sets. They do not need to converge, and if they do converge then they can do so at arbitrary slow rate. If they converge, however, then they converge to some γ-robust weakly asymptotically stable set A of the continuous time system.   □

It seems reasonable to expect that if we assume the existence of a unique minimal weakly asymptotically stable (or attracting) set, then some of the results in Section 6.2 might be reproducible. This, however, seems to be a rather restrictive assumption and we will therefore not pursue this idea.

The same limitation as for the time discretization also holds for the subdivision algorithm from Section 6.3. We will, however, see in Chapter 7 that subdivision algorithms are applicable to problems in weak stability, namely to the computation of weak domains of attraction.

7 Domains of Attraction

In this chapter we will discuss domains of attraction for strongly and weakly attracting sets as well as reachable sets. We introduce and discuss suitable robustness properties for these sets and investigate their behavior under discretization and perturbation. In addition we formulate and discuss algorithms for their computation.

Domains of attraction (and their close relatives reachable sets, cf. Section 7.7) play an important role in the analysis of nonlinear dynamical systems. For systems without input domains of attraction were extensively investigated in the late 1960s, see for instance Zubov [129], Coleman [21], Wilson [125] and Bhatia [13] or the textbooks by Hahn [61] or Khalil [71]. For controlled and perturbed nonlinear systems one should in particular mention the monograph by Colonius and Kliemann [22] which presents an approach for the analysis of such systems where domains of attraction and reachable sets play a prominent role.

Many of the mentioned results for systems without input could be transferred for systems with (control or perturbation) input u, see the papers by Camilli, Wirth and the author [16, 17] and by Wirth and the author [58]. In particular, it was shown that Zubov's method for the characterization of domains of attraction can be extended to systems with input, a result that we will investigate in more detail in Section 7.2, below.

In order to simplify the presentation, throughout this chapter we assume that (2.8) or (2.24) holds, even without explicitly mentioning it.

7.1 Definitions and Basic Properties

Let us start by defining the objects we want to investigate.

Definition 7.1.1 (domain of attraction)

(i) Consider a strongly attracting set A. We call the set

$$\mathcal{D}(A) := \left\{ x \in \mathbb{R}^n \;\middle|\; \begin{array}{l} \text{there exists a function } \beta(t) \to 0 \text{ as } t \to \infty \\ \text{such that } \|\Phi(t, x_0, u)\|_A \leq \beta(t) \text{ for all } t \in \mathbb{T}^+, u \in \mathcal{U} \end{array} \right\}$$

the *(uniform) strong domain of attraction* of A.

(ii) Consider a weakly asymptotically stable set A. We call the set

$$\mathcal{D}(A) := \{x \in \mathbb{R}^n \,|\, \text{there exists } u^* \in \mathcal{U} \text{ with } \|\Phi(t, x_0, u^*)\|_A \to 0 \text{ as } t \to \infty\}$$

the *weak domain of attraction* of A.   □

These objects are different from the attracted neighborhoods B which we have considered in the earlier chapters since here we explicitly consider the maximal sets of points converging to A. We always have the inclusion $B \subset \mathcal{D}(A)$, however, the uniformity imposed on the convergence for $x \in B$ (cf. Definitions 3.1.3(i) and 4.1.4(i)) excludes equality of B and $\mathcal{D}(A)$ both in the weak and in the strong case, except when $B = \mathcal{D}(A) = \mathbb{R}^n$. Since the definition of asymptotically stable sets includes the existence of an attracted neighborhood B we can define

$$t(x, u) := \inf\{t \in \mathbb{T}^+ \,|\, \Phi(t, x, u) \in B\}$$

using the convention $\inf \emptyset = \infty$.

We can use this value for a different characterization of the maximal domain of attraction.

Lemma 7.1.2 Consider a strongly or weakly asymptotically stable set A with attracted neighborhood B.

(i) The strong domain of attraction $\mathcal{D}(A)$ satisfies the equality

$$\mathcal{D}(A) := \{x \in \mathbb{R}^n \,|\, \sup_{u \in \mathcal{U}} t(x, u) < \infty\}.$$

(ii) The weak domain of attraction $\mathcal{D}(A)$ satisfies the equality

$$\mathcal{D}(A) := \{x \in \mathbb{R}^n \,|\, \inf_{u \in \mathcal{U}} t(x, u) < \infty\}.$$

Proof: Follows directly from the definitions using the imposed global bounds (2.8) or (2.24).   □

Remark 7.1.3 For continuous time systems it was observed in [17] that for a strong domain of attraction $\mathcal{D}(A)$ its closure $\operatorname{cl} \mathcal{D}(A)$ coincides with the closure of the set

$$\tilde{\mathcal{D}}(A) := \{x \in \mathbb{R}^n \,|\, \|\Phi(t, x_0, u)\|_A \to 0 \text{ as } t \to \infty \text{ for all } u \in \mathcal{U}\}$$

which could be considered as a more natural definition of a domain of attraction. This observation, in turn, is based on a result by Sontag and Wang [106,

Lemma III.2], which states that if $\sup_{u \in \mathcal{U}}\{t(x, u)\} = \infty$, while $t(x, u) < \infty$ for every $u \in \mathcal{U}$, then in every neighborhood of x there exists a point y and a functions $u_y \in \mathcal{U}$ such that $t(y, u_y) = \infty$, hence $x \in \partial \widetilde{\mathcal{D}}(A)$.

For both continuous and discrete time systems it is easily seen that we obtain the equality $\mathcal{D}(A) = \widetilde{\mathcal{D}}(A)$ under the continuity assumption (2.13).   □

The following proposition shows that domains of attraction are always open sets.

Proposition 7.1.4 Consider a strongly or weakly asymptotically stable set A with attracted neighborhood B. Then the strong or weak domain of attraction $\mathcal{D}(A)$ is an open set.

Proof: We first consider the strong case: Let $x \in \mathcal{D}(A)$. Then by Lemma 7.1.2(i) and the definition of B there exists a time $t > 0$ such that $\Phi_{\mathcal{U}}(t, x) \subset$ int B. By continuous dependence on the initial value we obtain that $\Phi_{\mathcal{U}}(t, y) \subset$ int B for all y sufficiently close to x. Hence $\sup_{u \in \mathcal{U}} t(y, u) < \infty$ which again by Lemma 7.1.2(i) implies $y \in \mathcal{D}(A)$. The weak case follows analogously using Lemma 7.1.2(ii).   □

7.2 Zubov's Method

In this section we describe Zubov's method for strong and weak domains of attraction. Basically, this section consists of results obtained by Camilli, Wirth and the author in [16, 17, 58], with the only exception being Proposition 7.2.5 and the extensions to discrete time systems.

The original method of Zubov [129] states that for a system without input, i.e., a system of the form

$$\dot{x} = f(x),$$

where $f : \mathbb{R}^n \to \mathbb{R}^n$ is C^1 and has an asymptotically stable equilibrium x^*, there exists a function $g : \mathbb{R}^n \to \mathbb{R}_0^+$ with $g(x) = 0 \Leftrightarrow x = x^*$ such that the partial differential equation

$$Dv(x)f(x) + g(x)(1 - v(x))\sqrt{1 + \|f(x)\|^2} = 0 \tag{7.1}$$

has a unique solution $v : \mathbb{R}^n \to [0, 1]$ with $v(x^*) = 0$. This function characterizes the domain of attraction via $\mathcal{D}(\{x^*\}) = \{x \in \mathbb{R}^n \mid v(x) < 1\}$ and is a Lyapunov function on $\mathcal{D}(\{x^*\})$. Equation (7.1) is called Zubov's equation. Later, this statement was generalized to periodic orbits instead of equilibria, see Aulbach [7].

In order to generalize this statement to systems with control or perturbation and to arbitrary asymptotically stable sets we need to introduce some definitions. We start with continuous time systems.

Let A be a strongly or weakly asymptotically stable set with attracted neighborhood B for which we assume $d_H(B, A) < \infty$. (Actually, all statements in this section remain true also if we do not assume the c-boundedness of A which is included in our definitions of asymptotically stable sets.) The class $\mathcal{KL}$ function β characterizing the rate of attraction is supposed to satisfy

$$\beta(r, t) \leq a_1(a_2(r)e^{-t}) \tag{7.2}$$

for suitable class $\mathcal{K}_\infty$ functions a_1 and a_2, which is always possible by Lemma B.1.3. Now consider a function $g : \mathbb{R}^n \to \mathbb{R}_0^+$, the nonnegative, extended value functional $J : \mathbb{R}^n \times \mathcal{U} \to \mathbb{R} \cup \{+\infty\}$ given by

$$J(x, u) := \int_0^{+\infty} g(\varphi(t, x, u))dt \tag{7.3}$$

and the related optimal value functions

$$v^+(x) := \sup_{u \in \mathcal{U}} \{1 - e^{-J(x,u)}\} \tag{7.4}$$

and

$$v^-(x) := \inf_{u \in \mathcal{U}} \{1 - e^{-J(x,u)}\}. \tag{7.5}$$

The function g is supposed to be continuous and satisfies

(i) There exists a constant $C > 0$ such that $g(x) \leq Ca_1^{-1}(\|x\|_A)$ for all $x \in B$ and a_1 from (7.2).

(ii) $\inf_{x \notin B(\epsilon, A)} g(x) > 0$ for all $\epsilon > 0$ and $\inf_{x \notin B} g(x) =: g_0 > 0$. (7.6)

(iii) For every $R > 0$ there exists a constant L_R such that $\|g(x) - g(y)\| \leq L_R\|x - y\|$ for all $\|x\|, \|y\| \leq R$.

Since g is nonnegative it is immediate that $J(x, u) \geq 0$, $v^+(x) \in [0, 1]$ and $v^-(x) \in [0, 1]$ for all $x \in \mathbb{R}^n$ and all $u \in \mathcal{U}$. Furthermore, standard techniques from optimal control (see e.g. [8, Chapter III]) imply that v^+ and v^- satisfy the dynamic programming principle, i.e., for each $t > 0$ we have

$$v^+(x) = \sup_{u \in \mathcal{U}} \left\{(1 - G(x, t, u)) + G(x, t, u)v^+(\varphi(t, x, u))\right\} \tag{7.7}$$

and

$$v^-(x) = \inf_{u \in \mathcal{U}} \left\{(1 - G(x, t, u)) + G(x, t, u)v^-(\varphi(t, x, u))\right\} \tag{7.8}$$

with

$$G(x, t, u) := \exp\left(-\int_0^t g(\varphi(\tau, x, u))d\tau\right). \tag{7.9}$$

In the next proposition we investigate the relation between $\mathcal{D}(A)$, v^+ and v^- and the continuity of v^+ and v^-.

Proposition 7.2.1 If A is a strongly asymptotically stable set with strong domain of attraction $\mathcal{D}(A)$ and $g : \mathbb{R}^n \to \mathbb{R}_0^+$ satisfies (7.6) then the following properties hold.

(i) $v^+(x) < 1$ if and only if $x \in \mathcal{D}(A)$

(ii) $v^+(x) = 0$ if and only if $x \in A$

(iii) $v^+(x) \to 1$ for $x \to x_0 \in \partial\mathcal{D}(A)$ and for $\|x\|_A \to \infty$.

(iv) v^+ is continuous on $\mathbb{R}^n$

(v) v^+ is a robust Lyapunov function on $\mathcal{D}(A)$, i.e., for each $x \in \mathcal{D}(A) \setminus A$, each $t > 0$ and each $u \in \mathcal{U}$ it satisfies

$$v^+(\varphi(t,x,u)) - v^+(x) \le (1 - G(t,x,u))(v^+(\varphi(t,x,u)) - 1) < 0 \quad (7.10)$$

for G from (7.9).

If A is a weakly asymptotically stable set with weak domain of attraction $\mathcal{D}(A)$ and $g : \mathbb{R}^n \to \mathbb{R}_0^+$ satisfies (7.6) then the following properties hold.

(i) $v^-(x) < 1$ if and only if $x \in \mathcal{D}(A)$

(ii) $v^-(x) = 0$ if and only if $x \in A$

(iii) $v^-(x) \to 1$ for $x \to x_0 \in \partial\mathcal{D}(A)$ and for $\|x\|_A \to \infty$

(iv) v^- is continuous on $\mathbb{R}^n$

(v) v^- is a control Lyapunov function on $\mathcal{D}(A)$, i.e., for each $x \in \mathcal{D}(A)$, each $t \ge 0$ and each $\varepsilon > 0$ there exists $u \in \mathcal{U}$ such that it satisfies

$$v^-(\varphi(t,x,u)) - v^-(x) \le (1 - G(t,x,u))(v^-(\varphi(t,x,u)) - 1) + \varepsilon \quad (7.11)$$

for G from (7.9). In particular for each $x \in \mathcal{D}(A) \setminus A$ and each $t > 0$ there exists $u \in \mathcal{U}$ such that

$$v^-(\varphi(t,x,u)) < v^-(x).$$

Proof: We prove the weak case; for the (analogous) strong case we refer to [16, 17].

(i) We first show the property for $x \in B$. In this case we know the existence of $u \in \mathcal{U}$ such that

$$\|\varphi(t,x,u)\|_A \le a_1(a_2(r)e^{-t}).$$

From this we obtain that

$$J(x,u) = \int_0^\infty g(\varphi(t,x,u))dt \le \int_0^\infty Ca_1^{-1}(a_1(a_2(\|x\|)e^{-t}))dt = Ca_2(\|x\|)$$

implying that $v^-(x) \le 1 - e^{-J(x,u)} \le 1 - e^{-Ca_2(\|x\|)} < 1$.

For $x \in \mathcal{D}(A)$ and $x \notin B$ we find $u \in \mathcal{U}$ and $t > 0$ such that $\varphi(t, x, u) \in B$, hence $v^-(x) < 1$ follows from (7.8) and the case $x \in B$.

Now let $x \notin \mathcal{D}(A)$. Then we know that for all $u \in \mathcal{U}$ we obtain $\varphi(t, x, u) \notin B$ for all $t \geq 0$. Consequently $J(x, u) = \infty$ for all $u \in \mathcal{U}$ and hence $v^-(x) = 1$.

(ii) Immediate since $g \equiv 0$ on A, $g > 0$ on A^c and A is weakly forward invariant.

(iii) From (7.3) and (7.6)(ii) it is immediate that if $\inf_{u \in \mathcal{U}} t(x_n, u) \to \infty$ for some sequence of points x_n then $v^-(x_n) \to 1$. Hence it suffices to show that $\inf_{u \in \mathcal{U}} t(x_n, u) \to \infty$ if $x_n \to x_0 \in \partial \mathcal{D}(A)$ or $\|x_n\|_A \to \infty$. In order to see this property define $T_n = \inf_{u \in \mathcal{U}} \{t(x_n, u)\}$. If we assume that T_n is bounded for all $n \in \mathbb{N}$, we can find T such that for any n there is a control u_n with $\varphi(T, x_n, u_n) \in \text{int } B$. Then if $x_n \to x_0 \in \partial \mathcal{D}(A)$ by continuous dependence on the initial value we can conclude that $\varphi(T, x_0, u_n) \in \text{int } B$ for $n > 0$ sufficiently large which contradicts the assumption that $x_0 \in \partial \mathcal{D}(A)$ since $\mathcal{D}(A)$ is open. The assertion is clear for $\|x_n\|_A \to \infty$, as our assumption (2.8) excludes solutions exploding in backward time.

(iv) Because of (iii) we only need to show continuity on $\mathcal{D}(A)$. For this we introduce the auxiliary function $V : \mathcal{D}(A) \to \mathbb{R}_0^+$ given by

$$V(x) := \inf_{u \in \mathcal{U}} J(x, u), \text{ i.e., } v^-(x) = 1 - e^{-V(x)}.$$

As in the proof of property (i) we obtain $V(x_0) \leq C a_2(\|x\|)$ for $x \in B$. Note that for all $\varepsilon > 0$ there exists $r^* > 0$ such that $C a_2(\|x\|) \leq \varepsilon$ for all $x \in B$ with $\|x\| \leq r^*$. Now fix $\varepsilon > 0$ and let $r^* > 0$ satisfy this property. Let $x \in \mathcal{D}(A)$ be arbitrary. Then we find $u \in \mathcal{U}$ and $T > 0$ such that $\|\varphi(T, x, u)\|_A \leq r^*/2$ and $V(x) + \varepsilon > J(x, u)$. By continuous dependence on the initial value there is a neighborhood $W \subset \mathcal{D}(A)$ of x such that $\|\varphi(T, y, u)\|_A \leq r^*$ for all $y \in W$. We may assume that $\text{cl } W \subset \mathcal{D}(A)$ is compact, whence also $\mathcal{R}_T := \text{cl } \bigcup_{t \in [0, T]} \varphi_u(t, W)$ is compact. This implies in particular, that $V(y) \leq T \max\{g(x) \mid x \in \mathcal{R}_T\} + \varepsilon$ for all $y \in W$. From this bound of V on W and the fact that g is bounded away from zero on $\mathbb{R}^n \setminus B(r^*/2, A)$ it follows that there exists $T^* > 0$ such that whenever $y \in W$ and $u \in \mathcal{U}$ are such that $V(y) + \varepsilon > J(y, u)$ we have $\varphi(t, y, u) \in B(r^*/2, A)$ for all $t > T^*$. We may now choose a Lipschitz constant L_g for g on $\mathcal{R}_{T^*}$. Let $y, z \in W$ and $u \in \mathcal{U}$ be such that $V(y) + \varepsilon > J(y, u)$. Then we obtain

$$V(z) - V(y) < V(z) - \int_0^{+\infty} g(\varphi(t, y, u)) dt + \varepsilon$$

$$\leq \int_0^{T^*} |g(\varphi(t, z, u)) - g(\varphi(t, y, u))| \, dt + V(\varphi(T^*, z, u)) + C a_2(r^*) + \varepsilon$$

$$\leq L_g \int_0^{T^*} e^{Lt} \|z - y\| \, dt + V(\varphi(T^*, z, u)) + 2\varepsilon$$

If $\|z-y\|$ is small enough then $\varphi(T^*, z, u) \in B(r^*, A)$ so that $V(\varphi(T^*, z, u)) < Ca_2(r^*)$ and also the integral can be bounded by ε, so that the whole expression is bounded by 4ε. As this condition is symmetric in y, z this shows continuity of V. The function v is then continuous by definition.

(v) This is immediate from (7.8). $\square$

We can now state the main result on the generalization of Zubov's equation for perturbed and controlled systems. For the definition of viscosity solutions see Appendix A.

Theorem 7.2.2 Consider a system (2.1) satisfying (2.8).

(i) Let A be a strongly attracting set with attracted neighborhood B, rate of attraction β and strong domain of attraction $\mathcal{D}(A)$. Let $g : \mathbb{R}^n \to \mathbb{R}_0^+$ satisfy (7.6). Then there exists a unique continuous and bounded viscosity solution v of the partial differential equation

$$\inf_{u \in U} \{-Dv(x)f(x, u) - (1 - v(x))g(x)\} = 0, \quad x \in \mathbb{R}^n \qquad (7.12)$$

with $v(x) = 0$ for all $x \in A$. This functions coincides with v^+ from (7.4). In particular, it characterizes the domain of attraction via $\mathcal{D}(A) := \{x \in \mathbb{R}^n \,|\, v(x) < 1\}$ and is a robust Lyapunov function on $\mathcal{D}(A)$ in the sense of (7.10).

(ii) Let A be a weakly attracting set with attracted neighborhood B, rate of attraction β and weak domain of attraction $\mathcal{D}(A)$. Let $g : \mathbb{R}^n \to \mathbb{R}_0^+$ satisfy (7.6). Then there exists a unique continuous and bounded viscosity solution v of the partial differential equation

$$\sup_{u \in U} \{-Dv(x)f(x, u) - (1 - v(x))g(x)\} = 0, \quad x \in \mathbb{R}^n \qquad (7.13)$$

with $v(x) = 0$ for all $x \in A$. This functions coincides with v^- from (7.5). In particular, it characterizes the domain of attraction via $\mathcal{D}(A) := \{x \in \mathbb{R}^n \,|\, v(x) < 1\}$ and is a control Lyapunov function on $\mathcal{D}(A)$ in the sense of (7.11).

Proof: We show (ii); the strong case (i) follows by similar arguments. By standard viscosity solution arguments similar to the ones used in the proofs of Propositions 3.5.6 and 4.5.6 one shows that v^- is a viscosity solution of (7.13).

Conversely, from the optimality principles in [108, Theorem 3.2 (i) and (ii)] one concludes that any continuous and bounded viscosity solution v of (7.13) satisfies

$$v(x) = \inf_{u \in \mathcal{U}} \sup_{t \geq 0} \{(1 - G(x, t, u)) + G(x, t, u)v(\varphi(t, x, u))\} \qquad (7.14)$$

and

$$v(x) = \inf_{u \in \mathcal{U}} \inf_{t \geq 0} \{(1 - G(x,t,u)) + G(x,t,u)v(\varphi(t,x,u))\}. \qquad (7.15)$$

We pick some $x \in \mathbb{R}^n$ and distinguish two cases:

$v^-(x) < 1$: First note that the inequality $1 - e^{-J(x,u)} < 1$ holds if and only if $\|\varphi(t,x,u)\|_A \to 0$. Hence for any $0 < \varepsilon < 1 - v^-(x)$ and any $u_\varepsilon \in \mathcal{U}$ satisfying

$$1 - e^{-J(x,u_\varepsilon)} \leq v^-(x) + \varepsilon$$

we can conclude $\|\varphi(t,x,u_\varepsilon)\|_A \to 0$. Thus, since v is continuous and satisfies $v|_A \equiv 0$ we obtain $v(\varphi(t,x,u_\varepsilon)) \to 0$. Hence by (7.15) we obtain

$$v(x) \leq \lim_{t \to \infty} 1 - G(x,t,u_\varepsilon) + G(x,t,u_\varepsilon)v(\varphi(t,x,u_\varepsilon))$$

$$= \lim_{t \to \infty} 1 - G(x,t,u_\varepsilon) = 1 - e^{-J(x,u_\varepsilon)} \leq v^-(x) + \varepsilon$$

and consequently $v(x) \leq v^-(x)$. For the converse inequality we may assume $v(x) < 1$ because otherwise $v(x) \geq v^-(x)$ is immediate. Fix $0 < \varepsilon < 1 - v(x)$ and choose a control $u_\varepsilon \in \mathcal{U}$ such that the inf in (7.14) is attained up to ε, i.e.,

$$v(x) + \varepsilon \geq \sup_{t \geq 0} \{(1 - G(x,t,u_\varepsilon)) + G(x,t,u_\varepsilon)v(\varphi(t,x,u_\varepsilon))\}.$$

Then by (7.8) for each $t \geq 0$ we obtain

$$v^-(x) \leq 1 - G(x,t,u_\varepsilon) + G(x,t,u_\varepsilon)v^-(\varphi(t,x,u_\varepsilon))$$

and thus for each $t \geq 0$ we have

$$v^-(x) - v(x) \leq G(x,t,u_\varepsilon)(v^-(\varphi(t,x,u_\varepsilon) - v(\varphi(t,x,u_\varepsilon))) + \varepsilon.$$

Now, again, we have two diferent cases: either we find a sequence $t_i \to \infty$ such that $\|\varphi(t_i,x,u_\varepsilon)\|_A \to 0$ as $t_i \to \infty$, which implies $v^-(\varphi(t_i,x,u_\varepsilon) \to 0$ and $v(\varphi(t_i,x,u_\varepsilon) \to 0$, hence $v^-(x) - v(x) \leq \varepsilon$. If no such sequence $t_i \to \infty$ exists we can conclude that there exists $\eta, T > 0$ such that $\|\varphi(t,x,u_\varepsilon)\|_A > \eta$ for all $t \geq T$. By the properties of g and the definition of G this implies $G(x,t,u_\varepsilon) \to 0$ and since v and v^- are bounded we again obtain $v^-(x) - v(x) \leq \varepsilon$. Since $\varepsilon > 0$ was arbitrary this shows $v^-(x) \leq v(x)$, and hence equality.

$v^-(x) = 1$: In this case $x \notin \mathcal{D}(A)$ holds, hence there exists $\eta > 0$ such that $\|\varphi(t,x,u)\|_A \geq \eta$ for all $t \geq 0$ and all $u \in \mathcal{U}$, which in turn implies $G(x,t,u) \to 0$ as $t \to \infty$, even uniformly in $u \in \mathcal{U}$. Hence for each $u \in \mathcal{U}$ we obtain

$$1 - G(x,t,u) + G(x,t,u)v(\varphi(t,x,u)) = 1 - G(x,t,u) \to 1$$

uniformly in u as $t \to \infty$ because v is bounded. Hence (7.14) implies $v(x) \geq 1$ and (7.15) implies $v(x) \leq 1$, thus $v(x) = 1 = v^-(x)$. $\square$

Remark 7.2.3 (i) In [15, 17, 58] several more general results are proved. In particular we obtain a comparison principle for semicontinuous viscosity sub- and supersolutions of (7.12) and (7.13) and we allow g to depend also on u.

(ii) In general a classical (i.e., C^1) solution to (7.12) and (7.13) does not exist. What can be shown for the strong case (see [15]) is that under mild conditions on g the solution of (7.12) is globally Lipschitz continuous and for suitable (but in general unknown) g the solution to (7.12) is even smooth outside a neighborhood of $\partial \mathcal{D}(A)$. For the weak case these properties are not known and we conjecture that they are false. In this case, however, we can at least obtain Lipschitz continuous functions which solve (7.13) outside a neighborhood of A, cf. Proposition 7.2.5, below.   □

Before we turn to this result we briefly want to state the discrete time version of Zubov's method. Of course, here the PDEs (7.12) and (7.13) do not make sense, we can, however, define v^+ and v^- and obtain an analogous result to Proposition 7.2.1.

For a discrete time system Φ_h^0 of type (2.17) we define the following nonnegative, extended value functional $J_n : \mathbb{R}^n \times \mathcal{U} \to \mathbb{R} \cup \{+\infty\}$ given by

$$J_h(x, u) := \sum_{i=0}^{\infty} hg(\Phi_h^0(hi, x, u))dt \tag{7.16}$$

and the related optimal value functions

$$v_h^+(x) := \sup_{u \in \mathcal{U}} \{1 - e^{-J_h(x,u)}\} \tag{7.17}$$

and

$$v_h^-(x) := \inf_{u \in \mathcal{U}} \{1 - e^{-J_h(x,u)}\}. \tag{7.18}$$

Again using standard optimal control techniques (see e.g. [8, Chapter III]) one sees that these functions satisfy the dynamic programming principles

$$v_h^+(x) = \sup_{u \in \mathcal{U}} \left\{ (1 - G_h(x, t, u)) + G_h(x, t, u)v_h^+(\Phi_h^0(t, x, u)) \right\}, \quad t \in h\mathbb{N} \tag{7.19}$$

and

$$v_h^-(x) = \inf_{u \in \mathcal{U}} \left\{ (1 - G_h(x, t, u)) + G_h(x, t, u)v_h^-(\Phi_h^0(t, x, u)) \right\}, \quad t \in h\mathbb{N}, \tag{7.20}$$

with

$$G_h(x, t, u) := \exp\left(-\sum_{i=0}^{t/h-1} hg(\Phi_h^0(hi, x, u))d\tau \right), \quad t \in h\mathbb{N} \tag{7.21}$$

The proof of the following proposition is completely analogous to the continuous time case in Proposition 7.2.1.

Proposition 7.2.4 If A is a strongly asymptotically stable set with strong domain of attraction $\mathcal{D}(A)$ and $g : \mathbb{R}^n \to \mathbb{R}_0^+$ satisfies (7.6) then the following properties hold.

(i) $v_h^+(x) < 1$ if and only if $x \in \mathcal{D}(A)$

(ii) $v_h^+(x) = 0$ if and only if $x \in A$

(iii) $v_h^+(x) \to 1$ for $x \to x_0 \in \partial\mathcal{D}(A)$ and for $\|x\|_A \to \infty$

(iv) v_h^+ is continuous on $\mathbb{R}^n$

(v) v_h^+ is a robust Lyapunov function on $\mathcal{D}(A)$, i.e., for each $x \in \mathcal{D}(A) \setminus A$, each $t \in h\mathbb{N}$ and each $u \in \mathcal{U}$ it satisfies

$$v_h^+(\Phi_h^0(t,x,u)) - v_h^+(x) \le (1 - G_h(t,x,u))(v_h^+(\Phi_h^0(t,x,u)) - 1) < 0 \tag{7.22}$$

for G from (7.9).

If A is a weakly asymptotically stable set with weak domain of attraction $\mathcal{D}(A)$ and $g : \mathbb{R}^n \to \mathbb{R}_0^+$ satisfies (7.6) then the following properties hold.

(i) $v_h^-(x) < 1$ if and only if $x \in \mathcal{D}(A)$

(ii) $v_h^-(x) = 0$ if and only if $x \in A$

(iii) $v_h^-(x) \to 1$ for $x \to x_0 \in \partial\mathcal{D}(A)$ and for $\|x\|_A \to \infty$

(iv) v_h^- is continuous on $\mathbb{R}^n$

(v) v_h^- is a control Lyapunov function on $\mathcal{D}(A)$, i.e., for each $x \in \mathcal{D}(A)$, each $t \in h\mathbb{N}$ and each $\varepsilon > 0$ there exists $u \in \mathcal{U}$ such that it satisfies

$$v_h^-(\Phi_h^0(t,x,u)) - v_h^-(x) \le (1 - G_h(t,x,u))(v_h^-(\Phi_h^0(t,x,u)) - 1) + \varepsilon \tag{7.23}$$

for G_H from (7.21). In particular for each $x \in \mathcal{D}(A) \setminus A$ and each $t \in \mathbb{T}^+$ there exists $u \in \mathcal{U}$ such that

$$v_h^-(\Phi_h^0(t,x,u)) < v_h^-(x).$$

An inspection of the proof of the Lipschitz continuity for v^+ in [16, Proposition 4.2] reveals that also v_h^+ is Lipschitz under suitable (mild) conditions on g.

As already mentioned, in general we do not expect that Zubov's equation for weak domains of attraction (7.13) has Lipschitz continuous solutions. Since the Lipschitz property of v^- will, however, turn out to be crucial in what follows, we need the following result. We formulate it for continuous time systems, an analogous version holds for discrete time systems (of course, without property (iii)).

Proposition 7.2.5 Let A be a weakly attracting set with attracted neighborhood B, rate of attraction β and weak domain of attraction $\mathcal{D}(A)$. Let $g : \mathbb{R}^n \to \mathbb{R}_0^+$ satisfy (7.6). Then for each closed neighborhood $\mathcal{B}(r, A)$ of A with $\mathcal{B}(r, A) \subset \mathcal{D}(A)$ there exists a function $\tilde{v}^- : \mathbb{R}^n \to [0, 1]$ with the following properties.

(i) $\tilde{v}^-$ is continuous on $\mathbb{R}^n$ and locally Lipschitz on $\mathcal{D}(A)$

(ii) $A = \{x \in \mathbb{R}^n \,|\, \tilde{v}^-(x) = 0\}$ and $\mathcal{D}(A) = \{x \in \mathbb{R}^n \,|\, \tilde{v}^-(x) < 1\}$

(iii) $\tilde{v}^-(x)$ is a viscosity solution of (7.13) on $\mathbb{R}^n \setminus \mathcal{B}(r, A)$

(iv) $\tilde{v}^-$ is a control Lyapunov function on $\mathcal{D}(A) \setminus \mathcal{B}(r, A)$ in the following sense: for each $x \in \mathcal{D}(A)$, each $t \geq 0$ and each $\varepsilon > 0$ there exists a $u^* \in \mathcal{U}$ such that either $\varphi(\tau, x, u^*) \in \mathcal{B}(r, A)$ for some $\tau \in [0, t]$ or the inequality

$$\tilde{v}^-(\varphi(t, x, u^*)) - \tilde{v}^-(x) \leq (1 - G(t, x, u^*))(\tilde{v}^-(\varphi(t, x, u^*)) - 1) + \varepsilon \quad (7.24)$$

holds for G from (7.9).

Proof: Consider a Lipschitz continuous function $\rho : \mathbb{R}^n \to [0, 1]$ with $\rho(x) = 1$ if $x \in \mathcal{B}(r/2, A)$ and $\rho(x) = 0$ if $x \notin \mathcal{B}(r, A)$. We define a new right hand side $\tilde{f} : \mathbb{R}^n \times \tilde{U} \to \mathbb{R}^n$ by

$$\tilde{f}(x, \tilde{u}) := f(x, u) + a\rho(x)(d - f(x, u)) \quad (7.25)$$

where $\tilde{u} = (u, a, d) \in \tilde{U} := U \times [0, 1] \times D$ with $D := \{x \in \mathbb{R}^n \,|\, \|x\| \leq 1\}$. Note that $f(x, u) = \tilde{f}(x, u, a, d)$ for all $(u, a, d) \in \tilde{U}$ and all $x \notin \mathcal{B}(r, A)$. We denote the trajectories of the f–system by $\varphi(t, x, u)$ and those of the $\tilde{f}$–system (7.25) by $\tilde{\varphi}(t, x, \tilde{u})$. Clearly, A remains weakly asymptotically stable for (7.25) and if $\mathcal{B}(r, A) \subset \mathcal{D}(A)$ then also the domain of attraction is the same. Now pick some function g satisfying (7.6) and consider the function

$$\tilde{v}^-(x) := \inf_{\tilde{u} \in \tilde{\mathcal{U}}} \{1 - e^{-\tilde{J}(x, u)}\}$$

from (7.5) defined for $\tilde{J}$ based on $\tilde{\varphi}$, i.e.,

$$\tilde{J}(x, \tilde{u}) := \int_0^{+\infty} g(\tilde{\varphi}(t, x, \tilde{u}))dt,$$

which is the solution of (7.13) for $\tilde{f}$. Then the properties (ii)–(iv) are immediate from Proposition 7.2.1. It remains to show the Lipschitz property. We show this for the auxiliary function $\tilde{V} : \mathcal{D}(A) \to \mathbb{R}_0^+$ given by

$$\tilde{V}(x) := \inf_{\tilde{u} \in \tilde{\mathcal{U}}} \tilde{J}(x, \tilde{u}).$$

Since $\tilde{v}^- = 1 - e^{-\tilde{V}}$ this implies the claimed Lipschitz property of $\tilde{v}^-$.

In order to prove Lipschitz continuity of $\tilde{V}$ first observe that for all $x \in \mathcal{D}(A)$, all $t \geq 0$ and all $\tilde{u} \in \tilde{\mathcal{U}}$ for which $\tilde{\varphi}(t, x, u) \in \mathcal{D}(A)$ we have the inequality

$$\tilde{V}(x) \leq \int_0^t g(\tilde{\varphi}(\tau, x, \tilde{u}))d\tau + \tilde{V}(\tilde{\varphi}(t, x, \tilde{u})). \tag{7.26}$$

Now for each $x \in \mathcal{B}(r/4, A)$ we can conclude that $\mathcal{B}(r/4, x) \subset \mathcal{B}(r/2, A)$. Then the construction of $\tilde{f}$ implies that for all $y \in \mathcal{B}(r/4, x)$ we can use the (constant) control

$$\tilde{u}_{x,y} = \left(u, 1, \frac{y - x}{\|y - x\|} \right) \in \tilde{U}, \quad u \in U \text{ arbitrary}$$

to obtain that

$$\tilde{\varphi}(t_{x,y}, x, \tilde{u}_{x,y}) = y \text{ for } t_{x,y} = \|x - y\|.$$

Since g is bounded on $\mathcal{B}(r/2, A)$ by some suitable constant M_g and this trajectory does not leave the convex set $\mathcal{B}(r/4, x) \subset \mathcal{B}(r/2, A)$, from (7.26) we obtain $\tilde{V}(x) \leq \tilde{V}(y) + \|x - y\|M_g$ which by symmetry implies

$$\|\tilde{V}(x) - \tilde{V}(y)\| \leq \|x - y\|M_g,$$

i.e., Lipschitz continuity of $\tilde{V}$ with constant M_g on each ball $\mathcal{B}(r/4, x)$ for all $x \in \mathcal{B}(r/4, A)$.

Now we proceed similar to the proof of Proposition 7.2.1(iv): Let $x \in \mathcal{D}(A)$ and $\varepsilon \in (0, 1]$ be arbitrary. Then we find $\tilde{u}^* \in \tilde{\mathcal{U}}$ and $T > 0$ such that $\|\tilde{\varphi}(T, x, \tilde{u}^*)\|_A < r/4$ and $\tilde{V}(x) + \varepsilon > \tilde{J}(x, \tilde{u})$. By continuous dependence on the initial value (which holds uniformly for all $\tilde{u} \in \tilde{\mathcal{U}}$) there is a neighborhood $W \subset \mathcal{D}(A)$ of x such that

$$\|\tilde{\varphi}(t, y, \tilde{u}) - \tilde{\varphi}(t, z, \tilde{u})\| < r/4 \tag{7.27}$$

for all $y, z \in W$, all $t \in [0, T]$ and all $\tilde{u} \in \tilde{\mathcal{U}}$. We may assume that $\mathrm{cl}\, W \subset \mathcal{D}(A)$ is compact, thus also $\mathcal{R}_T := \mathrm{cl} \bigcup_{t \in [0,T]} \tilde{\varphi}_{\tilde{u}}(t, W)$ is compact. Since by choice of $\tilde{u}^*$ and (7.27) we know that $\tilde{\varphi}(t, y, \tilde{u}^*) \in \mathcal{B}(r/2, A)$, this implies in particular, that $\tilde{V}(y) \leq T \max\{g(x) \mid x \in \mathcal{R}_T\} + M_{\mathcal{B}}$ for all $y \in W$ where $M_{\mathcal{B}} := \sup_{x \in \mathcal{B}(r/2, A)} \tilde{V}(x) \leq M_g r/2$. From this bound of $\tilde{V}$ on W and the fact that g is bounded away from zero on $\mathbb{R}^n \setminus \mathcal{B}(r/4, A)$ it follows that there exists $T^* > 0$ such that whenever $y \in W$ and $\tilde{u} \in \tilde{\mathcal{U}}$ are such that $\tilde{J}(y, \tilde{u}) < \tilde{V}(y) + \varepsilon$ we have $\tilde{\varphi}(t, y, \tilde{u}) \in \mathcal{B}(r/4, A)$ for all $t \geq T^*$. Note that since ε is bounded from above the time T^* is independent of ε. By making W smaller, if necessary, we may assume that (7.27) holds for all $t \in [0, T^*]$ instead of $t \in [0, T]$. We may now choose a Lipschitz constant L_g for g on

$\mathcal{R}_{T^*}$. Let $y, z \in W$ and $\tilde{u}^* \in \tilde{\mathcal{U}}$ be such that $\tilde{V}(y) + \varepsilon > \tilde{J}(y, \tilde{u}^*)$. Then we obtain

$$
\begin{aligned}
\tilde{V}(z) - \tilde{V}(y) &< \tilde{V}(z) - \int_0^{+\infty} g(\tilde{\varphi}(t, y, \tilde{u}^*))dt + \varepsilon \\
&\leq \int_0^{T^*} |g(\tilde{\varphi}(t, z, \tilde{u}^*)) - g(\tilde{\varphi}(t, y, \tilde{u}^*))| \, dt \\
&\quad + \tilde{V}(\tilde{\varphi}(T^*, z, \tilde{u}^*)) - \tilde{V}(\tilde{\varphi}(T^*, z, \tilde{u}^*)) + \varepsilon \\
&\leq L_g \int_0^{T^*} e^{Lt} \|z - y\| \, dt + M_g e^{LT^*} \|z - y\| + \varepsilon \\
&= \left(\frac{L_g}{L}(e^{LT^*} - 1) + M_g e^{LT^*} \right) \|z - y\| + \varepsilon
\end{aligned}
$$

using that $\tilde{\varphi}(T^*, y, \tilde{u}^*) \in \mathcal{B}(r/4, A)$ and $\tilde{\varphi}(T^*, z, \tilde{u}^*) \in \mathcal{B}(r/4, \tilde{\varphi}(T^*, y, \tilde{u}^*))$, which implies that the local Lipschitz estimate for $\tilde{V}$ on $\mathcal{B}(r/2, A)$ holds. Since $\varepsilon > 0$ was arbitrary, T^* does not depend on ε and this estimate is symmetric in y and z this shows the Lipschitz continuity of $\tilde{V}$.   □

7.3 Robustness for Domains of Attraction

In order to be able to make statements about numerical approximations of domains of attractions we will now introduce a robustness concept for these objects. We will use a suitable modification of the concepts we have developed for attracting sets. To illustrate this procedure we first explain the relation between domains of attraction and attracting sets.

Proposition 7.3.1 Consider a continuous time system (2.1) and consider a strongly (or weakly) asymptotically stable set A with bounded domain of attraction $\mathcal{D}(A)$. Then $\mathcal{D}(A)^c$ is a strongly (or weakly) asymptotically stable set for the time reversed system. In particular it is ISDS (or wISDS) for the inflated time reversed system for suitable rate and gains.

Sketch of Proof: We sketch the proof for the strong case, the weak case being similar: By Theorem 7.2.2(i) we find a solution v^+ of Zubov's equation (7.12) which is a continuous robust Lyapunov function for A with $v^+ \equiv 1$ on $\mathcal{D}(A)^c$, i.e., v^+ satisfies

$$
v^+(\varphi(t, x, u)) \leq \mu(v^+(x), t)
$$

for all $x \in \mathcal{D}(A)$, all $u \in \mathcal{U}$ and some suitable class $\mathcal{KLD}$ function μ.

A straightforward computation shows that $V := 1 - v^+$ is a robust Lyapunov function for $\mathcal{D}(A)^c$ for the time reversed system, which implies the strong asymptotic stability.   □

It is essentially the ISDS (or wISDS) property of $\mathcal{D}(A)^c$ for the inflated time reversed system that we want to use in what follows. However, since working with time reversed systems it is in general not convenient (or even impossible, e.g., for discrete time systems which are not invertible) we will formulate this property directly in terms of the original system. The following definition gives such a formulation which we state in terms of Lyapunov functions.

Definition 7.3.2 (dynamical robustness of domains of attraction)

Let Φ denote the inflated system of type (2.10) or (2.26) corresponding to some unperturbed system Φ^0 of type (2.1) or type (2.17), respectively.

(i) Consider a strongly attracting set A for Φ^0 with domain of attraction $\mathcal{D}(A)$. We call $\mathcal{D}(A)$ *dynamically robust* with rate μ of class $\mathcal{KLD}$ and gains σ, γ of class $\mathcal{K}_\infty$ on an (open) *robustness neighborhood* B of $\mathcal{D}(A)^c$ if there exists a bounded function $V : \mathbb{R}^n \to \mathbb{R}_0^+$ which satisfies

$$V(x) \geq \|x\|_{\mathcal{D}(A)^c} \text{ for all } x \in B,$$

$$V(x) \leq \sigma(\|x\|_{\mathcal{D}(A)^c}) \text{ for all } x \in B$$

$$V(x) \equiv a \text{ for all } x \notin B \quad \text{and} \quad V(x) < a \text{ for all } x \in B$$

for some constant $a > 0$ and

$$V(\Phi(t, x, u, w)) \geq \min\{\mu(V(x), -t), a\}$$

for all $x \in B$, all $u \in \mathcal{U}$ and all $w \in \mathcal{W}$, with $\gamma(\|w(\tau)\|) \leq \mu(V(x), -\tau)$ for almost all $\tau \in [0, t]$ for continuous time systems or $\gamma(\|w\|_{[hi, h(i+1)]}) \leq \mu(V(x), -hi)$ for all $i \in \mathbb{N}_0$ with $h(i+1) \leq t$ for discrete time systems.

(ii) Consider a weakly attracting set A for Φ^0 with domain of attraction $\mathcal{D}(A)$. We call $\mathcal{D}(A)$ *dynamically robust* (with respect to some set of perturbation strategies $\mathcal{P}$) with rate μ of class $\mathcal{KLD}$ and gains σ, γ of class $\mathcal{K}_\infty$ on an (open) *robustness neighborhood* B of $\mathcal{D}(A)^c$ if there exists a bounded function $V : B \to \mathbb{R}_0^+$ which satisfies

$$V(x) \geq \|x\|_{\mathcal{D}(A)^c} \text{ for all } x \in B,$$

$$V(x) \leq \sigma(\|x\|_{\mathcal{D}(A)^c}) \text{ for all } x \in B$$

$$V(x) \equiv a \text{ for all } x \notin B \quad \text{and} \quad V(x) < a \text{ for all } x \in B$$

for some constant $a > 0$ and

$$V(\Phi(t, x, u, p[u])) \geq \min\{\mu(V(x), -t), a\} \text{ weakly for all } t \in \mathbb{T}^+$$

for all $x \in B$ and all $p \in \mathcal{P}$, with $\gamma(\|p[u](\tau)\|) \leq \mu(V(x), -\tau)$ for all $u \in \mathcal{U}$ and almost all $\tau \in [0, t]$ for continuous time systems or $\gamma(\|p[u]\|_{[hi, h(i+1)]}) \leq \mu(V(x), -hi)$ for all $u \in \mathcal{U}$ and all $i \in \mathbb{N}_0$ with $h(i+1) \leq t$ for discrete time systems. □

We have decided to formulate this definition directly in terms of Lyapunov functions since these will be the crucial tools in the proofs to follow. It seems reasonable to expect that one could start with a comparison function formulation analogous to the ISDS or wISDS property and then obtain the Lyapunov function from Definition 7.3.2 similar to the procedure in the Theorems 3.5.3 and 4.5.4. In a way, this could be considered as the more systematic and "clean" way to introduce robustness for domains of attraction. Nevertheless, since we want to avoid a lengthy repetition of all the arguments from the first Chapters and since we will show below that dynamical robustness in the sense of Definition 7.3.2 is a typical property of domains of attraction we consider the chosen "shortcut" as justified.

Clearly, if the complement of a domain of attraction of some strongly attracting set is ISDS for the time reversed system then it is dynamically stable with the same rate and gains. This is easily seen by using the ISDS Lyapunov function from Theorem 3.5.3(iii) for which the verification of Definition 7.3.2(i) is straightforward. Indeed, as mentioned before, this was the main motivation for Definition 7.3.2. Note, however, that for weakly attracting sets it is not clear whether this implication is valid since here under time reversal it is not a priori clear which control function has to be used for the opposite time direction. This reversal is always possible if the wISDS Lyapunov function is continuous and we consider perturbations from $\mathcal{P}^0$ since in this case we can use the infinitesimal characterization of V via the Hamilton-Jacobi-Isaacs equation and conclude the growth of V by arguments similar to Theorem A.2.1.

It turns out that dynamical robustness is an inherent property of domains of attraction, at least when these are bounded, as shown in the following theorem.

Proposition 7.3.3 Consider a (strongly or weakly) asymptotically stable set A with bounded domain of attraction $\mathcal{D}(A)$. Then for each $\delta > 0$ there exists a robustness neighborhood B of $\mathcal{D}(A)^c$ with $\mathrm{cl}\, B \cap A = \emptyset$ on which the domain of attraction is dynamically robust for the inflated system with perturbations from $\mathcal{P} = \mathcal{P}^\delta$ for suitable rate μ of class $\mathcal{KLD}$ and gains σ and γ of class $\mathcal{K}_\infty$.

Proof: We show the assertion for weakly asymptotically stable sets and for continuous time systems; the other cases follow similarly with the obvious modifications.

Consider the locally Lipschitz Lyapunov function $\tilde{v}^-$ from Theorem 7.2.5 for some $r > 0$ such that $\mathcal{B}(r, A) \subset \mathcal{D}(A)$, choose $C \in (0, 1)$ such that the superlevel set $\{x \in \mathbb{R}^n \,|\, \tilde{v}^-(x) \geq C\}$ does not intersect $\mathcal{B}(r, A)$, and set $B = \{x \in \mathbb{R}^n \,|\, \tilde{v}^-(x) > C\}$.

We set $\tilde{V}(x) = 1 - \tilde{v}^-(x)$ for $x \in B$ and $\tilde{V}(x) = 1 - C$ for $x \notin B$. Then the local Lipschitz continuity of $\tilde{v}^-$ on $\mathcal{D}(A)$ implies the existence of constants

$L_r > 0$ for $r > 0$ such that

$$\tilde{V}(y) \geq \tilde{V}(x) - L_r \|x - y\| \tag{7.28}$$

for all $x \in \mathbb{R}^n$ with $\tilde{V}(x) \geq r$. Furthermore, using that g is bounded from below on B by some positive constant M we obtain from the optimality principle for $\tilde{v}^-$ that $\tilde{V}$ satisfies

$$\sup_{u \in \mathcal{U}} \tilde{V}(\varphi(t, x, u)) \geq \min\{e^{Mt} \tilde{V}(x), C\}.$$

Now fix $x \in B$ and $t \in [0, \delta]$. Then we find a $u^* \in \mathcal{U}$ such that

$$\tilde{V}(\varphi(t, x, u^*)) \geq \min\{e^{Mt/2} \tilde{V}(x), C\}.$$

Now consider $p \in \mathcal{P}_\varepsilon^\delta$. Then by Lemma 4.7.1(i) and estimate (7.28) we can conclude that either $\varphi(t, x, u^*, p[u^*]) \notin B$ or

$$\tilde{V}(\varphi(t, x, u^*, p[u^*])) \geq e^{Mt/2} \tilde{V}(x) - L_r \varepsilon(e^{Lt} - 1)/L \geq (1 + Mt/2)\tilde{V}(x) - L_r \varepsilon C_1 t$$

for some suitable constant $C_1 > 0$ depending on L and δ, where L is the Lipschitz constant of f from (2.8).

Now we choose $C_2 > 0$ such that $(1 + Mt/4) \geq e^{C_2 t}$ for all $t \in [0, \delta]$ and $\tilde{\gamma}$ of class $\mathcal{K}_\infty$ such that

$$\tilde{\gamma}^{-1}(r) \leq \frac{Mr}{4 L_r C_1 e^{C_2 \delta}}.$$

Then for all $p \in \mathcal{P}$ with $\tilde{\gamma}(\|p[u](\tau)\|) \leq e^{C_2 \delta} \tilde{V}(x)$ for almost all $\tau \in [0, t]$ and all $u \in \mathcal{U}$ we find a $u' \in \mathcal{U}$ such that either $\varphi(t, x, u', p[u']) \notin B$ or

$$\tilde{V}(\varphi(t, x, u', p[u'])) \geq Mt/4 \tilde{V}(x) \geq e^{C_2 t} \tilde{V}(x).$$

Now pick a class $\mathcal{K}_\infty$ function ρ such that

$$V(x) := \rho(\tilde{V}(x)) \geq \|x\|_{\mathcal{D}(A)^c} \text{ for all } x \in B.$$

We set $\mu(r, t) := \rho(e^{C_2 t} \rho^{-1}(V(x)))$ and $\gamma(r) = \rho(\tilde{\gamma}(r))$. Then for all $p \in \mathcal{P}$ satisfying $\gamma(\|p[u](\tau)\|) \leq \mu(V(x), \tau)$ for almost all $\tau \in [0, t]$ and all $u \in \mathcal{U}$ we find a $u'' \in \mathcal{U}$ such that

$$V(\varphi(t, x, u'', p[u''])) \geq \min\{\mu(V(x), t), \rho(C)\}.$$

By induction we can extend this estimate for arbitrary $t > 0$ which shows the assertion for some suitable class $\mathcal{K}_\infty$ function σ.   □

Due to the fact that we will use robustness properties of $\mathcal{D}(A)^c$ we will typically end up with discretization error estimates for the distance between $\mathcal{D}(A)^c$ and the complement of its approximation, which—in general—do not allow the derivation of estimates for the distance between $\mathcal{D}(A)$ and its approximation itself, cf. Remark 2.3.3. This is an inherent consequence from the fact that we use the (natural) robustness of $\mathcal{D}(A)^c$ here.

We end this section by showing that for the computation of domains of attraction we can entirely work on the discrete time level even if we start from a continuous time system. This is an analogous result to Lemma 6.2.6 for strong attractors.

Lemma 7.3.4 Consider a continuous time system φ of type (2.1) and let A be a (weakly or strongly) asymptotically stable set for φ. Then for each $h > 0$ the domains of attraction of A coincide for φ and for its time-h map φ^h.

Proof: We show the assertion for a strongly asymptotically stable set, for weakly asymptotically stable sets it follows similarly. Denote the respective domains of attraction by $\mathcal{D}_0(A)$ and $\mathcal{D}_h(A)$. Clearly, $\mathcal{D}_0(A) \subseteq \mathcal{D}_h(A)$, since if x is such that $\varphi(t, x, u)$ converges to A uniformly in u then also $\varphi^h(t, x, u)$ does so. Conversely, let $x \in \mathcal{D}_h(A)$. Then $\varphi^h(t, x, u)$ converges to A uniformly for all u. Hence we reach the attracted neighborhood B of A for some time $t^* \in \mathbb{T}^+$ bounded independently of u, and from there we know that $\varphi(t, \varphi^h(t^*, x, u), u(t^* + \cdot))$ converges to A uniformly in u. Hence $x \in \mathcal{D}_0(A)$, and consequently $\mathcal{D}_0(A) \subseteq \mathcal{D}_h(A)$ which finishes the proof. $\square$

In other words, it is always sufficient to look for the domains of attraction for the time-h map of φ, even for arbitrary large h.

7.4 Domains of Attraction under One–Step Discretizations

In this section we will analyze how a domain of attraction changes under a numerical one–step discretization of an ordinary differential equation. Here and in the following sections we will develop our results for weak domains of attraction; all results, however, hold accordingly and with analogous proofs for the strong case.

As already mentioned, we will state the results assuming the global bounds (2.8). For bounded domains of attraction we can use standard cutoff techniques in order to ensure these global bounds under the local assumption (2.7). Note, however, that the following statements also apply to unbounded domains of attraction, in which case the global bounds (2.8) cannot be deduced from their local counterparts (2.7). Here one could weaken the global bounds using suitable uniform Lipschitz and boundedness conditions with respect to the (unbounded) domain of attraction as they have been used, e.g., in [70] for unbounded weakly attracting sets.

It will turn out that the inherent dynamical robustness of domains of attraction from Definition 7.3.2 is not sufficient for a convergence result with respect to the Hausdorff distance. In fact, we encounter exactly the same problems as we have seen for attractors under discretizations (cf. Theorem 6.2.7), namely in general we only obtain an estimate for the "dist" as stated in the following theorem.

Theorem 7.4.1 Consider a system φ of type (2.1) satisfying (2.8) and a numerical one step approximation $\widetilde{\Phi}_h$ of order $q \in \mathbb{N}$. Assume that φ has a weakly attracting set A whose domain of attraction $\mathcal{D}(A)$ is dynamically robust (with respect to $\mathcal{P} = \mathcal{P}^\delta$ for some $\delta > 0$) with robustness gain γ of class $\mathcal{K}_\infty$. Then for all $h > 0$ sufficiently small the discrete time system $\widetilde{\Phi}_h$ has a weakly attracting set A_h close to A with domain of attraction $\mathcal{D}_h(A_h)$ satisfying
$$\text{dist}(\mathcal{D}_h(A_h)^c, \mathcal{D}(A)^c) \leq \gamma(e^{Lh} ch^q).$$
In particular, if there exists a sequence of time–steps $h_n \to 0$ and a set $\mathcal{D}$ with $d_H(\mathcal{D}_{h_n}(A_{h_n}), \mathcal{D}) \to 0$ as $h_n \to 0$, then we obtain $\mathcal{D}(A) \subseteq \mathcal{D}$.

Proof: From Lemma 4.8.4 and Lemma 7.3.4 it is easily seen that $\mathcal{D}(A)$ is a dynamically robust domain of attraction with gain $\gamma(e^{Lh} \cdot)$ for the inflated time-h map φ^h.

By Theorem 6.4.1 we obtain the existence of A_h. Let B_h be an attracted neighborhood. For $h > 0$ sufficiently small we may assume that B_h is so large that the robustness neighborhood B on which we have the the dynamical robustness satisfies $B \supset B_h^c$, which implies that the Lyapunov function V characterizing the dynamical robustness satisfies the implication $V(x) = a$ $\Rightarrow x \in B_h$. Clearly, all initial values $x \in \mathbb{R}^n$ which can be controlled to B_h lie in $\mathcal{D}_h(A_h)$. By Lemma 5.1.4 we obtain that $\widetilde{\Phi}_h$ is $(ch^q, 1)$-embedded in the inflated time-h map. Hence we can conclude that for all $x \in B_h^c$ with $V(x) \geq ch^q$ and all $t \geq 0$ we obtain the existence of a $u^* \in \mathcal{U}$ such that
$$V(\widetilde{\Phi}_h(t, x, u^*)) \geq \min\{\mu(-t, V(x)), a\}$$
which implies that $V(\widetilde{\Phi}_h(t, x, u^*)) = a$ and thus $\widetilde{\Phi}_h(t, x, u^*) \in B_h$ for some $t \in \mathbb{T}^+$. Hence
$$\{x \in \mathbb{R}^n \mid x \in B_h \text{ or } V(x) \geq e^{Lh} ch^q\} \subseteq \mathcal{D}_h(A_h)$$
which implies
$$\mathcal{D}_h(A_h)^c \subseteq \{x \in \mathbb{R}^n \mid V(x) \leq e^{Lh} ch^q\}.$$
By the bounds on V implies the asserted inequality.

The inclusion $\mathcal{D}(A) \subseteq \mathcal{D}$ follows, since by Lemma 2.3.6(iv) we obtain the equality $\mathcal{D} = \text{Lim sup}_{h_n \to 0} \mathcal{D}_{h_n}(A_{h_n})$. By Lemma 2.3.6(ii) this set contains the open set $\mathcal{D}(A)$. $\square$

Remark 7.4.2 (i) This result does not imply an estimate for the distance $\text{dist}(\mathcal{D}(A), \mathcal{D}_h(A_h))$, cf. Remark 2.3.3. If $\mathcal{D}(A)$ is bounded then by Lemma 2.3.2(v) we can, however, conclude that $\text{dist}(\mathcal{D}(A), \mathcal{D}_h(A_h)) \to 0$ as $h \to 0$.

(ii) The rate of attraction to A (or, alternatively, the robustness of A) which was used in Chapter 6 does not appear in this result, apart from the fact that it justifies the assumption on the existence of the sets A_{h_n}. Note however, that this rate of attraction determines the upper bound for h for which the result is valid.

(iii) Note that under suitable conditions on the numerical one–step scheme we can also obtain the result for $\delta = 0$, cf. Remark 6.4.6.   □

We can give a necessary and sufficient condition in terms of rates of attraction to a neighborhood of A, which is in the same spirit as the corresponding result for strong attractors in Theorem 6.2.8.

Theorem 7.4.3 Consider a system φ of type (2.1) satisfying (2.8) and a numerical one step approximation $\widetilde{\Phi}_h$. Consider a sequence $h_n \to 0$ and assume that the numerical discrete time systems $\widetilde{\Phi}_{h_n}$ have weakly attracting sets A_{h_n} converging to some weakly attracting set A of the continuous time system. Let $\mathcal{D}_{h_n}(A_{h_n})$ and $\mathcal{D}(A)$ denote the corresponding domains of attraction and assume that there exists an open set $\mathcal{D}$ containing A and satisfying $d_H(\mathcal{D}, \mathcal{D}_{h_n}(A_{h_n})) \to 0$ and $d_H(\mathcal{D}^c, \mathcal{D}_{h_n}(A_{h_n})^c) \to 0$ as $n \to \infty$.

Then $\mathcal{D} = \mathcal{D}(A)$ holds if and only if there exist an attracted neighborhood B_0 of A with $d_{\min}(\mathcal{D}(A), B_0) > 0$, a number $N \in \mathbb{N}$ and sets $\widetilde{\mathcal{D}}_{h_n}$ satisfying $d_H(\widetilde{\mathcal{D}}_{h_n}^c, \mathcal{D}_{h_n}(A_{h_n})^c) \to 0$ such that for each $\varepsilon > 0$ there exists a time $T > 0$ with the property that for all h_n, $n \geq N$, and all $x \in \widetilde{\mathcal{D}}_{h_n}$ with $\|x\|_{\widetilde{\mathcal{D}}_{h_n}^c} \geq \varepsilon$ there exist $u_{h_n} \in \mathcal{U}$ with

$$\widetilde{\Phi}_{h_n}(t_{h_n}, x, u_{h_n}) \in B_0$$

for some $t_{h_n} \in [0, T]$.

Proof: Assume $\mathcal{D} = \mathcal{D}(A)$. By Proposition 7.3.3 we can assume dynamical robustness of $\mathcal{D}(A)$, i.e., in particular the existence of a Lyapunov function V characterizing the dynamical robustness with attraction rate μ and gains σ and γ, which is defined on some robustness neighborhood B satisfying $B \supset B_0^c$ for some attracted neighborhood B_0 of A. Then, as in the proof of Theorem 7.4.1, one easily sees that defining the sets

$$\widetilde{\mathcal{D}}_{h_n} := \{x \in \mathbb{R}^n \,|\, x \in B_0 \text{ or } V(x) \geq e^{Lh} c h_n{}^q\}$$

for each $x \in \widetilde{\mathcal{D}}_{h_n}$ we obtain the existence of u such that

$$V(\widetilde{\Phi}_{h_n}(t, x, u)) \geq \min\{\mu(V(x), -t), a\}$$

where $V(x) = a$ implies $x \in B_0$. Hence for all $x \in \widetilde{\mathcal{D}}_{h_n}$ with $V(x) \geq \varepsilon$ we obtain that $\widetilde{\Phi}_{h_n}(t, x, u) \in B_0$ for some $t \in [0, T]$ with $T > 0$ chosen such that $\mu(\varepsilon, -T) \geq a$. Since from the bounds on V we obtain the implication $\|x\|_{\widetilde{\mathcal{D}}_{h_n}^c} \geq \varepsilon \Rightarrow V(x) \geq \varepsilon$ this shows the claim, since the estimate $d_H(\widetilde{\mathcal{D}}_{h_n}^c, \mathcal{D}(A)^c) \leq \sigma(e^{Lh} ch^q)$, which follows from the bounds on V, and the convergence $d_H(\mathcal{D}_{h_n}(A_{h_n})^c, \mathcal{D}(A)^c) \to 0$ imply the asserted property $d_H(\widetilde{\mathcal{D}}_{h_n}^c, \mathcal{D}_{h_n}(A_{h_n})^c) \to 0$ by triangle inequality.

Conversely, assume the existence of the $\widetilde{\mathcal{D}}_{h_n}$ with the asserted properties. From Theorem 7.4.1 we already know that $\mathcal{D}(A) \subseteq \mathcal{D}$. It remains to show $\mathcal{D}(A) \supseteq \mathcal{D}$. For this let $x \in \mathcal{D}$. By the assumption on B_0 we obtain the existence of a neighborhood $B_1 := \mathcal{B}(\varepsilon, B_0)$ which is contained in $\mathcal{D}(A)$. Clearly, $\mathcal{D}(A)$ is exactly the set of points that can be controlled to B_1. Now by Lemma 2.3.2(v) we know that for each $x \in \mathcal{D}$ there exists an $N \in \mathbb{N}$ such that $x \in \widetilde{\mathcal{D}}_{h_n}$ for all $n \geq N$. For N sufficiently large we can assume that $\|x\|_{\widetilde{\mathcal{D}}_{h_n}}^c \geq \varepsilon$ for all $n \geq N$ and some $\varepsilon > 0$. Hence there exists a $T > 0$ such that for all numerical systems $\widetilde{\Phi}_{h_n}$, $n \geq N$, there exist $u_{h_n} \in \mathcal{U}$ and $t_{h_n} \in [0, T]$ with

$$\widetilde{\Phi}_{h_n}(t_{h_n}, x, u_{h_n}) \in B_0.$$

By Lemma 5.1.2 we hence obtain that $\varphi(t_{h_n}, x, u_{h_n}) \in B_1$ for all n sufficiently large which shows $x \in \mathcal{D}(A)$. $\square$

Simplifying a little bit, this seemingly complicated theorem just states that if the rate of convergence to A of the trajectories starting in sufficiently large subsets of $\mathcal{D}_{h_n}(A_{h_n})$ does not depend on the time step h_n, then we obtain convergence.

Remark 7.4.4 For simplicity we have decided to formulate the rate of attraction via an ε–T formalism. Alternatively, one could also use comparison functions to estimate $\|\widetilde{\Phi}(t, x, u)\|_{B_0}$ which would have to depend on t and $\|x\|_{\widetilde{\mathcal{D}}_{h_n}^c}$. $\square$

In the next theorem we will use a suitable "structural stability" condition on the domain of attraction guaranteeing convergence without looking at rates of attraction. We will base our condition on the simple observation that the complement of a weak (strong) domain of attraction is a strongly (weakly) forward invariant set. For this recall the robustness conditions for forward invariant sets from Definitions 3.2.1 and 4.2.1.

Theorem 7.4.5 Consider a system φ of type (2.1) satisfying (2.8) and a numerical one step approximation $\widetilde{\Phi}_h$ of order $q \in \mathbb{N}$. Assume that φ has a weakly attracting set A whose domain of attraction $\mathcal{D}(A)$ is dynamically robust (with respect to $\mathcal{P} = \mathcal{P}^\delta$ for some $\delta > 0$) with robustness gain γ of class $\mathcal{K}_\infty$. Assume furthermore that $\mathcal{D}(A)^c$ is a direct robust strongly forward invariant set with gain $\tilde\gamma$ for the corresponding inflated system. Then for all $h > 0$ sufficiently small the discrete time system $\tilde\Phi_h$ has a weakly attracting set A_h close to A with domain of attraction $\mathcal{D}_h(A_h)$ satisfying

$$\operatorname{dist}(\mathcal{D}_h(A_h)^c, \mathcal{D}(A)^c) \leq \gamma(e^{Lh} ch^q) \tag{7.29}$$

and

$$\operatorname{dist}(\mathcal{D}(A)^c, \mathcal{D}_h(A_h)^c) \leq \tilde\gamma(e^{Lh} ch^q). \tag{7.30}$$

In particular, we obtain the Hausdorff distance estimate

$$d_H(\mathcal{D}_h(A_h)^c, \mathcal{D}(A)^c) \leq \max\{\gamma(e^{Lh} ch^q), \tilde\gamma(e^{Lh} ch^q)\}. \tag{7.31}$$

If the robust forward invariance of $\mathcal{D}(A)^c$ is inverse instead of direct then we obtain

$$\operatorname{dist}(\mathcal{D}_h(A_h), \mathcal{D}(A)) \leq \tilde\gamma(e^{Lh} ch^q) \tag{7.32}$$

instead of (7.30) and (7.31). In this case, if either $\mathcal{D}(A)$ or $\mathcal{D}(A)^c$ is bounded then we obtain

$$d_H(\mathcal{D}(A), \mathcal{D}_h(A_h)) \to 0 \text{ as } h \to 0.$$

Proof: Inequality (7.29) follows from Theorem 7.4.1. For the proof of inequality (7.30) observe that by Lemma 4.8.4 the direct robust forward invariance condition also holds for the inflated time-h map with gain $\tilde\gamma(e^{Lh} \cdot)$. Hence we obtain the existence of a strongly forward invariant set C_h for $\tilde\Phi_h$ with $d_H(C_h, \mathcal{D}(A)^c) \leq \tilde\gamma(e^{Lh} ch^q)$. Clearly, if $h > 0$ is sufficiently small then $C_h \cap A_h = \emptyset$, hence also $C_h \cap \mathcal{D}_h(A_h) = \emptyset$ which implies that $C_h \subseteq \mathcal{D}_h(A_h)^c$ and consequently

$$\operatorname{dist}(\mathcal{D}(A)^c, \mathcal{D}_h(A_h)^c) \leq \operatorname{dist}(\mathcal{D}(A)^c, C_h) \leq d_H(C_h, \mathcal{D}(A)^c) \leq \tilde\gamma(e^{Lh} ch^q)$$

which shows the desired inequality and thus also (7.31).

The case of inverse robustness follows analogously.

If $\mathcal{D}(A)$ or $\mathcal{D}(A)^c$ is bounded and we have inverse robustness then we can apply Lemma 2.3.2(v) to inequality (7.29), which yields $\operatorname{dist}(\mathcal{D}(A), \mathcal{D}_h(A_h)) \to 0$ as $h \to 0$. Together with estimate (7.32) we obtain the desired convergence for d_H. $\square$

Remark 7.4.6 Both direct and inverse robustness of the strong forward invariance of $\mathcal{D}(A)^c$ are restrictive assumptions, which are not satisfied in general, and a priori it is not clear whether direct or inverse robustness is the more restrictive condition.

In any case, inverse robustness of $\mathcal{D}(A)^c$ is not less natural than direct robustness; maybe it is even more natural since it corresponds to robustness conditions on $\mathcal{D}(A)$. If, for instance, $\mathcal{D}(A)$ happens to be a strongly attracting set for the time–reversed system, then—using the inherent robustness of strongly attracting sets—we can find α–strongly attracting sets $\mathcal{D}_\alpha$ for the inflated system with $d_H(\mathcal{D}(A), \mathcal{D}_\alpha) \leq \gamma(\alpha)$ for some suitable gain γ, whose complements $\mathcal{D}_\alpha^c$ then are strongly forward invariant for the original system. In this case, the inverse robustness is satisfied. Actually, we can even impose weaker conditions on $\mathcal{D}(A)$ which guarantee inverse robustness of the forward invariance of $\mathcal{D}(A)^c$. We will come back to this topic when discussing reachable sets in Section 7.7. $\square$

7.5 Subdivision Algorithm

In this section we want to formulate a subdivision algorithm for domains of attractions and analyze its performance. We start by briefly describing an algorithm by Szolnoki [115, 116] which computes viability kernels (see, e.g., [5]), i.e., maximal weakly forward invariant subsets of some given compact set Ω.

Algorithm 7.5.1 Let Φ_h^0 be a discrete time system of type (2.17). Consider a compact set Ω, a cell covering $\mathbf{Q}^0$ of Ω with P^0 cells Q_i^0, $i = 1, \ldots, P^0$, and a cell–cell space discretization $\widehat{\Phi}_h^0$ on $\mathbf{Q}^0$ with accuracy $\Delta_i^0 \leq h\varepsilon^0$ for all $i = 1, \ldots, P^0$. Let $A^0 = \Omega$, $j = 0$ and proceed iteratively:

(1) **(Selection Step)**
    Let $A_0^{j+1} := A^j$, $l := 0$ and compute iteratively A_{l+1}^{j+1} as the union of all cells Q_i^j, $i = 1, \ldots, P^j$ for which there exists $\bar{u} \in \mathcal{U}(Q_i^j)$ such that

$$\widehat{\Phi}_h(Q_i^j, \bar{u}) \subseteq A_l^{j+1},$$

repeat this computation until $A_{l+1}^{j+1} = A_l^{j+1}$ and set $A^{j+1} = A_l^{j+1}$.

(2) **(Refinement Step)**
    Consider a new cell covering $\mathbf{Q}^{j+1}$ of Ω with P^{j+1} cells Q_i^{j+1}, $i = 1, \ldots, P^{j+1}$, satisfying $C_{\mathbf{Q}^j} \subset C_{\mathbf{Q}^{j+1}}$ and a new cell–cell space discretization $\widehat{\Phi}_h^{j+1}$ on $\mathbf{Q}^{j+1}$ with accuracy $\Delta_i^{j+1} \leq h\varepsilon^{j+1}$ for all $i = 1, \ldots, P^{j+1}$ for some $\varepsilon^{j+1} < \varepsilon^j$.

(3) Set $j := j + 1$ and continue with (1).

$\square$

It is not too difficult to see that for rigorous discretizations $\widehat{\Phi}_h^j$ the set $\bigcap_{j \geq 0} A^j$ is the maximal weakly forward invariant subset of Ω, i.e., its viability kernel

We refer to [118] for a detailed analysis in which also different versions of the selection step (1) are considered, cf. Remark 6.3.7. Under certain conditions these viability kernels have a strong relation to domains of attractions, see [117] and Remark 7.5.7, below.

Algorithm 7.5.1 in connection with ideas from a numerical approximation to Zubov's equation (cf. Remark 7.6.2, below) served as a motivation for the development of the following subdivision technique for the computation of domains of attraction.

Algorithm 7.5.2 Let Φ_h^0 be a discrete time system of type (2.17). Consider a compact set Ω, a cell covering $\mathbf{Q}^0$ of Ω with P^0 cells Q_i^0, $i = 1, \ldots, P^0$, and a cell-cell space discretization $\widehat{\Phi}_h^0$ on $\mathbf{Q}^0$ with accuracy $\Delta_i^0 \leq h\varepsilon^0$ for all $i = 1, \ldots, P^0$, and let $j = 0$. To each cell Q_i^j associate a status $s(Q_i^j)$ which can take the values in (inside), pin (partially inside), und (undefined) and out (outside) and define the sets

$$\Omega_{in}^j = \bigcup_{i,\, s(Q_i^j)=in} Q_i^j, \quad \Omega_{pin}^j = \bigcup_{i,\, s(Q_i^j)=pin} Q_i^j \cup \Omega_{in}^j$$

and

$$\Omega_{out}^j = \bigcup_{i,\, s(Q_i^j)=out} Q_i^j \cup \Omega^c.$$

Consider some target set $\mathcal{S} \subset \operatorname{int}\Omega$ and set $s(Q_i^0) := in$ for all $Q_i \subset \mathcal{S}$, and $s(Q_i^0) := und$ else. Set $D^0 := \Omega_{in}^0$, $C^0 := \Omega_{out}^0$ and proceed iteratively

(1) **(Selection Step)**
   For all Q_i^j with $s(Q_i^j) = und$ or $s(Q_i^j) = pin$ set

   $s(Q_i^j) := pin$, if there exists $\bar{u} \in \mathcal{U}(Q_i^j)$ with $\widehat{\Phi}_h^j(Q_i^j, \bar{u}) \cap \Omega_{pin}^j \neq \emptyset$

   $s(Q_i^j) := in$, if there exists $\bar{u} \in \mathcal{U}(Q_i^j)$ with $\widehat{\Phi}_h^j(Q_i^j, \bar{u}) \subseteq \Omega_{in}^j$

   $s(Q_i^j) := out$, if $\widehat{\Phi}_h^j(Q_i^j, \bar{u}) \subseteq \Omega_{out}^j$ for all $\bar{u} \in \mathcal{U}(Q_i^j)$,

   where we update the sets Ω_{in}^j, Ω_{pin}^j and Ω_{out}^j after each new assignment. If some $s(Q_i^j)$ changed its value during this computation repeat this step.

(2) **(Status Update Step)**
   Set $s(Q_i^j) := out$ for all Q_i^j with $s(Q_i^j) = und$.
   Set $s(Q_i^j) := und$ for all Q_i^j with $s(Q_i^j) = pin$.
   Update the sets Ω_{in}^j, Ω_{pin}^j and Ω_{out}^j and set
   $D^{j+1} := \Omega_{in}^j$, $C^{j+1} := \Omega_{out}^j$ and $E^{j+1} := \Omega \setminus (D^{j+1} \cup C^{j+1})$.

(3) **(Refinement Step)**
   Consider a new cell covering $\mathbf{Q}^{j+1}$ of Ω with P^{j+1} cells Q_i^j, $i =$

$1, \ldots, P^{j+1}$, satisfying $\mathcal{C}_{\mathbf{Q}^j} \subset \mathcal{C}_{\mathbf{Q}^{j+1}}$ and a new cell–cell space discretization $\widehat{\Phi}_h^{j+1}$ on $\mathbf{Q}^{j+1}$ with accuracy $\Delta_i^{j+1} \leq h\varepsilon^{j+1}$ for all $i = 1, \ldots, P^{j+1}$ for some $\varepsilon^{j+1} < \varepsilon^j$. Set $s(Q_{i'}^{j+1}) = s(Q_i^j)$ for all $Q_{i'}^{j+1} \subset Q_i^j$, $j := j+1$ and continue with Step (1).

□

Remark 7.5.3 Note that in practice only the cells lying in E^{j+1} have to be refined in Step (3) since the cells in $D^{j+1} = \Omega_{in}^j$ and $C^{j+1} = \Omega_{out}^j$ remain unchanged in the future iterations.   □

Remark 7.5.4 We have formulated Algorithm 7.5.2 without using the iterates $\widehat{\Phi}_h(t, \cdot, \cdot)$, $t = ih$, of $\widehat{\Phi}_h$ as defined by (5.20), because we wanted to present a directly implementable version. Using these iterates, however, we can simplify the description of the algorithm. Denoting by $\mathbf{U}$ the set of all sequences $\bar{\mathbf{u}} = (\bar{u}_j)_{j \in \mathbb{N}_0}$ with $\bar{u}_j \in \mathcal{U}(\Omega)$, we can replace steps (1) and (2) by the following single step (1').

(1') **(Selection Step)**
    Compute the sets

$$C^{j+1} = \{Q_i^j \in \mathbf{Q}^j \mid \widehat{\Phi}_h(t, x, \bar{\mathbf{u}}) \cap D^j = \emptyset \text{ for all } t \in \mathbb{T}^+ \text{ and all } \bar{\mathbf{u}} \in \mathbf{U}\}, (7.33)$$

$$D^{j+1} = \{Q_i^j \in \mathbf{Q}^j \mid \text{there exist } \bar{\mathbf{u}} \in \mathbf{U}, t \in \mathbb{T}^+ \text{ with } \widehat{\Phi}_h(t, Q_i, \bar{\mathbf{u}}) \subseteq D^j\} (7.34)$$

    and set $E^{j+1} := \Omega \setminus (D^{j+1} \cup C^{j+1})$.

A straightforward induction shows that step (1') indeed computes the same sets D^{j+1} and C^{j+1} as the steps (1) and (2) in the original description of Algorithm 7.5.2.   □

The following theorem shows the convergence of this method when we use rigorous space discretizations.

Theorem 7.5.5 Consider a discrete time system Φ_h^0 of type (2.21) and let A be a weakly attracting set with domain of attraction $\mathcal{D}(A)$. Let $S \subset \mathcal{D}(A)$ be a neighborhood of A, let $\Omega \subset \mathbb{R}^n$ be a compact set containing $\mathcal{D}(A)$ and consider Algorithm 7.5.2 with rigorous space discretizations $\widehat{\Phi}_h^j$. Then the inclusions $D^{j+1} \subseteq \mathcal{D}(A) \subseteq D^{j+1} \cup E^{j+1}$, $C^{j+1} \subseteq \mathcal{D}(A)^c \subseteq C^{j+1} \cup E^{j+1}$ and $\partial \mathcal{D}(A) \subseteq E^{j+1}$ hold. Furthermore for all $j \in \mathbb{N}_0$ with ε^j sufficiently small we have the following estimates:

(i) If $\mathcal{D}(A)$ is dynamically robust for some robustness gain γ of class $\mathcal{K}_\infty$, then

$$d_H(C^{j+1} \cup E^{j+1}, \mathcal{D}(A)^c) \leq \gamma(\varepsilon^j) + h\varepsilon^j.$$

(ii) If $\mathcal{D}(A)^c$ is a direct robustly forward invariant set for some gain $\tilde{\gamma}$ of class $\mathcal{K}_\infty$, then

$$d_H(C^{j+1}, \mathcal{D}(A)^c) \leq \tilde{\gamma}(\varepsilon^j) + h\varepsilon^j.$$

If the robust forward invariance is inverse then we obtain

$$d_H(D^{j+1} \cup E^{j+1}, \mathcal{D}(A)) \leq \tilde{\gamma}(\varepsilon^j) + h\varepsilon^j.$$

(iii) If $\mathcal{D}(A)$ is dynamically robust and $\mathcal{D}(A)^c$ is inversely γ-robustly forward invariant, both with gain γ of class $\mathcal{K}_\infty$, then

$$d_H(E^{j+1}, \partial\mathcal{D}(A)) \leq \gamma(\varepsilon^j) + h\varepsilon^j.$$

Proof: Recall the representations (7.34) and (7.33) for D^{j+1} and E^{j+1}. Then by the embedding property from Remark 5.3.5(iii) and the rigorous discretization one easily sees by induction over j that $D^{j+1} \subseteq \mathcal{D}(A)$ and $C^{j+1} \subseteq \mathcal{D}(A)^c$. This implies all of the stated inclusions.

In order to see estimate (i) consider the Lyapunov function V characterizing the dynamical robustness for some robustness neighborhood B of $\mathcal{D}(A)^c$. For $\varepsilon^j > 0$ sufficiently small by Theorem 6.4.7 we can assume that S contains an attracting set $\hat{A}$ for $\hat{\Phi}_h^j$ with attracted neighborhood $\hat{B}$ containing B^c. Note that since $S \subseteq D^j$ for all $j \geq 0$ by (7.34) this implies $\hat{B} \subset D^{j+1}$. For any closed set $C \subset \mathbb{R}^n$ define $V^m(C) := \min\{V(x)\,|\,x \in C\}$. Then by the embedding property from Remark 5.3.5(iii) for each cell $Q_i^j \subset \mathcal{D}(A)$, $Q_i^j \not\subseteq \hat{B}$ with $V^m(Q_i^j) \geq \gamma(\varepsilon^j)$ we obtain the existence of $\bar{u} \in U$ such that

$$V^m(\hat{\Phi}_h(t, Q_i^j, \bar{u})) \geq \min\{\mu(V^m(Q_i^j), -t), a\},$$

which implies $\hat{\Phi}_h(t, Q_i^j, \bar{u}) \subseteq B^c \subseteq \hat{B} \subseteq D^j$ for t sufficiently large, thus $Q_i^j \subset D^{j+1}$. Since $V(x) \geq \|x\|_{\mathcal{D}(A)^c}$ this implies that all cells Q_i^j with $\|x\|_{\{\mathcal{D}(A)^c)\}} \geq \gamma(\varepsilon^j)$ for all $x \in Q_i^j$ are contained in D^{j+1}, thus, since $\mathrm{diam}(\mathbf{Q}^j) \leq h\varepsilon^j$, all cells with $\mathrm{dist}(Q_i^j, \mathcal{D}(A)^c) \geq \gamma(\varepsilon^j) + h\varepsilon^j$ lie in D^{j+1}, which gives (i).

We now prove (ii) for the inverse case, the direct one being similar. For $\varepsilon^j > 0$ consider the ε^j-strongly forward invariant set C_{ε^j} satisfying $d_H(C_{\varepsilon^j}^c, \mathcal{D}(A)) \leq \tilde{\gamma}(\varepsilon^j)$. Note that for ε^j sufficiently small this implies $C_{\varepsilon^j} \cap \mathcal{D}(A) = \emptyset$ since otherwise $\mathcal{D}(A) \subseteq C_{\varepsilon^j}$ contradicting this distance estimate. In particular, since $D^j \subseteq \mathcal{D}(A)$ for all $j \geq 0$ we obtain $C_{\varepsilon^j} \cap D^j = \emptyset$ for all $j \geq 0$.

Now each cell $Q_i^j \subset \mathcal{D}(A)^c$ with $\|x\|_{\mathcal{D}(A)} \geq \tilde{\gamma}(\varepsilon^j)$ for all $x \in Q_i^j$ is contained in C_{ε^j}, thus each cell Q_i^j intersecting

$$\{x \in \mathcal{D}(A)^c\,|\,\|x\|_{\mathcal{D}(A)} \geq \gamma(\varepsilon^j) + \mathrm{diam}(\mathbf{Q}^j)\}$$

is contained in C_{ε^j}. Hence, if we consider the largest set $\widehat{C} \in \mathcal{C}_{\mathbf{Q}^j}$ with $\widehat{C} \subseteq C_{\varepsilon^j}$ we obtain

$$d_H(\widehat{C} \cup \Omega^c, \mathcal{D}(A)^c) \leq \gamma(\varepsilon^j) + \mathrm{diam}(\mathbf{Q}^j) \leq \gamma(\varepsilon^j) + h\varepsilon^j$$

and by the embedding we have

$$\widehat{\Phi}_h^j(\widehat{C}, \bar{u}) \subset C_{\varepsilon^j}$$

for all $\bar{u} \in \mathcal{U}(\Omega)$. Since $\widehat{\Phi}_h^j(\widehat{C}, \bar{u}) \cap \Omega \in \mathcal{C}_{\mathbf{Q}^j}$ we can conclude that

$$\widehat{\Phi}_h^j(\widehat{C}, \bar{u}) \subset \widehat{C} \cup \Omega^c$$

for all $\bar{u} \in \mathcal{U}(\Omega)$. Since $C_{\varepsilon^j} \cap D^j = \emptyset$ by (7.33) we obtain $\widehat{C} \subset C^{j+1}$ which shows (ii).

In order to prove estimate (iii) observe that from (i) and (ii) we obtain

$$E^{j+1} \subseteq \mathcal{B}(\gamma(\varepsilon^j) + h\varepsilon^j, \mathcal{D}(A)) \cap \mathcal{B}(\gamma(\varepsilon^j) + h\varepsilon^j, \mathcal{D}(A)^c) = \mathcal{B}(\gamma(\varepsilon^j) + h\varepsilon^j, \partial\mathcal{D}(A))$$

which implies (iii).   □

Remark 7.5.6 (i) The amount of cells needed in this approximation is determined by the size of the sets E^{j+1}, since these are the sets that need to be refined during the iteration. Hence we can expect the algorithm to be efficient if the forward invariance of $\mathcal{D}(A)^c$ is inversely robust, since in this case the E^{j+1} shrink down to $\partial\mathcal{D}(A)$.

Even without this robustness the convergence $d_H(\mathrm{int}\, D^j, \mathcal{D}(A)) \to 0$ is ensured by Lemma 2.3.2(v), since $\mathcal{D}(A)$ is open and bounded and Theorem 7.5.5 gives

$$d_H(\mathcal{D}(A), \mathrm{int}\, D^{j+1}) = \mathrm{dist}(\mathcal{D}(A), \mathrm{int}\, D^{j+1})$$

and

$$\mathrm{dist}((\mathrm{int}\, D^{j+1})^c, \mathcal{D}(A)^c) = \mathrm{dist}(C^{j+1} \cup E^{j+1}, \mathcal{D}(A)^c) \to 0.$$

(ii) It follows immediately from the proof that the distance estimates from Theorem 7.5.5 remain valid for non–rigorous discretizations provided the inclusions $D^{j+1} \subseteq \mathcal{D}(A) \subseteq D^{j+1} \cup E^{j+1}$, $C^{j+1} \subseteq \mathcal{D}(A)^c \subseteq C^{j+1} \cup E^{j+1}$ and $\partial\mathcal{D}(A) \subseteq E^{j+1}$ hold. Since the proof only relies on the embedding of $\widehat{\Phi}_h$ in the inflated system, in this case Property (ii) of Definition 5.3.4 is not needed.   □

Remark 7.5.7 It was already mentioned that also the viability kernels which are computed by Algorithm 7.5.1 have a relation to domains of attraction. In the nicest case Ω contains exactly one weakly asymptotically stable set A and all trajectories starting outside $\mathcal{D}(A)$ (i.e., those which cannot be controlled to A) eventually leave Ω. In this case $\mathcal{D}(A)$ coincides with the viability kernel of Ω, and consequently Algorithms 7.5.1 and 7.5.2 compute the same objects. Then the main difference between these algorithms is that in Algorithm 7.5.1 each cell in A^j (which approximates $\mathcal{D}(A)$) is refined while in Algorithm 7.5.2 only the cells in E^j are refined. Under the robust invariance condition E^j approximates $\partial\mathcal{D}(A)$, hence this set is considerably smaller than A^j, and consequently also the number of cells needed in the approximation can be expected to be much smaller.   □

The discussion in Section 5.3 shows that a rigorous space discretization is not so easy to implement and expensive to evaluate in practice, furthermore numerical experiments show that also non rigorous discretizations often show reasonable results. We would therefore like to have at least some mathematical result telling us what we can expect in the non rigorous case. This is possible if we replace the condition for a rigorous discretization by the following weaker one.

Definition 7.5.8 (inner error for space discretizations)

Consider a cell covering $\mathbf{Q}$ of some compact set $\Omega \subset \mathbb{R}^n$, a discrete time system Φ_h^0 of type (2.1) and a cell–cell space discretization $\widehat{\Phi}_h$ of Φ_h^0 on $\mathbf{Q}$ according to Definition 5.3.4.

Then we say that $\widehat{\Phi}_h$ has *inner error* ε for some $\varepsilon > 0$ if condition (ii) of Definition 5.3.4 holds for $\min\{\varepsilon, \Delta_i\}$ instead of Δ_i.   □

Note that a rigorous discretization corresponds to an inner error $\varepsilon = 0$. While a rigorous discretization based on (5.23) needs some clever strategy to cover the whole image $\Phi_h^0(Q_i, \bar{u})$ a discretization with small inner error "only" needs sufficiently many test points such that (5.23) is satisfied with $r = \varepsilon$.

Let us state the result of Algorithm 7.5.2 for non rigorous discretizations satisfying Definition 7.5.8.

Theorem 7.5.9 Consider a discrete time system Φ_h^0 of type (2.21) and let A be a weakly attracting set with domain of attraction $\mathcal{D}(A)$. Let $\mathcal{S} \subset \mathcal{D}(A)$ be a neighborhood of A, let $\Omega \subset \mathbb{R}^n$ be a compact set containing $\mathcal{D}(A)$ and consider Algorithm 7.5.2 using space discretizations $\widehat{\Phi}_h^j$ with inner error $h\varepsilon$ for all $j = 0, 1, \dots$. Assume that $\mathcal{D}(A)$ is dynamically robust and $\mathcal{D}(A)^c$ is inversely γ-robustly strongly forward invariant, both with gain γ. Then for $\varepsilon > 0$ sufficiently small and all $j \in \mathbb{N}_0$ with ε^j sufficiently small we obtain

$$\mathrm{dist}(C^{j+1} \cup E^{j+1}, \mathcal{D}(A)^c) \le r^j, \tag{7.35}$$

$$\mathrm{dist}(D^{j+1} \cup E^{j+1}, \mathcal{D}(A)) \le r^j \tag{7.36}$$

and

$$\mathrm{dist}(E^{j+1}, \partial\mathcal{D}(A)) \le r^j \tag{7.37}$$

for $r^j = \max\{\gamma(\varepsilon),\, \gamma(\varepsilon^j)\} + h\varepsilon^j$.

If the robust invariance is direct we obtain (7.35) and

$$d_H(\mathcal{D}(A)^c, C^{j+1}) \leq r^j. \tag{7.38}$$

In this case no statement about $\text{dist}(E^{j+1}, \partial\mathcal{D}(A))$ is possible.

Proof: We show the assertion for the inverse case, the direct one follows similarly. Again we use the representations (7.34) and (7.33) for D^{j+1} and C^{j+1}.

We first show the existence of sets $C_\varepsilon \subseteq \mathcal{D}(A)^c$ and $D_\varepsilon \subseteq \mathcal{D}(A)$ with $C^j \subseteq D_\varepsilon^c$ and $D^j \subseteq C_\varepsilon^c$ for all $j > 0$ and which satisfy the estimates

$$d_H(C_\varepsilon^c, \mathcal{D}(A)) \leq \gamma(\varepsilon) \ \text{ and } \ d_H(D_\varepsilon^c, \mathcal{D}(A)^c) \leq \gamma(\varepsilon).$$

This implies

$$\text{dist}(D^j, \mathcal{D}(A)) \leq \gamma(\varepsilon) \ \text{ and } \ \text{dist}(C^j, \mathcal{D}(A)^c) \leq \gamma(\varepsilon). \tag{7.39}$$

We start showing the existence of C_ε. The assumption on the inner error of the space discretization implies that for each $x \in Q_i^j$ and each $\bar{u} \in \mathcal{U}(Q_i)$ there exists $u \in \mathcal{U}$ and $p \in \mathcal{P}_\varepsilon^h$ such that

$$\Phi_h^0(x, u, p[u]) \in \hat{\Phi}_h(Q_i^j, \bar{u}). \tag{7.40}$$

Now the inverse robustness assumption for $\mathcal{D}(A)^c$ implies the existence of an ε–strongly forward invariant set C_ε with

$$d_H(C_\varepsilon^c, \mathcal{D}(A)) \leq \gamma(\varepsilon).$$

Note that for ε sufficiently small this implies $C_\varepsilon \cap \mathcal{D}(A) = \emptyset$ since otherwise $\mathcal{D}(A) \subseteq C_\varepsilon$ contradicting this distance estimate.

The distance yields that each point $x \in \mathbb{R}^n$ with $\|x\|_{\mathcal{D}(A)} \geq \gamma(\varepsilon)$ lies in C_ε which implies that

$$\Phi_h^0(t, x, u, p[u]) \in C_\varepsilon$$

for all $u \in \mathcal{U}$, $p \in \mathcal{P}_\varepsilon^h$, $t \geq 0$ and all $x \in C_\varepsilon$. Thus for each cell Q_i^j intersecting $B(\gamma(\varepsilon), \mathcal{D}(A))^c$ we obtain

$$\hat{\Phi}_h(Q_i^j, \bar{u}) \cap C_\varepsilon \neq \emptyset$$

for each $\bar{u} \in \mathcal{U}(\Omega)$. Hence a simple induction over j using (7.34) implies that $Q_i^j \cap D^{j+1} = \emptyset$ for all $j \geq 0$, i.e., the assertion.

Similarly, setting $D_\varepsilon := \{x \in \mathbb{R}^n \mid V(x) \geq \varepsilon\}$ for the dynamical robustness Lyapunov function V one obtains the existence of D_ε with the asserted properties.

Now we show inequality (7.35). Consider a cell $Q_i^j \subseteq C^{j+1} \cup E^{j+1}$. If $Q_i^j \subseteq C^{j+1}$, then by (7.39) we can conclude $\text{dist}(Q_i^j, \mathcal{D}(A)^c) \leq \gamma(\varepsilon)$. If $Q_i^j \subseteq E^{j+1}$, then as in the proof of Theorem 7.5.5(i) we obtain that $\text{dist}(Q_i^j, \mathcal{D}(A)^c) \geq \gamma(\varepsilon^j) + h\varepsilon^j$ cannot hold, because all these sets must be contained in D^{j+1}. Thus we obtain $\text{dist}(Q_i^j, \mathcal{D}(A)^c) \leq \gamma(\varepsilon^j) + h\varepsilon^j$ which shows (7.35).

For inequality (7.36) consider a cell $Q_i^j \subseteq D^{j+1} \cup E^{j+1}$. Then we either have $Q_i^j \subseteq D^{j+1}$ and thus by (7.39) we obtain $\text{dist}(Q_i^j, \mathcal{D}(A)) \leq \gamma(\varepsilon)$. Otherwise we have $Q_i^j \subseteq E^{j+1}$ and we proceed as in the proof of Theorem 7.5.5(ii) observing that in this proof for ε^j sufficiently small we can without loss of generality choose $C_{\varepsilon^j} \subseteq C_\varepsilon$ for C_ε from above, which implies $D^j \cap C_{\varepsilon^j} = \emptyset$. Thus we obtain the estimate $\text{dist}(Q_i^j, \mathcal{D}(A)) \leq \max\{\gamma(\varepsilon^j), \gamma(\varepsilon)\} + h\varepsilon^j$ which shows (7.36).

Finally, inequality (7.37) is immediate from (7.35) and (7.36). $\qquad\qquad\square$

In other words, when using a non rigorous discretization with some inner accuracy ε we still get an approximation for the domain of attraction whose accuracy is determined by ε and ε^j and the robustness gain γ.

Remark 7.5.10 Note that the estimates provided by Theorem 7.5.9 under the assumption of inverse robustness of $\mathcal{D}(A)^c$ in general do not provide an estimate of the Hausdorff distance between the respective sets, cf. Remark 2.3.3. This problem was observed before, e.g. in the approximation of reachable sets by Häckl [60], where a so called outer ball condition was proposed to overcome this difficulty.

By Lemma 2.3.2(v) (analogous to Remark 7.5.6(i)) we can only conclude the convergence

$$d_H(\text{int } D^{j+1}, \mathcal{D}(A)) \to 0$$

for $r^j \to 0$ (i.e., when both ε and ε^j tend to 0) although we cannot control the convergence rate of this expression. Note, however, that after each change of ε (which is the initial inner accuracy) the whole algorithm has to be restarted.

$\qquad\qquad\square$

7.6 Numerical Approximation of Zubov's Method

In this section we will describe a numerical method for the approximation of domains of attraction which is directly obtained from Zubov's equation. For systems without input Zubov's method has been used by a number of authors for the construction of numerical schemes for the computation of domains of attraction, see, e.g., [1, 72, 123].

Our method is based on the application of a (now) standard two–step approximation scheme for Hamilton–Jacobi type equations to (7.13) which goes back to Capuzzo Dolcetta [18] and Falcone [32], see also the survey paper [19] and Appendix A in [8]. Starting from a continuous time system we first fix a time step $h > 0$ and consider some numerical approximation Φ_h^0 of φ^h, satisfying

$$\|\Phi_h^0(x, u) - \varphi^h(h, x, u)\| \le h c_{time}$$

for some constant $c_{time} > 0$. This could be a one step approximation or an iterate of some one step approximation, cf. Remark 5.3.11. When we start from a discrete time system, of course, no time discretization is necessary.

Again we consider a compact set $\Omega \subset \mathbb{R}^n$ which—for simplicity of exposition—is assumed to be a union of finitely many simplices (e.g., triangles in $\mathbb{R}^2$). Then we can consider a simplicial grid Γ with nodes z_j covering Ω. Note that the grid induces a cell covering $\mathbf{Q}$ of Ω. On Γ we can consider the space P_Γ of continuous functions which are linear on each simplex Q_i. Each of these functions is uniquely determined by its values in the nodes z_j. Alternatively, one could use a rectangular grid as described in [46, 56], since the statements in this section are independent on the underlying interpolation rule as long as each function in P_Γ has the property that the maximum on each cell is attained in one of the grid nodes lying in this cell. Now we are looking for the function satisfying

$$\hat{v}(z_j) = T_h(\hat{v})(z_j) \tag{7.41}$$

for all nodes z_j of the grid Γ where the operator T_h is given by

$$T_h(v)(x) = \begin{cases} 1, & x \notin \text{int}\,\Omega \\ 0, & x \in A \\ \min_{u \in \mathcal{U}}\{(1 - hg(x))v(\Phi_h^0(x, u)) + hg(x)\}, & \text{else} \end{cases}$$

Note that the term $hg(x)$ corresponds to a first order approximation of the integral

$$\int_0^h g(\varphi^h(t, x, u))dt.$$

Here one could also use a higher order approximation, cf., e.g., [33, 56].

Since the restriction of T_h to P_Γ is a contraction (with contraction rate $1 - \min_{z_j \notin A} hg(z_j) < 1$) we obtain a unique solution $\hat{v}$ to (7.41). There are, however, examples, where even for finer and finer grids and $h \to 0$ this solution v_h does not converge to the solution v of Zubov's equation (7.13), cf. [15] for an example where we even have $\hat{v} \equiv 1$ on all Q_i not intersecting A. The reason for this lies in the fact that g vanishes on A and consequently the contraction rate tends to 1 for finer and finer grids, i.e., we have a singularity at ∂A.

It was shown in [15] that we can avoid this undesirable behavior by regularizing (7.13) before applying the discretization. To this end consider some parameter $\varepsilon > 0$, define

$$g^\varepsilon(x) := \max\{g(x), \varepsilon\}$$

and replace (7.13) by

$$\inf_{u \in U} \{-Dv(x)f(x, u) - g(x) + v(x)g^\varepsilon(x)\} = 0 \qquad x \in \mathbb{R}^n. \qquad (7.42)$$

It was shown in [15] that the (unique) solution v^ε of (7.42) uniformly converges to v, furthermore for all $\varepsilon > 0$ sufficiently small we obtain $\mathcal{D}(A) = \{x \,|\, v^\varepsilon(x) < 1\}$, and v^ε is a Lyapunov function outside the domain where the regularization changes g. (Actually the results in [15] apply to the strong case, i.e. to (7.12). All proofs, however, easily carry over to the weak case.)

Applying the above discretization procedure to (7.42) we end up with the unique solution of

$$\hat{v}(z_j) = T_h^\varepsilon(\hat{v})(z_j) \qquad (7.43)$$

for all nodes z_j of the grid Γ where the operator T_h^ε is given by

$$T_h^\varepsilon(v)(x) = \begin{cases} 1, & x \notin \text{int}\,\Omega \\ 0, & x \in A \\ \min_{u \in U}\{(1 - hg^\varepsilon(x))v(\Phi_h^0(x, u)) + hg(x)\}, & \text{else} \end{cases}$$

Note that now the contraction rate is $1 - h\varepsilon$ and hence no longer depends on the choice of the grid. It is easily seen that the unique bounded solution $\hat{v}$ of (7.43) has values in $[0, 1]$, because otherwise (7.43) immediately implies unboundedness of $\hat{v}$.

For this approximation of (7.42), using the error analysis for this scheme from [32], it is possible to show that $\hat{v}$ converges to v^ε as h and the cell size tends to 0, see [15]. The problem with this result, however, is twofold: First, the estimated rate of convergence depends on ε, i.e., the straightforward adaptation of the analysis from [32] to (7.42) gives an estimate of the order $c_{time}^\varepsilon + h^\varepsilon + \text{diam}(\mathbf{Q})^\varepsilon/h$ (for certain HJB–equations this was improved in [34] to $c_{time}^\varepsilon + h^\varepsilon + \text{diam}(\mathbf{Q})^\varepsilon/h^{\varepsilon/2}$, but the applicability of this result to (7.42) has not yet been checked rigorously). Secondly, and more importantly, these estimates give a bound for the L_∞ error between $\hat{v}$ and v^ε which does not imply any a priori estimate for the distance of the level sets characterizing $\mathcal{D}(A)$. Using the techniques from [14] one can at least ensure the convergence of suitable sublevel sets of $\hat{v}$ to $\mathcal{D}(A)$ but no distance estimate can be deduced. (Actually, it seems possible to overcome this difficulty since essentially the distance between the level sets depends on how fast $v(x)$ converges to 1 as x approaches $\partial\mathcal{D}(A)$ which in turn depends on how fast the trajectories can be steered away from $\partial\mathcal{D}(A)$. Nevertheless, we expect that this detour via the L_∞ error produces very conservative estimates.)

What we want to do in what follows is to use the robustness concepts which we already utilized in the analysis of the subdivision algorithm for the analysis of the solution $\hat{v}$ from (7.43).

In order to do this we will impose a "compatibility condition" on the relation between the accuracy of the scheme and the regularization parameter ε. For this we consider the cell–cell discretization $\widehat{\Phi}_h$ from (5.21) and Remark 5.3.6 corresponding to Φ_h^0 on $\mathbf{Q}$, which we interpret as a (total) discretization of φ^h, cf. Remark 5.3.10. (The necessary changes for the case when we start from a discrete time system Φ_h not related to some continuous time system should be obvious). We set $B_{reg} = \{x \in \mathbb{R}^n \,|\, g^\varepsilon(x) = \varepsilon\}$, and assume that the system has a weakly attracting set A with $\mathcal{D}(A) \subset \Omega$, where $\mathcal{D}(A)$ is dynamically robust on some robustness neighborhood B. Then we impose the following condition on the discretization and regularization.

> There exists an attracting set $\widehat{A} \in \mathcal{C}_{\mathbf{Q}}$ for $\widehat{\Phi}_h$ with $\widehat{A} \subset B_{reg}$
> and attracted neighborhood $\widehat{B}$ containing B^c (7.44)

Note that under the assumption on A and $\mathcal{D}(A)$ Theorem 6.4.7 ensures that this condition is satisfied if $\widehat{\Phi}_h$ is sufficiently accurate, i.e., if the error in time c_{time} and the discretization width in space diam$(\mathbf{Q})$ are sufficiently small. Under this condition it is immediate from (7.43) that $\hat{v}(x) < 1$ on B^c.

The following theorem analyzes the performance of this approximation for grids with equidistant cellwidth $hc_{space} > 0$.

Theorem 7.6.1 Consider a system φ of type (2.1), a weakly asymptotically stable set A for φ and a compact set Ω containing $\mathcal{D}(A)$. Assume that $\mathcal{D}(A)$ is dynamically robust on some robustness neighborhood B with gain γ of class $\mathcal{K}_\infty$, and that $\mathcal{D}(A)^c$ is directly robustly forward invariant with same gain γ. Consider the solution $\hat{v}$ of (7.43) for some $\varepsilon > 0$ and assume that $B_{reg} \subset \mathcal{D}(A)$. Consider an equidistant grid with diam$(\mathbf{Q}) \le hc_{space}$ for some constant $c_{space} > 0$ and a time step $h > 0$. Then if $c_{space} > 0$ is sufficiently small and (7.44) is satisfied with $B_0 = B^c$ the set $\widehat{\mathcal{D}} := \{x \in \Omega \,|\, \hat{v}(x) < 1\}$ satisfies

$$d_H(\widehat{\mathcal{D}}^c, \mathcal{D}(A)^c) \le \gamma\left((1 + Lh)(c_{time} + c_{space})\right).$$

If $\mathcal{D}(A)^c$ is inversely robustly forward invariant (instead of direct), then we obtain the estimates

$$\text{dist}(\widehat{\mathcal{D}}^c, \mathcal{D}(A)^c) \le \gamma\left((1 + Lh)(c_{time} + c_{space})\right),$$

and

$$\text{dist}(\widehat{\mathcal{D}}, \mathcal{D}(A)) \le \gamma\left((1 + Lh)(c_{time} + c_{space})\right).$$

Proof: Let $\overline{\Phi}_h$ denote the point–cell discretization related to Φ_h^0 given by (5.21) and let $\widehat{\Phi}_h$ denote the corresponding cell–cell discretization from Remark 5.3.6. Then it follows that

$$d_H(\varphi^h(h, Q_i, \bar{u}, 0), \widehat{\Phi}_h(Q_i, \bar{u})) \leq hc_{time} + hc_{space}. \tag{7.45}$$

Applying one step of the Subdivision Algorithm 7.5.2 for this map with $\mathcal{S} = B_0$ we obtain the sets D^1 and C^1. Clearly, $\hat{v} = 1$ on C^1 and $\hat{v} < 1$ on D^1, hence $D^1 \subseteq \hat{\mathcal{D}}$ and $C^1 \subseteq \hat{\mathcal{D}}^c$. Thus the desired inequalities follow from Theorem 7.5.9. $\qquad\qquad\qquad\qquad\qquad\qquad\qquad\qquad\qquad\qquad\qquad\square$

Obviously, a grid with equidistant cell size is not an efficient choice for this approximation. In fact, using the same arguments as in the proof of Theorem 6.1.6 (see also Theorem 6.4.7) one can show that if the corresponding cell–cell space discretization has an accuracy which is less than $h \max\{\gamma^{-1}(\text{dist}(Q_i, \partial\mathcal{D}(A)), c_{space}\}$ then we obtain the same estimates as in Theorem 7.6.1, implying that we only need fine cells "close to $\partial\mathcal{D}(A)$".

This explains the good results for this method using the adaptive grids based on the a posteriori error estimates developed in [47] (see, e.g., the numerical experiment in [15]), which lead to a grid being essentially refined around $\partial\mathcal{D}(A)$. The "theoretical problem" with these error estimates is that again they are designed to minimize the L_∞ error and hence do not allow a direct development of estimates for the distance between the computed sets. Up to now we could not find an efficient a posteriori refinement criterion providing or using information about these distances. We conjecture, however, that a possible coupling between a cell covering from the subdivision Algorithm 7.5.2 and the discrete PDE operator (7.43) on the grid induced by this covering could lead into this direction.

It should finally be mentioned that the numerical scheme for the computation of strong domains of attraction based on the maximum time optimal control problem, which was developed by Falcone, Wirth and the author [35] can also be analyzed by the techniques used in this Section.

Remark 7.6.2 The approximation to Zubov's equation described in this section served as a motivation for the development of the Subdivision Algorithm 7.5.2. In fact, the original idea behind Algorithm 7.5.2 was to simultaneously compute the set on which $\hat{v} \equiv 1$ (i.e., D^j) and the set on which $\hat{v} \to 0$ as the function g uniformly tends to 0 (i.e., C^j), without having to compute $\hat{v}$ explicitly. $\qquad\qquad\qquad\qquad\qquad\qquad\qquad\qquad\qquad\square$

7.7 Reachable Sets

A concept very related to (asymptotic) domains of attraction is that of (finite time) reachable sets. These sets play an important role in the analysis of nonlinear control systems. In this last section of this chapter we will show how these sets relate to domains of attractions and how we can apply the techniques from the last sections to these sets. Furthermore, we will investigate how the robust invariance condition for domains of attraction translates to reachable sets and how it relates to a robustness condition well known in the geometric study of nonlinear control systems.

For some subset $B \subset \mathbb{R}^n$ the reachable set is defined according to the following definition.

Definition 7.7.1 (reachable set)

The *reachable set* for some subset $B \subset \mathbb{R}^n$ is defined by

$$\mathcal{R}(B) := \{y \in \mathbb{R}^n \,|\, \text{there exist } u \in \mathcal{U},\, x \in B \text{ and } t \geq 0 \text{ with } \varphi(t, x, u) = y\}.$$

If $B = \{x\}$ then we also write $\mathcal{R}(x)$ instead of $\mathcal{R}(\{x\})$. $\square$

Note that this definition is different from the one given on Page 17 because the one here uses the unperturbed system.

Besides being useful by themselves, reachable sets can in particular be used to define control sets (sometimes also called controllability sets), which are subsets of the spate space where the system is completely controllable.

Definition 7.7.2 (control set)

A subset $D \subset \mathbb{R}^n$ is called a *control set* if it satisfies the following properties.

(i)   The set D is weakly forward invariant.

(ii) The inclusion $D \subseteq \mathcal{R}(x)$ holds for all $x \in D$.

(iii) The set D is maximal (with respect to set inclusion) with properties (i) and (ii).

$\square$

For more information and a huge number of applications of control sets we refer to [22]. Here we only remark that under a local accessibility condition (cf. [22, Assumption (3.1.2)]) we can steer any point $x \in \text{int}\, D$ to any other point $y \in \text{int}\, D$ in finite time (note that the definition of D only demands approximate controllability). Furthermore, for all $x \in \text{int}\, D$ the interior of D is uniquely determined by the reachable sets $\mathcal{R}(x)$ for the original and the corresponding time reversed system.

The usefulness of reachable and control sets in the analysis of controlled and perturbed systems together with the fact that an explicit calculation of reachable sets is hardly ever possible has lead to the development of several numerical methods during the last decade. Häckl [60] (see also [59] for a description of an early version of the algorithm) constructs an approximation from the inside by small convex sets. The subdivision Algorithm 7.5.1 by

Szolnoki [115, 116, 118], which computes viability kernels can also be used for the computation of control sets (and their domains of attraction) using results about the relation between viability kernels, domains of attractions, reachable and control sets from [117]. Dang and Maler [28] suggest an approximation by polyhedra where again convex hulls play an important role. All these methods have in common that (explicitly or implicitly) certain robustness condition of $\mathcal{R}(x)$ with respect to small numerical errors are imposed. More precisely, in Häckl's algorithm this condition is needed for the convergence result, while in Szolnoki's algorithm it is needed to establish the connection between viability kernels and control sets, see [117]. The conditions used by these authors are formulated via (ε, T)-chains which we will discuss below. Dang and Maler just assume that small local errors do not lead to wrong solutions or "accept them as a sad fact of life, as do all engineers who use simulation methods" (an attitude which—by the way—makes many of the results developed in this book obsolete. I hope that the reader who made it so far in this monograph is less fatalistic and rather prefers to search for the happy facts).

In the remainder of this section we want to clarify the relation between reachable sets and the domains of attraction we considered in the preceding sections, sketch how the algorithms from the preceding sections can be used for the computation of reachable sets and show how the robustness condition via (ε, T)-chains relates to our robust invariance condition.

The following proposition shows the relation of reachable sets to domains of attraction.

Proposition 7.7.3 Consider a locally accessible control system given by some vectorfield f. Let D be a control set with nonvoid interior and consider some point $x^* \in \operatorname{int} D$. Then the following properties hold:

(i) For each $\varepsilon > 0$ with $B(\varepsilon, x^*) \subset \operatorname{int} D$ the equality $\mathcal{R}(x^*) = \mathcal{R}(B(\varepsilon, x^*))$ holds.

(ii) For each $\varepsilon > 0$ with $B(\varepsilon, x^*) \subset \operatorname{int} D$ there exists a control system with right hand side $\tilde{f} : \mathbb{R}^n \times U \times [0, 1] \to \mathbb{R}^n$ satisfying $\tilde{f}(x, u, \tilde{u}) = -f(x, u)$ for all $x \notin B(\varepsilon, x^*)$ and all $\tilde{u} \in [0, 1]$, such that $A = \{x^*\}$ is a weakly asymptotically stable set for $\tilde{f}$ with $\mathcal{D}(A) = \mathcal{R}(x^*)$.

Proof: (i) The assertion is immediate from the fact that under the local accessibility assumption any point $y \in B(\varepsilon, x)$ can be reached from x^* in finite time.

(ii) Consider a Lipschitz continuous function $\rho : \mathbb{R}^n \to [0, 1]$ with $\rho(x) = 1$ if $x \in B(\varepsilon/2, x^*)$ and $\rho(x) = 0$ if $x \notin B(\varepsilon, x^*)$. We define

$$\tilde{f}(x, u, \tilde{u}) := -f(x, u) + \tilde{u}\rho(x)(f(x, u) - (x - x^*)).$$

Clearly, x^* is weakly asymptotically stable with attracted neighborhood $\mathcal{B}(\varepsilon/2, x^*)$ using the constant control $(u, \tilde{u}) = (u, 1)$ where $u \in U$ is arbitrary. The assertion about the domain of attraction follows easily from the fact that $\mathcal{R}(x^*) = \mathcal{R}(\mathcal{B}(\varepsilon/2, x^*)) = \mathcal{R}(\mathcal{B}(\varepsilon, x^*))$. $\qquad\square$

Thus we can apply all results for domains of attraction to reachable sets $\mathcal{R}(x)$ for $x \in D$ for some control set D. In particular, we can apply the algorithms of the previous sections in order to obtain a numerical approximation of reachable sets.

When using the subdivision Algorithm 7.5.2 then we can even work directly with $-f$ instead of $\tilde{f}$, provided we are careful enough in the choice of the target set $\mathcal{S}$ and the initial cell covering in Algorithm 7.5.2: We choose a neighborhood $\mathcal{S} \subset D$ of x^* and an initial cell covering $\mathbf{Q}$ such that there exists $C \in \mathcal{C}_\mathbf{Q}$ and a neighborhood $\mathcal{B}(\varepsilon, x^*)$ satisfying

$$\mathcal{B}(\varepsilon, x^*) \subset C \subset \mathcal{S}.$$

Constructing $\widehat{\Phi}_h$ from $\tilde{f}$ for this ε then results in a map which coincides with the respective cell–cell discretization of $-f$ outside C. Since, however, the restriction $\widehat{\Phi}_h|_C$ is never used in Algorithm 7.5.2 we can simply work with a discretization of $-f$.

In the numerical approximation of Zubov's equation this trick does not work, since here we really need an asymptotically stable set A to start with. Nevertheless, also here one could avoid the explicit use of $\tilde{f}$ by a suitable change of the fixed point operator T_h^ε in a small neighborhood of x^*, but we do not want to go into these technical details here.

Let us finally discuss what the robust invariance condition on $\mathcal{D}(A)^c$ means in the context of reachable sets. For this we explain the concept of chain reachable sets as introduced in Häckl [60].

Definition 7.7.4 ((ε, T)-chains and chain reachable sets)

(i) Let $\varepsilon, T > 0$. An (ε, T)-chain ζ is a sequence of points $x_0, \ldots, x_k$, $k \in \mathbb{N}$ together with sequences of times $t_0 \geq T$, $t_1 \geq T$, $\ldots$, $t_{k-2} > T$, $t_{k-1} > 0$ and control functions $u_0, \ldots, u_{k-1} \in \mathcal{U}$ satisfying

$$\|\varphi(t_i, x_i, u_i) - x_{i+1}\| \leq \varepsilon \text{ for all } i = 0, \ldots, k-1.$$

Note that the last time t_{k-1} is required only to be positive. We write $\zeta_0 = x_0$ and $\zeta_{end} = x_k$.

(ii) Let $\varepsilon, T > 0$ and $x \in \mathbb{R}^n$. The (ε, T)-chain reachable set of x is defined by

$$\mathcal{R}_{(\varepsilon, T)}(x) := \{y \in \mathbb{R}^n \mid \text{there exists an } (\varepsilon, T)\text{-chain } \zeta \text{ with } \zeta_0 = x, \zeta_{end} = y\}.$$

(iii) Let $x \in \mathbb{R}^n$. The chain reachable set of x is defined by

$$\mathcal{R}_c(x) := \bigcap_{\varepsilon > 0,\, T > 0} \mathcal{R}_{(\varepsilon,T)}(x).$$

□

For many theoretical as well as numerical considerations and statements the assumption $\operatorname{cl}\mathcal{R}(x) = \mathcal{R}_c(x)$ or variants thereof are a crucial ingredient; in particular assumptions of this kind are made for the numerical approximations of Häckl and Szolnoki mentioned above. In order to show the relation to our robust invariance condition, observe that strong forward invariance of $\mathcal{D}(A)^c$ is equivalent to the fact that no trajectory can move from $\mathcal{D}(A)^c$ to $\mathcal{D}(A)$. For the time reversed system this translates to strong forward invariance of $\mathcal{R}(x)$. Consequently, inverse robust forward invariance of $\mathcal{D}(A)^c$ translates to direct robust forward invariance of $\mathcal{R}(x)$. The following proposition shows that this, in turn, is equivalent to the identity $\mathcal{R}_c(x) = \operatorname{cl}\mathcal{R}(x)$.

Proposition 7.7.5 Assume that the reachable set $\mathcal{R}(x)$ of some point x is bounded. Then $\mathcal{R}(x)$ is a direct robustly strongly forward invariant set for the inflated system with some gain γ of class $\mathcal{KL}$ and perturbations from $\mathcal{W}_{\alpha_0}$ for some $\alpha_0 > 0$ if and only if the identity $\operatorname{cl}\mathcal{R}(x) = \mathcal{R}_c(x)$ holds.

Proof: Observe that by Lemma 3.7.1(iii) for all $\alpha > 0$ there exist $\varepsilon > 0$ sufficiently small and $T > 0$ sufficiently large such that for any (ε, T)-chain ζ there exists $u \in \mathcal{U}$ and $w \in \mathcal{W}_\alpha$ with $\|\varphi(T(\zeta), \zeta_0, u, w) - \zeta_{end}\| \le \varepsilon$ (the difference ε is due to the fact that the trajectory time before the last jump can be arbitrarily small, in which case it cannot be reproduced by the inflated system). Conversely, by Lemma 3.7.1(i) for all $\varepsilon > 0$ and all $T > 0$ there exists $\alpha > 0$ sufficiently small such that for each $u \in \mathcal{U}$ and each $w \in \mathcal{W}_\alpha$ there exists an (ε, T)-chain ζ with $\zeta_{end} = \varphi(T(\zeta), \zeta_0, u, w)$.

Now assume direct robust forward invariance. Then by the considerations from above, for any $\delta > 0$ there exists $\varepsilon > 0$ such that any (ε, T)-chain ζ satisfies $\|\zeta_{end}\|_{\mathcal{R}(x)} \le \delta$, hence the limit for $\varepsilon \to 0$ of the endpoints of these chains has to lie in $\operatorname{cl}\mathcal{R}(x)$, thus $\mathcal{R}_c(x) = \operatorname{cl}\mathcal{R}(x)$.

Conversely, assume $\mathcal{R}_c(c) = \operatorname{cl}\mathcal{R}(x)$. Then $\bigcap_{\varepsilon \ge 0, T \ge 0} \mathcal{R}_{(\varepsilon,T)}(x) = \operatorname{cl}\mathcal{R}(x)$. Observing that for for any two monotone sequences $\varepsilon_k \to 0$ and $T_k \to \infty$ the equality

$$\bigcap_{\varepsilon \ge 0,\, T \ge 0} \mathcal{R}_{(\varepsilon,T)}(x) = \operatorname{Lim\,sup}_{\varepsilon_k \to 0,\, T_k \to \infty} \mathcal{R}_{(\varepsilon_k, T_k)}(x)$$

holds, by Lemma 2.3.5 (using the compactness of $\operatorname{cl}\mathcal{R}(x)$) we can conclude that

$$d_H(\mathcal{R}_{(\varepsilon_k, 1/\varepsilon_k)}(x), \operatorname{cl}\mathcal{R}(x)) \to 0 \text{ as } \varepsilon_k \to 0.$$

By the considerations from above we obtain that for each $\varepsilon_k > 0$ there exists $\alpha > 0$ such that

$$\bigcup_{t\geq 0} \varphi_\alpha(t,x) \subset \mathcal{R}_{(\varepsilon_k,1/\varepsilon_k)}(x),$$

which implies

$$d_H\left(\bigcup_{t\geq 0} \varphi_\alpha(t,x), \mathcal{R}(x)\right) \to 0$$

as $\alpha \to 0$ and consequently there exist strongly α-invariant sets C_α with the property that $d_H(C_\alpha, \mathcal{R}(x)) \to 0$ as $\alpha \to 0$. This implies the existence of a robustness gain γ. □

In other words, our robust invariance condition is nothing else than the "chain robustness condition" well known in the (theoretical and numerical) analysis of reachable and control sets, cf. [22, 60, 115, 116]. Note that this condition is satisfied generically for families of systems satisfying a so called inner pair condition This property is proved in [22, Theorem 4.5.7] for control sets, the same arguments apply, however, to reachable sets. It should be noted that in Häckl [60, Lemma 1.3.5] it was already observed that strong exponential attraction of $\mathcal{R}(x)$ implies robust invariance with linear robustness gain γ. The case of general attraction rates, however, was not considered there.

As far as domains of attraction are concerned, it is immediate that one can define "chain domains of attraction" $\mathcal{D}_c(A)$ similar to chain reachable sets. Just as in Proposition 7.7.5 one can then obtain equivalence between inverse robust strong invariance of $\mathcal{D}(A)^c$ and the identity $\operatorname{cl}\mathcal{D}(A) = \mathcal{D}_c(A)$.

Appendices

A Viscosity Solutions

In this appendix we will briefly review the concept of viscosity solutions and discuss those properties that we need for treating the ISDS and wISDS Lyapunov functions.

The theory of viscosity solutions started in the early 1980's with the papers by Crandall and Lions [24, 25], Crandall, Evans and Lions [23] and the monograph by Lions [89]. A comprehensive up–to–date overview (especially in connection with optimal control problems) can be found in the monograph by Bardi and Capuzzo Dolcetta [8].

Viscosity solutions provide a powerful solution concept for those partial differential equations (and also partial differential inequalities) which in general do not admit smooth classical solutions. In particular (but of course not exclusively) this applies to first order equations such as those appearing in the characterization of Lyapunov functions and in Zubov's method, both of which play a vital role in this book.

A.1 Definition

In this section we give the definition of viscosity solutions via test functions and via sub- and superdifferentials. Here for some function $g : O \to \mathbb{R}$ with $O \subseteq \mathbb{R}^n$ we use the definitions

$$\mathrm{argmin}_O \, g := \{x \in O \,|\, g \text{ attains a local minimum in } x\}$$

and

$$\mathrm{argmax}_O \, g := \{x \in O \,|\, g \text{ attains a local maximum in } x\}.$$

Definition A.1.1 (viscosity solutions, sub- and supersolutions)

Consider an open subset O of $\mathbb{R}^n$ and a continuous function $H : O \times \mathbb{R} \times \mathbb{R}^n \to \mathbb{R}$.

(i) A lower semicontinuous function $V : O \to \mathbb{R}$ is called a *viscosity supersolution* of the equation

$$H(x, V, DV) = 0 \qquad x \in O \qquad\qquad (A.1)$$

if for all $\phi \in C^1(O)$ and $x \in \mathrm{argmin}_O (V - \phi)$ we have

$$H(x, V(x), D\phi(x)) \geq 0.$$

(ii) An upper semicontinuous function $U : O \to \mathbb{R}$ is called a *viscosity sub-solution* of equation (A.1) if for all $\phi \in C^1(O)$ and $x \in \mathrm{argmax}_O (U - \phi)$ we have

$$H(x, U(x), D\phi(x)) \leq 0.$$

(iii) A continuous function $V : O \to \mathbb{R}$ is called a *viscosity solution* of equation (A.1) if V is a viscosity supersolution and a viscosity subsolution of (A.1). □

Remark A.1.2 It is not difficult to see (cf. [8, Lemma II.1.7]) that the set of derivatives $D\phi(x)$ for $x \in \mathrm{argmin}_O(V - \phi)$ coincides with the set

$$D^- V(x) := \{ p \in \mathbb{R}^n \,|\, V(x) - V(y) - p(x - y) \leq o(\|x - y\|) \text{ for all } y \in \mathbb{R}^n \}$$

and that the set of derivatives $D\phi(x)$ for $x \in \mathrm{argmax}_O (U - \phi)$ equals

$$D^+ U(x) := \{ p \in \mathbb{R}^n \,|\, U(x) - U(y) - p(x - y) \geq -o(\|x - y\|) \text{ for all } y \in \mathbb{R}^n \},$$

where in both cases p is interpreted as row vector, i.e., $p(x - y)$ is the inner product in $\mathbb{R}^n$.

Hence, one can alternatively define viscosity solutions via the sets D^- and D^+, the so called *viscosity sub- and superdifferentials*. Note that if a function $V : O \to \mathbb{R}$ is differentiable in some $x \in O$ the equality $D^+ V(x) = D^- V(x) = \{DV(x)\}$ follows, hence for smooth functions viscosity solutions coincide with classical solutions. □

A.2 Optimality Principles

In this section we will formulate the so called optimality principles. Loosely speaking, for classical solutions to differential equations the formulation as a differential equation or as an integral equation is essentially equivalent. The same turns out to be true for viscosity solutions. This was observed already in the early days of viscosity solutions (cf. the monograph by Lions [89]), recently these results were refined by Soravia [107, 108, 109]. We will cite the corresponding result for continuous supersolutions related to "sup inf" equations, i.e., for Hamilton–Jacobi–Isaacs equations related to differential games.

For the definition of $\mathcal{P}^0$ see Definition 2.1.1. Throughout this appendix we assume that f satisfies the usual assumptions from Chapter 2 and that W is compact.

Theorem A.2.1 Let V be a continuous supersolution of

$$\sup_{u \in U} \inf_{w \in W} \{-f(x, u, w)DV(x) - h(x, u, w)\} \geq 0$$

on some open set $O \subset \mathbb{R}^n$, where $h : \mathbb{R}^n \times U \times W$ is a bounded and continuous function which is Lipschitz in x uniformly in u and w. Then V satisfies

$$V(x) = \sup_{p \in \mathcal{P}^0} \inf_{u \in \mathcal{U}} \sup_{t \in [0, \tau(x, u, p[u]))} \left\{ \int_0^t h(\varphi(\tau, x, u, p[u]), u(\tau), p[u](\tau))d\tau \right.$$

$$\left. + V(\varphi(t, x, u, p[u])) \right\}$$

where $\tau(x, u, p[u]) := \inf\{\tau \geq 0 \,|\, \varphi(\tau, x, u, p[u]) \notin O\}$.

Proof: See [107, Section 4]. $\qquad\qquad\qquad\qquad\qquad\qquad\qquad\qquad\qquad$ □

Remark A.2.2 (i) If we only have "$\sup_u$" or "$\inf_w$" (instead of "$\sup_u \inf_w$") then one can also obtain results for lower semicontinuous V, where for the "$\sup_u$"–problem we need the continuity condition (2.13) (or use relaxed controls), see [108, 109]. Since in the case of strong attraction we can always ensure the existence of an "ε–optimal" continuous supersolution (see Theorem 3.5.7) we refrain from using these more complicated discontinuous versions here.

(ii) Statements for discontinuous supersolutions for the "$\sup_u \inf_w$"–problem might be useful (cf. the discussion after Theorem 4.5.5), and it does not seem impossible that some results in this direction could be obtained. However, since we were not able to find suitable references in the literature and since the derivation of such results is beyond the scope of this appendix we leave this question open. $\qquad\qquad\qquad\qquad\qquad\qquad\qquad\qquad\qquad\qquad\qquad$ □

The next theorem states a consequence of the preceding one for the inequalities we need for ISDS and wISDS Lyapunov functions.

Theorem A.2.3 Let $g : \mathbb{R}_0^+ \to \mathbb{R}_0^+$ be a continuous function which is Lipschitz on $\mathbb{R}^+$ and satisfies $g(r) = 0$ if and only if $r = 0$. Let $V : \mathbb{R}^n \to \mathbb{R}_0^+$ be a continuous viscosity supersolution of the equation

$$\sup_{u \in U} \inf_{w \in W} \{-DV(x)f(x, u) - g(V(x))\} \geq 0$$

on some open set $O \subset \mathbb{R}^n$. Then V satisfies

$$\inf_{u \in \mathcal{U}} V(\varphi(t, x, u, p[u])) \leq \mu(V(x), t)$$

for all $x \in O$, $p \in \mathcal{P}^0$ and all $t \geq 0$ with $\varphi(\tau, x, u, p[u]) \in O$ for all $\tau \in [0, t]$ and all $u \in \mathcal{U}$, where μ is the unique solution of the initial value problem

$$\frac{d}{dt}\mu(r,t) = -g(\mu(r,t)), \quad \mu(r,0) = r.$$

Proof: Fix $x \in O$, $t > 0$ and $p \in \mathcal{P}$. Then by the assumption on f there exists a compact set $K \subset O$ such that $\varphi(\tau, x, u, p[u]) \in K$ for all $u \in \mathcal{U}$ and all $\tau \in [0, t]$. For any $\varepsilon > 0$ we can approximate $g(V(x))$ by some Lipschitz continuous function $h(x)$ such that

$$h(x) \leq g(V(x)) \text{ for all } x \in K \quad \text{and} \quad \sup_{x \in K} |h(x) - g(V(x))| \leq \varepsilon.$$

Now Theorem A.2.1 yields

$$V(x) \geq \inf_{u \in \mathcal{U}} V(\varphi(t, x, u, p[u])) + \int_0^t h(\varphi(\tau, x, u, p[u]))d\tau.$$

Letting $\varepsilon \to 0$ we obtain

$$V(x) \geq \inf_{u \in \mathcal{U}} V(\varphi(t, x, u, p[u])) + \int_0^t g(V(\varphi(\tau, x, u, p[u])))d\tau.$$

Now we set $V^*(t) := \inf_{u \in \mathcal{U}} V(\varphi(t, x, u, p[u]))$. Then for each time $t_0 \in [0, t]$ we can choose a sequence of control functions $u_n \in \mathcal{U}$ with

$$V(\varphi(t_0, x, u_n, p[u_n])) \leq V^*(t_0) + \varepsilon_n$$

for some sequence $\varepsilon_n \to 0$. Fix $t_0 > 0$ and abbreviate $x_n := \varphi(t_0, x, u_n, p[u_n])$. For $t_1 \in (t_0, t]$ and all $u \in \mathcal{U}$ we obtain

$$V^*(t_1) \leq V(\varphi(t_1, x, u_n \&_{t_0} u, p[u_n \&_{t_0} u])) = V(\varphi(t_1 - t_0, x_n, u, \tilde{p}[u]))$$

for $\tilde{p}[u](\tau) = p[u_n \&_{t_0} u](t_0 + \tau)$. Since we know that

$$V(x_n) \geq \inf_{u \in \mathcal{U}} V(\varphi(t_1 - t_0, x_n, u, \tilde{p}[u])) + \int_{t_1 - t_0}^{t_1} g(V(\varphi(\tau, x_n, u, \tilde{p}[u])))d\tau$$

this implies

$$V(x_n) \geq \inf_{u \in \mathcal{U}} V^*(t_1) + \int_{t_1 - t_0}^{t_1} g(V(\varphi(\tau, x_n, u, \tilde{p}[u])))d\tau$$

and since $V(x_n) \to V^*(t_0)$ we finally obtain that the upper Dini derivative satisfies

$$\limsup_{t_1 \searrow t_0} \frac{V^*(t_1) - V^*(t_0)}{t_1 - t_0} \leq -g(V^*(t_0)).$$

Hence by a standard argument (see, e.g., [84, Theorem 1.2.1]) these solutions are less or equal than the exact solutions $\mu(V(x), t)$ of

$$\frac{d}{dt}\mu(r,t) = -g(\mu(r,t)).$$

and the assertion follows. $\square$

Let us state two consequences of this theorem.

Corollary A.2.4 Let g be a function satisfying the assumptions of Theorem A.2.3, let $O \subset \mathbb{R}^n$ be an open set and let $V : O \to \mathbb{R}_0^+$ be a continuous supersolution of the equation

$$\inf_{u \in U}\{-DV(x)f(x,u) - g(V(x))\} \geq 0$$

on some open ring $R = \{x \in \mathbb{R}^n \,|\, a < V(x) < b\} \subset O$. Then V satisfies

$$V(\varphi(t,x,u)) \leq \max\{\mu(V(x),t),a\}$$

for all $x \in R$, $u \in \mathcal{U}$ and all $t \geq 0$ with $\varphi(\tau,x,u) \in O$ for all $\tau \in [0,t]$.

Proof: Fix $t > 0$ and $u \in \mathcal{U}$. As long as the solution remains in R, Theorem A.2.3 (applied with $U = \{0\}$ and $W = U$) implies $V(\varphi(t,x,u)) \leq \mu(V(x),t)$. Now consider the minimal time $t^* \in [0,t]$ such that $x^* = \varphi(t^*,x,u) \in \partial R$, i.e. $V(x^*) = a$ or $V(x^*) = b$. Since μ is monotone decreasing in t (by positivity of g) and $x \in R$ we have $V(\varphi(t^*,x,u)) < \mu(b,t^*) < b$ hence $V(x^*) = a$. Now assume that there exists $T \in (t^*,t]$ with $V(\varphi(T,x,u)) > a$. By continuity we can assume $V(\varphi(\tau,x,u)) < b$ for all $\tau \in [t^*,T]$ (otherwise we can choose a smaller $T > t^*$). Then there exists a time $\tilde{t} \in [t^*,T]$ such that $V(\varphi(\tilde{t},x,u)) = a$ and $V(\varphi(\tau,x,u)) > a$ for all $\tau \in (\tilde{t},T]$. This implies $\varphi(\tau,x,u) \in R$ for all $\tau \in (\tilde{t},T]$ and again by Theorem A.2.3 we obtain

$$V(\varphi(T,x,u)) \leq \inf_{\tau \in (\tilde{t},T]}\mu(V(\varphi(\tau,x,u)),T-\tau) \leq a$$

contradicting the choice of T. $\square$

Corollary A.2.5 Let g be a function satisfying the assumptions of Theorem A.2.3, let $O \subset \mathbb{R}^n$ be an open set and let $V : \text{cl}\,O \to \mathbb{R}_0^+$ be a continuous supersolution of the equation

$$\sup_{u \in U}\inf_{w \in W}\{-DV(x)f(x,u,w) - g(V(x))\} \geq 0$$

on some open ring $R = \{x \in \mathbb{R}^n \,|\, a < V(x) < b\} \subset O$. Assume that R is bounded and that $V(x) > b$ for all $x \in \partial O$. Then for each $x \in O$ with $V(x) < b$, each $t > 0$, each $p \in \mathcal{P}^0$ and each $\varepsilon > 0$ there exists a $u^* \in \mathcal{U}$ such that $\varphi(t,x,u^*,p[u^*]) \in O$ and

$$V(\varphi(t,x,u^*,p[u^*])) \leq \max\{\mu(V(x),t),a\} + \varepsilon.$$

Proof: Fix $p \in \mathcal{P}$, $\varepsilon > 0$ and $\varepsilon_0 \in (0, \varepsilon]$ and consider the rings $R_\varepsilon := \{x \in \mathbb{R}^n \,|\, a + \varepsilon < V(x) < b - \varepsilon\}$ and $R_{\varepsilon_0} := \{x \in \mathbb{R}^n \,|\, a + \varepsilon_0 < V(x) < b - \varepsilon_0\}$. Since cl R is compact and f is bounded we find a time $t_{\varepsilon_0} > 0$ such that every solution $\varphi(\tau, x, u, p[u])$ for $x \in R_{\varepsilon_0}$ stays inside R for all $\tau \in [0, t_\varepsilon]$. Thus by Theorem A.2.3 we obtain

$$\inf_{u \in \mathcal{U}} V(\varphi(t, x, u, p[u])) \leq \mu(V(x), t)$$

for all $t \in [0, t_{\varepsilon_0}]$ and all $x \in R_{\varepsilon_0}$. Now by induction for each $t > 0$ and each $x \in R_\varepsilon$ we obtain the existence of a $u' \in \mathcal{U}$ such that either

$$V(\varphi(t, x, u', p[u'])) \leq \mu(V(x), t) + \varepsilon$$

or $V(\varphi(\tau, x, u', p[u'])) \leq a + \varepsilon_0$ for some $\tau \in [0, t]$.

For $x \in O$ with $V(x) \leq a + \varepsilon$ we can use an arbitrary control $u \in \mathcal{U}$ until we reach R_ε (the assumption $V(x) > b$ for all $x \in \partial O$ implies that each trajectory must cross R_ε before leaving O) and then proceed as above. Together this yields the assertion for all $x \in O$ with $V(x) < b - \varepsilon_0$. Since $\varepsilon_0 \in (0, \varepsilon]$ was arbitrary we obtain the assertion for all $x \in O$ with $V(x) < b$. $\square$

B Comparison Functions

In the definitions of asymptotic stability and robustness concepts we extensively use the concept of comparison functions. There are two main reasons why we prefer these functions to the (qualitatively equivalent) concept of the ε–δ formalism: First, it is notationally convenient and often more intuitive, and secondly, it automatically leads to a natural concept of a convergence rate or a robustness gain. Historically, this concept apparently goes back to Hahn [61] in the 1960's; recently these functions gained new popularity especially in the field of nonlinear mathematical control theory, see for instance the references in Section 3.2. In this appendix we will define a number of classes of comparison functions and investigate some properties of these functions.

B.1 Definition

In this section we first define the classes of functions which we need and then cite a useful Lemma for class $\mathcal{KL}$ functions along with a simple consequence for our newly introduced class of class $\mathcal{KLD}$ functions.

Definition B.1.1 (class $\mathcal{K}$, $\mathcal{K}_\infty$ and $\mathcal{L}$ functions)

A continuous function $\sigma : \mathbb{R}_0^+ \to \mathbb{R}_0^+$ is called of class $\mathcal{K}$ if it is strictly increasing and satisfies $\sigma(0) = 0$. It is called of class $\mathcal{K}_\infty$ if, in addition, it is unbounded.

A continuous function $\rho : \mathbb{R}_0^+ \to \mathbb{R}_0^+$ is called of class $\mathcal{L}$ if it is strictly decreasing and satisfies $\lim_{r \to \infty} \rho(r) = 0$. □

Definition B.1.2 (class $\mathcal{KL}$ and $\mathcal{KLD}$ functions)

A continuous function $\beta : \mathbb{R}_0^+ \times \mathbb{R}_0^+ \to \mathbb{R}_0^+$ is called of class $\mathcal{KL}$ if it is of class $\mathcal{K}$ in the first and of class $\mathcal{L}$ in the second argument.

A continuous function $\mu : \mathbb{R}_0^+ \times \mathbb{R} \to \mathbb{R}_0^+$ is called of class $\mathcal{KLD}$ if its restriction to $\mathbb{R}_0^+ \times \mathbb{R}_0^+$ is of class $\mathcal{KL}$ and, in addition, it satisfies

$$\mu(r, 0) = r \text{ and } \mu(\mu(r,t), s) = \mu(r, t+s) \text{ for all } r \geq 0, \ s, t \in \mathbb{R}.$$

□

While the letters $\mathcal{K}$ and $\mathcal{L}$ have no obvious meaning (apart from the saying that these were the first letters Hahn found on his keyboard when introducing this concept) the letter $\mathcal{D}$ stands for "dynamical" since any class $\mathcal{KLD}$ function μ defines a dynamical system on $\mathbb{R}_0^+$.

The following Lemma (sometimes referred to as "Sontag's $\mathcal{KL}$-Lemma") shows a fundamental property of class $\mathcal{KL}$ functions.

Lemma B.1.3 For any class $\mathcal{KL}$ function β there exist two class $\mathcal{K}_\infty$ functions a_1 and a_2 such that

$$\beta(r,t) \le a_1(a_2(r)e^{-t}) \text{ for all } r, t \ge 0.$$

Proof: See Sontag [103, Proposition 7].   □

We can use this lemma to obtain a relation between class $\mathcal{KL}$ and class $\mathcal{KLD}$ functions.

Lemma B.1.4 For any class $\mathcal{KL}$ function β there exist a class $\mathcal{KLD}$ function μ and a class $\mathcal{K}_\infty$ function σ such that

$$\beta(r,t) \le \mu(\sigma(r),t) \text{ for all } r, t \ge 0.$$

Proof: From Lemma B.1.3 one obtains the existence of two class $\mathcal{K}_\infty$ functions a_1 and a_2 such that

$$\beta(r,t) \le a_1(a_2(r)e^{-t}) \text{ for all } r, t \ge 0.$$

Now the assertion follows by setting $\mu(r,t) := a_1(a_1^{-1}(r)e^{-t})$ and $\sigma(r) := a_1(a_2(r))$.   □

Remark B.1.5 The asymptotic stability definition via class $\mathcal{KL}$ functions (as used, e.g., in Definition 3.1.3) is equivalent to the more usual definition via ε–δ relations. More precisely, for any function $a : \mathbb{R}_0^+ \times \mathbb{R}_0^+ \to \mathbb{R}_0^+$ satisfying the two properties

(i)   for all $\varepsilon > 0$ there exists $\delta > 0$ such that if $r \le \delta$ then $a(r,t) < \varepsilon$ for all $t \ge 0$

(ii) for all $\varepsilon > 0$ and for all $R > 0$ there exists $T > 0$ such that $a(r,t) < \varepsilon$ for all $0 \le r \le R$ and for all $t \ge T$

there exists a class $\mathcal{KL}$ function β with $a(r,t) \le \beta(r,t)$ for all $r, t \ge 0$.

This fact was already implicitly used in Hahn's book [61]; in this form it is stated (but not proved) in Albertini and Sontag [2, Lemma 4.1] and proved (but not explicitly stated) in Lin, Sontag and Wang [88, Section 3].   □

B.2 Approximation by Smooth Functions

We now show that—if we are willing to slightly increase all of these comparison functions in a suitable way—we can restrict ourselves to smooth comparison functions. We start with the following lemma on class $\mathcal{K}_\infty$ functions.

Lemma B.2.1 Consider a function σ of class $\mathcal{K}_\infty$ and a bounded and continuous function $\rho : \mathbb{R}_0^+ \to \mathbb{R}_0^+$ with $\rho(0) = 0$ and $\rho(r) > 0$ for all $r > 0$. Then there exists a function $\tilde\sigma$ of class $\mathcal{K}_\infty$ which is smooth on $\mathbb{R}^+$ and satisfies

$$|\sigma(r) - \tilde\sigma(r)| \le \rho(\sigma(r)) \text{ for all } r \ge 0$$

and $\frac{d}{dr}\tilde\sigma(r) > 0$ for all $r > 0$.

Proof: Without loss of generality we can assume that ρ is globally Lipschitz with constant $L \le 1/2$, otherwise we can replace it by its inf-convolution

$$\tilde\rho(r) := \inf_{s \in \mathbb{R}_0^+} \left\{ \rho(s) + \frac{|r - s|}{2} \right\}.$$

Then both $h_+(s) := s + \rho(s)$ and $h_-(s) := s - \rho(s)$ are of class $\mathcal{K}_\infty$, and hence $h(s) := \min\{h_+(s), h_-^{-1}(s)\}$ is a function of class $\mathcal{K}_\infty$ with $h(s) \le s + \rho(s)$, $h^{-1}(s) \ge s - \rho(s)$, and $h(s) > s$ for all $s > 0$. Now we define a two sided sequence s_i, $i \in \mathbb{Z}$ by $s_0 = 1$, $s_{i+1} = h(s_i)$ for $i \ge 0$ and $s_{i-1} = h^{-1}(s_i)$ for $i \le -1$. From the construction we obtain $s_{i+1} > s_i$. We claim that $s_i \to +\infty$ as $i \to +\infty$. Assume $s_i \to s^* < +\infty$ for $i \to +\infty$. Then by continuity we obtain that $s^* = h(s^*)$ which is impossible since $s^* > 0$ by choice of s_0 and $h(s) > s$ for all $s > 0$. A similar argument shows $s_i \to 0$ as $i \to -\infty$. Now we set $r_i := \sigma^{-1}(s_i)$ and define a piecewise linear function σ_l via

$$\sigma_l(r) = s_i + \frac{r - r_i}{r_{i+1} - r_i}(s_{i+1} - s_i)$$

for $r \in [r_i, r_{i+1}]$. Clearly, σ_l is continuous, of class $\mathcal{K}_\infty$, satisfies $\sigma_l(r_i) = \sigma(r_i)$ and $\frac{d}{dr}\sigma(r) > 0$ for all $r > 0$ with $r \ne r_i$ for $i \in \mathbb{Z}$. Furthermore, for each $r \in [r_i, r_{i+1}]$ we have that

$$\sigma_l(r) \le s_{i+1} = h(s_i) \le h(\sigma(r)) = \sigma(r) + \rho(\sigma(r))$$

and

$$\sigma_l(r) \ge s_i = h^{-1}(s_{i+1}) \ge h^{-1}(\sigma(r)) \ge \sigma(r) - \rho(\sigma(r))$$

which implies the asserted inequality for σ_l.

Hence we can obtain the desired $\tilde\sigma$ by a standard regularization of σ_l at the points r_i. $\qquad\square$

Lemma B.2.2 Consider a function σ of class $\mathcal{K}_\infty$ and a bounded and continuous function $\rho : \mathbb{R}_0^+ \to \mathbb{R}_0^+$ with $\rho(0) = 0$ and $\rho(r) > 0$ for all $r > 0$. Then there exists a function $\tilde{\sigma}$ of class $\mathcal{K}_\infty$ which is smooth on $\mathbb{R}^+$ and satisfies

$$\sigma(r) < \tilde{\sigma}(r) \leq \sigma(r) + \rho(\sigma(r)) \text{ for all } r > 0$$

and $\frac{d}{dr}\tilde{\sigma}(r) > 0$ for all $r > 0$.

Proof: As in the proof of Lemma B.2.1 we can assume that ρ is globally Lipschitz with constant $L \leq 1/2$. Then the function $h(s) := s + C\rho(s)$ is of class K_∞ for all $C \in [0, 1]$, and hence $\sigma_1(r) := \sigma(r) + \rho(\sigma(r))/2$ is of class $\mathcal{K}_\infty$. Setting $\rho_1(r) := \rho(\sigma(\sigma_1^{-1}(r)))/3$ (i.e. $\sigma_1(r) + \rho_1(\sigma_1(r)) = \sigma(r) + 5\rho(\sigma(r))/6$ and $\sigma_1(r) - \rho_1(\sigma_1(r)) = \sigma(r) + \rho(\sigma(r))/6$) we can apply Lemma B.2.1 to $\sigma_1(r)$ and $\rho_1(r)$. This yields

$$\sigma(r) + \rho(\sigma(r))/6 \leq \tilde{\sigma}(r) \leq \sigma(r) + 5\rho(\sigma(r))/6$$

and thus the assertion. □

Proposition B.2.3 Consider functions σ, γ of class $\mathcal{K}_\infty$ and μ of class $\mathcal{KLD}$. Consider furthermore a bounded and continuous function $\rho : \mathbb{R}_0^+ \to \mathbb{R}_0^+$ with $\rho(0) = 0$ and $\rho(r) > 0$ for all $r > 0$. Then there exist functions $\tilde{\sigma}, \tilde{\gamma}$ of class $\mathcal{K}_\infty$ and $\tilde{\mu}$ of class $\mathcal{KLD}$ which are smooth on $\mathbb{R}^+$ or $\mathbb{R}^+ \times \mathbb{R}$, respectively, and satisfy

$$\tilde{\mu}(r, t) \leq \mu(r, t) + \rho(\mu(r, t)) \tag{B.1}$$
$$\mu(\sigma(r), t) < \tilde{\mu}(\tilde{\sigma}(r), t) \tag{B.2}$$
$$\mu(\gamma(r), t) < \tilde{\mu}(\tilde{\gamma}(r), t) \tag{B.3}$$
$$\sigma(r) < \tilde{\sigma}(r) \leq \sigma(r) + \rho(\sigma(r)) \tag{B.4}$$
$$\gamma(r) < \tilde{\gamma}(r) \leq \gamma(r) + \rho(\gamma(r)) \tag{B.5}$$

for all $r > 0$ and $t \geq 0$, as well as $\frac{d}{dt}\big|_{t=0}\tilde{\mu}(r, t) < 0$, $\frac{d}{dr}\tilde{\sigma}(r) > 0$ and $\frac{d}{dr}\tilde{\gamma}(r) > 0$ for all $r > 0$.

Proof: We set $a(s) = \mu(1, -\ln s)$. Then a is of class $\mathcal{K}_\infty$ and a straightforward calculation shows that $\mu(r, t) = a(a^{-1}(r)e^{-t})$ for all $r, t \geq 0$. Setting $b(s) = a^{-1}(\sigma(s))$ and $c(s) = a^{-1}(\gamma(s))$ we obtain functions of class $\mathcal{K}_\infty$ satisfying

$$\mu(\sigma(r), t) = a(b(r)e^{-t}) \text{ and } \mu(\gamma(r), t) = a(c(r)e^{-t}).$$

Now by Lemma B.2.2 we can approximate a by some function $\tilde{a}$ which is smooth on $\mathbb{R}^+$ and satisfies

$$a(r) \leq \tilde{a}(r) \leq a(r) + \rho(a(r))/2 \text{ for all } r \geq 0$$

and $\frac{d}{dr}\tilde{a}(r) > 0$ for all $r > 0$. Now for all $r > 0$ we have $\tilde{a}(r) < a(r) + \rho(a(r))$ which implies $r < \tilde{a}^{-1}(a(r) + \rho(a(r)))$ and thus $\rho_1(r) := \tilde{a}^{-1}(a(r) + \rho(a(r))) -$

$r > 0$ for $r > 0$. Hence we can apply Lemma B.2.2 to b and c giving functions $\tilde{b}$ and $\tilde{c}$ which are smooth on $\mathbb{R}^+$ and satisfy $b(r) \leq \tilde{b}(r) \leq b(r) + \rho_1(b(r))$ and $c(r) \leq \tilde{c}(r) \leq c(r) + \rho_1(c(r))$ for all $r \geq 0$ and $\frac{d}{dr}\tilde{b}(r) > 0$ and $\frac{d}{dr}\tilde{c}(r) > 0$ for all $r > 0$. The choice of ρ_1 implies

$$a(b(r)) < \tilde{a}(\tilde{b}(r)) \leq \tilde{a}(b(r) + \tilde{a}^{-1}(a(b(r)) + \rho(a(b(r)r))) - b(r))$$
$$= (a(b(r)) + \rho(a(b(r)r))$$

for all $r > 0$ and similarly for $a(c(r))$. Furthermore we have

$$a(b(r)e^{-t}) < \tilde{a}(\tilde{b}(r)e^{-t}) \text{ and } a(c(r)e^{-t}) < \tilde{a}(\tilde{c}(r)e^{-t})$$

for all $r > 0$ and $t \geq 0$, as well as

$$\tilde{a}(\tilde{a}^{-1}(r)e^{-t}) \leq \tilde{a}(a^{-1}(r)e^{-t}) \leq a(a^{-1}(r)e^{-t}) + \rho(a(a^{-1}(r)e^{-t})).$$

Hence the assertion follows setting $\tilde{\mu}(r, t) := \tilde{a}(\tilde{a}^{-1}(r)e^{-t})$, $\tilde{\sigma}(r) = \tilde{a}(\tilde{b}(r))$ and $\tilde{\gamma}(r) = \tilde{a}(\tilde{c}(r))$. □

C Numerical Examples

In this last appendix we want to illustrate the convergence results and the performance of the algorithms from Chapter 7. The main purpose of this appendix is indeed to illustrate some examples rather than to present implementational tricks on high–end computers. Clearly, we believe that any proposed numerical algorithm should eventually be applied to real world problems (the last example in Section C.2 leads in this direction), however, creating highly efficient implementations of mathematical or numerical algorithms is a serious problem at its own right, which is beyond the scope of this more conceptionally oriented monograph.

In Section C.1 we start with the subdivision Algorithm 7.5.2, which we first apply to two simple test problems for which the exact solution is known, verifying in particular the estimates from Theorem 7.5.5. In Section C.2 we show some other examples for which the exact solution is not known.

Finally, in Section C.3 we turn to the numerical solution of Zubov's equation (7.43) and illustrate Theorem 7.6.1 using again the test examples from Section C.1.

C.1 Subdivision Algorithm: Test Examples

In this section we apply Algorithm 7.5.2 to two simple test problems (without input) for which the domains of attraction as well as their robustness gains are explicitly computable. We show the convergence of the algorithm depending on the chosen time step h and compare the results to the theoretically expected values from Theorem 7.5.5.

We consider the following two differential equations in $\mathbb{R}^2$.

$$\dot{x} = \begin{pmatrix} 0 & 1 \\ -1 & 0 \end{pmatrix} x + \rho_1(\|x\|)x, \quad \rho_1(r) = r - 1 \qquad \text{(C.1)}$$

and

$$\dot{x} = \begin{pmatrix} 0 & 1 \\ -1 & 0 \end{pmatrix} x + \rho_2(\|x\|)x, \quad \rho_2(r) = \begin{cases} r-1, & r \le 1 \\ (r-1)^2, & r > 1 \end{cases} \tag{C.2}$$

Figure C.1 shows solution trajectories $x(t)$ for these equations with initial values $(0, 1.1)^T$, $(0, 1)^T$ and $(0, 0.9)^T$ for $t \in [0, 20]$. The difference between these systems is that for trajectories of system (C.2) with initial value x_0 satisfying $\|x_0\| > 1$ the norm $\|\varphi(t, x_0)\|$ grows slower than for the trajectories for system (C.1) as long as $\|\varphi(t, x_0)\|$ is close to 1.

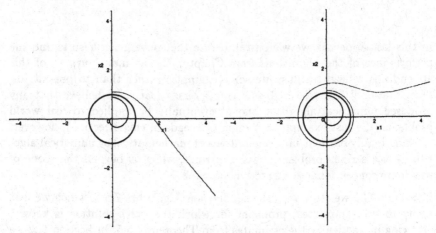

Fig. C.1. Solution trajectories for systems (C.1) (left) and (C.2) (right)

In polar coordinates these equations read

$$\dot{\theta} = 1, \quad \dot{r} = \rho_i(r),$$

which reveals (as Figure C.1 indicates) that for both examples the open unit disk $D^1 = \{x \in \mathbb{R}^2 \,|\, \|x\| < 1\}$ is the domain of attraction $\mathcal{D}(A)$ of the attracting set $A = \{0\}$. Taking the function

$$V(x) = \begin{cases} \|x\| - 1, & x \in D^1 \\ 0, & x \notin D^1 \end{cases}$$

one sees that for any $\eta > 1$ the class $\mathcal{K}_\infty$ function $\gamma_\eta(r) = \eta r$ is a robustness gain for which $\mathcal{D}(A)$ is dynamically robust.

Furthermore, one easily verifies that for system (C.1) the set $\mathcal{D}(A)^c$ is inversely robustly invariant for $\gamma^1(r) = r$, while for system (C.2) it is inversely robustly invariant for $\gamma^2(r) = \sqrt{r}$.

Hence from Theorem 7.5.5 we can conclude that for a rigorous space discretization (or if the inclusions from Remark 7.5.6(ii) hold) we can expect the estimates

$$d_H(E^{j+1}, \partial \mathcal{D}(A)) \leq \varepsilon^j + h\varepsilon^j \tag{C.3}$$

for system (C.1) and

$$d_H(E^{j+1}, \partial \mathcal{D}(A)) \leq \sqrt{\varepsilon^j} + h\varepsilon^j \tag{C.4}$$

for system (C.2), provided $\varepsilon^j > 0$ is sufficiently small.

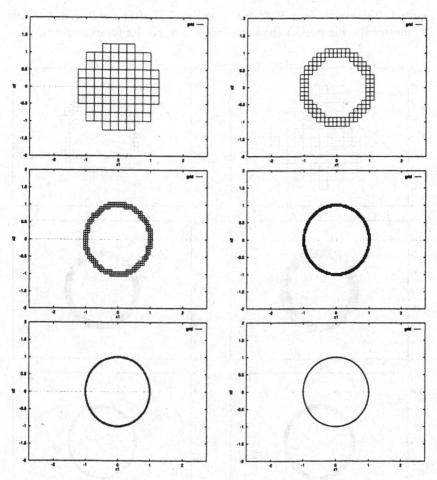

Fig. C.2. Results for system (C.1)

We test our algorithm using the time–1 map of the solution trajectories (i.e., $h = 1$), which are computed using a highly accurate extrapolation method (see [110, Section 7.2.4]) such that by Theorem 7.4.5 we can neglect the error due to time discretization.

For our tests we use the space discretization constructed according to the description on Page 132, where the test points are chosen to be the vertices of the rectangular cells, i.e., 4 test points per cell. The computational domain is chosen to be $\Omega = [-2, 2]^2$, the initial grid consists of 16×16 uniform rectangles and the set S was chosen as the ball around the origin with radius 0.25. In each refinement step the elements of E^j are refined by subdividing the rectangles in one coordinate direction, i.e., in x_1–direction in the first step, in x_2–direction in the second, in x_1–direction in the third, etc.

Figure C.2 shows the sets E^{j+1} for example (C.1) after $j = 0, 2, 4, 6, 8$ and 10 refinements; Figure C.3 shows the respective results for example (C.2).

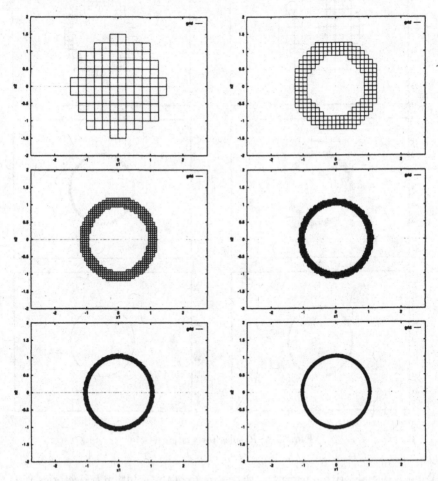

Fig. C.3. Results for system (C.2) after 0, 2, 4, 6, 8 and 10 refinements

It is easily seen that the inclusions from Remark 7.5.6(ii) hold, hence we expect the estimates (C.3) and (C.4) to hold for $\varepsilon^j = \operatorname{diam}(Q^j)$, provided this value is sufficiently small. Figures C.4 shows the real error depicted as points and the expected error from the estimates (C.3) and (C.4) as solid lines. Indeed, the expected error estimates are satisfied provided the computation is sufficiently accurate.

In addition, we have repeated the same calculations for the time–0.1 map, i.e., for $h = 0.1$. Here by Theorem 7.5.5 the convergence is expected to be much slower due to the fact that now we obtain $\varepsilon^j = 10\operatorname{diam}(\mathbf{Q}^j)$. Figure C.5 shows the expected and real results for this case which show that this is exactly what happens.

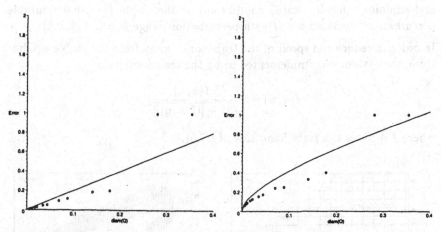

Fig. C.4. Real and expected errors for Systems (C.1) and (C.2) with $h = 1$

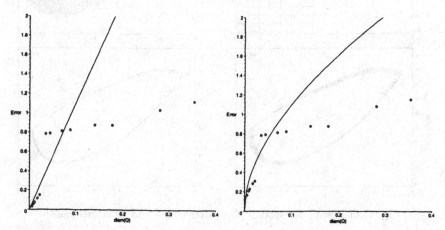

Fig. C.5. Real and expected errors for Systems (C.1) and (C.2) with $h = 0.1$

C.2 Subdivision Algorithm: Further Examples

The next example we want to consider is a model from [43], where it was introduced without input. Here we have added an additional perturbation term. It is given by

$$\begin{aligned}
\dot{x}_1 &= -x_1 + x_2 \\
\dot{x}_2 &= x_1/10 - 2x_2 - x_1^2 + (u - 1/10)x_1^3
\end{aligned} \tag{C.5}$$

and exhibits a locally stable equilibrium at the origin for all measurable perturbation functions $u \in \mathcal{U}$ with perturbation range $U = [-0.4, 0.4]$.

In order to reduce the speed of the trajectories away from the stable equilibrium, the system was implemented using the transformation

$$\tilde{f}(x, u) = \frac{5\, f(x, u)}{\sqrt{25 + \|f(x, u)\|^2}},$$

where f denotes the right hand side of (C.5).

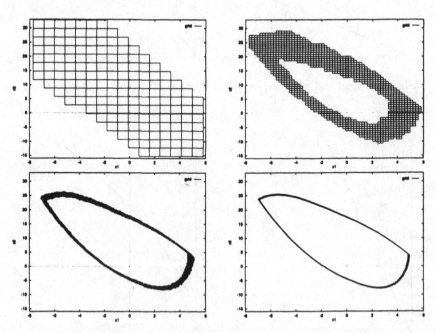

Fig. C.6. Results for system (C.5) after 0, 4, 8 and 12 refinements

For the time discretization (i.e., the computation of $\widetilde{\Phi}_h$) we have used the Euler scheme where, however, we have replaced $x + h\tilde{f}(x, u)$ by the extrapolation method from [110, Section 7.2.4]. This implies that while for a given measurable function $u \in \mathcal{U}$ we can only expect a first order approximation, for a piecewise constant u with $u_{[hi,h(i+1))} \equiv const$ we get a much more accurate approximation. The reason for this choice lies in the (numerical) observation that $\mathcal{D}(\{0\})$ does not change if we replace the measurable perturbations $u \in \mathcal{U}$ by the restricted class of perturbations which are piecewise constant on intervals of length h with $h \in (0,1]$ (for $h \gg 1$ this is no longer true). This even remains true if we only use the two extremal control values $u = -0.4$ and $u = 0.4$ for the computation of $\widetilde{\Phi}_h$.

Figure C.6 shows the results of the algorithm computing the strong domain of attraction $\mathcal{D}(\{0\})$ for $h = 1$, with $\Omega = [-7.71687, 5.81038] \times [-15.2562, 32.7901]$, and $\widetilde{\Phi}$ constructed as in the first two examples, i.e., with 4 test points per cell located at the vertices of the rectangular cells. The set S is chosen as an ellipse around the origin with radii $r_1 = 0.754286$ (in x_1–direction) and $r_2 = 2.27528$ (in x_2–direction). The initial cell covering again consists of 16×16 rectangular cells.

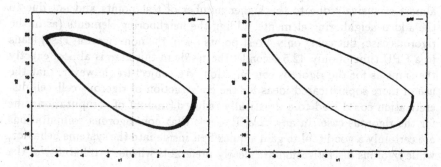

Fig. C.7. Rigorous and non–rigorous result for system (C.5) after 12 iterations

For this example we have also performed a computation using a rigorous discretization, following the ideas described in Section 5.3: For each cell in the image of a test–point under some control value we have added one layer of neighboring cells. Denoting the width and height of the rectangular domain Ω by $(\Delta x_1, \Delta x_2)$ and taking into account that we first refine by subdividing in the x_1–direction one sees that each rectangular cell either has width and height $(\Delta x_1/n, \Delta x_1/n)$ or $(\Delta x_1/(2n), \Delta x_1/n)$ for some $n \in \mathbb{N}$. Hence, using the norm $\|x\|_a = \max\{x_1/\Delta x_1, x_2/\Delta x_1\}$ in the first case or $\|x\|_b = \max\{2x_1/\Delta x_1, x_2/\Delta x_1\}$ in the second case, the diameter of each cell equals $1/n$ and one layer of neighboring elements gives exactly the ball of radius $1/n$ around the image in these norms. Hence, if we estimate $L_{\max} = \max\{1 + hL_a, 1 + hL_b\}$ for the Lipschitz constants $1 + hL_a$ and $1 + hL_b$ of the numerical time–h map $\widetilde{\Phi}_h$ in these norms, then by (5.22) we can conclude that the space discretization is rigorous if the distance of each point x to the closest test point (in the appropriate norm) is less than $1/nL_{\max}$. For a number of k^2 equidistributed test points the distance of an arbitrary point to the closest test point is easily seen to be less or equal $1/((k-1)2n)$, hence we obtain the condition $1/((k-1)2n \leq 1/nL_{max}$, which yields $k \geq L_{max}/2 + 1$. In our case we have estimated the Lipschitz constants for $h = 1$ numerically using 160000 points in Ω, which resulted in the estimate $L = 9.56$, i.e., $k^2 = 6^2 = 36$ test points in each cell. Figure C.7 shows the result of the corresponding computation after 12 iterations using the same initial grid as above, together with the final non–rigorous result from Figure C.6 above.

It turns out that in this example the non–rigorous computation apparently does not contain any errors, at least up to the accuracy of the rigorous computation.

In this case it is interesting to compare not only the results but also the CPU times needed for these computations, which were performed on a Pentium III CPU with 448.8 MHz under a SuSe LINUX system with 384 MB RAM. For the non–rigorous computation the CPU time was about 19.2 seconds while for its rigorous counterpart amounted to 900.1 seconds. This difference is almost exclusively due to the higher number of test points and not due to the added neighboring elements: adding the neighboring elements (as in the rigorous case) but using only 4 test points (as in the non–rigorous case) leads to a CPU time of only 42.5 seconds; the result in this case is almost exactly the same as for the rigorous computation. We conjecture, however, that the use of more sophisticated ideas for the construction of rigorous cell–cell discretization could lead to a drastically reduced amount of computation time for the rigorous case. In any case, it seems that non–rigorous computations are certainly a good tool to gain a quick first inside into the systems behavior, while rigorous discretization are clearly preferable whenever reliability of the result is needed.

The second example in this section is a three–dimensional model of a synchronous generator taken from [94]. It is given by the equations

$$\dot{x}_1 = x_2$$
$$\dot{x}_2 = -b_1 x_3 \sin x_1 - b_2 x_2 + P \qquad \text{(C.6)}$$
$$\dot{x}_3 = b_3 \cos x_1 - b_4 x_2 + E + u.$$

Using the parameters $b_1 = 34.29$, $b_2 = 0.0$, $b_3 = 0.149$, $b_4 = 0.3341$, $P = 28.22$ and $E = 0.2405$ this system exhibits a locally stable equilibrium at $x^* = (1.12, 0.0, 0.914)$. In [94] the feedback law $u(x) = a_1((x_1 - x_1^*)b_4 + x_2)$ with feedback gain $a_1 > 0$ was proposed in order to enlarge the domain of attraction of x^*. Here we show three computations, where the discrete time map $\widetilde{\Phi}_h$ is an approximation of the time–20 map of the system for constant control functions again computed by the extrapolation method from [110, Section 7.2.4]. Similar to the previous example, for the computations the system was slowed down by replacing its right hand side $f(x, u)$ by $f(x, u)/(1 + \|f(x, u)\|^2)$.

Here we have computed the weak domain of attraction $\mathcal{D}(\{x^*\})$ for three settings:

(i) for the uncontrolled system (i.e., $u \equiv 0$)

(ii) for constant feedback gain $a_1 = 0.45$

(iii) for time varying feedback gain $a_1(t) \in \{0, 0.15, 0.3, 0.45\}$, i.e., a_1 now plays the role of the control u in (2.21). Note that $a_1(t)$ here is the input of the discrete time system $\tilde{\Phi}_h$ for $h = 20$.

The corresponding weak domains of attraction are shown in Figure C.8(i)–(iii); the visualization was done with the graphics programming environment GRAPE developed at the Universities of Bonn and Freiburg (see www.iam.uni-bonn.de/sfb256/grape/). The computation was done on the domain $\Omega = [0, \pi/2] \times [-5, 5] \times [0, 3]$ with a starting grid consisting of $16 \times 16 \times 16 = 4096$ cubes. The set S here was chosen as the level set $V \leq 0.2$ of the Lyapunov function proposed in [94]

$$V(x) = x_2^2/2 + b_1 x_3 (\cos x_1^* - \cos x_1) + b_1 b_4 \tilde{x}_3^2/2b_3$$
$$+ b_1 a_1 \left(\cos x_1^* \tilde{x}_1 - (\sin x_1 - \sin x_1^*) + b_4/b_3 \left(\tilde{x}_1 \tilde{x}_3 + a_1 \tilde{x}_1^2/2\right)\right)$$

with $\tilde{x}_i = x_i - x_i^*$ and $a_1 = 0$ in (i) and $a_1 = 0.45$ in (ii) and (iii).

Note that in this example $\mathcal{D}(\{x^*\})$ is not completely contained in Ω, in which case the algorithm gives the set of those points in $\mathcal{D}(\{x^*\})$ which can be controlled to x^* without leaving Ω. Although this case is not completely covered by our theory (cf. Remark 5.3.8) the results look reasonable since the sets E^j apparently shrink down to a "thin" set.

The cell–cell map $\hat{\Phi}_h$ was constructed as in the previous examples using 8 test points per cube located in the vertices. The cases (i) and (ii) were computed with 12 subdivision steps (4 in each coordinate direction), while for case (iii) only 9 subdivision steps (3 in each coordinate direction) were needed.

C.3 Zubov's Method

In this last section we return to the two examples from Section C.1 and show some numerical approximations to Zubov's equation computed using the scheme (7.43). In particular, we illustrate Theorem 7.6.1 in order to show that the sublevel set $\hat{v} < 1$ is accurately computed even if the usual L_∞ error estimates for this class of schemes does not yield useful information.

In order to make this precise we consider Zubov's equation (here without input)

$$-Dv(x)f(x) - (1 - v(x))g(x) = 0$$

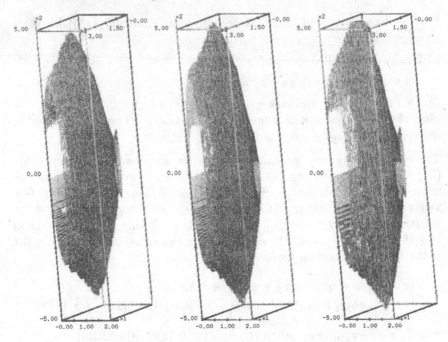

Fig. C.8. Results for Example C.6 (i), (ii) and (iii)

for the function $g(x) = \|x\|/1000$. With f denoting the right hand side of System (C.1) or system (C.2) (recall that $\mathcal{D}(\{0\}) = D^1$ for both examples), it is easily verified that for both systems the function

$$v(x) = \begin{cases} 1 - (1 - \|x\|)^{0.001}, & \|x\| < 1 \\ 1, & \|x\| \geq 1 \end{cases}$$

solves Zubov's equation for $x \notin \partial D^1$ in the classical sense. For $x \in \partial D^1$ one computes that the superdifferential satisfies $D^+(x) = \{\lambda x \mid \lambda \geq 0\}$ while the subdifferential $D^-(x)$ is empty. Since $p\,f(x) = 0$ for all $p \in D^+(x)$ and all $x \in \partial D^1$ we obtain the desired inequality

$$-p\,f(x) - (1 - v(x))g(x) = 0 \leq 0$$

from the definition of viscosity solutions, cf. Remark A.1.2.

For the numerical computation we have chosen the regularization parameter $\varepsilon = 1/10000$. Note that the L_∞ estimate from Section 7.2 does not yield any useful information for reasonably sized values of $\mathrm{diam}(\mathbf{Q})$, since $\mathrm{diam}(\mathbf{Q})^\varepsilon$ will only become small for a virtually impossibly large number of rectangles.

On the domain $\Omega = [-2, 2]^2$ and for time step $h = 0.1$ we have computed three approximations $\hat{v}$ for each of the examples (C.1) and (C.2), using equidistant space discretizations with $\mathrm{diam}(\mathbf{Q}) = 0.283$, 0.0707 and 0.0177. The Figures C.9 and C.10 show the respective results.

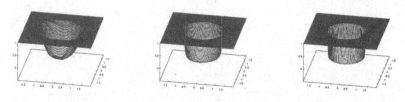

Fig. C.9. Approximation of v for Example C.1

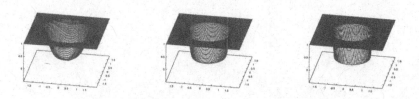

Fig. C.10. Approximation of v for Example C.2

The results clearly show that while the L_∞ error may certainly be large in neighborhood of ∂D^1, the Hausdorff–distance d_H between the sublevel set $\hat{v} < 1$ and D^1 obviously becomes small, which is exactly what was expected by Theorem 7.6.1.

Notation

$\mathbb{R}$	real numbers
$\mathbb{R}^+$	positive real numbers
$\mathbb{R}_0^+$	nonnegative real numbers: $\mathbb{R}_0^+ = \mathbb{R}^+ \cup \{0\}$
$\mathbb{Z}$	integer numbers
$\mathbb{N}$	natural numbers (without 0)
$\mathbb{N}_0$	$\mathbb{N} \cup \{0\}$
$h\mathbb{Z}$	integer multiples of $h > 0$: $h\mathbb{Z} = \{hk \mid k \in \mathbb{Z}\}$
$\mathbb{T}$	time axis: $\mathbb{T} = \mathbb{R}$ for continuous, $\mathbb{T} = h\mathbb{Z}$ for discrete time systems
$\mathbb{T}^+$	positive times: $\mathbb{T}^+ = \{t \in \mathbb{T} \mid t > 0\}$
$\mathbb{T}_0^+$	nonnegative times: $\mathbb{T}_0^+ = \mathbb{T}^+ \cup \{0\}$
$\|\cdot\|$	euclidean norm
$\|\cdot\|_A$	point–set euclidean distance, p. 22
$\mathrm{dist}(\cdot,\cdot)$	Hausdorff semidistance, p. 22
$d_H(\cdot,\cdot)$	Hausdorff distance, p. 22
$d_{\min}(\cdot,\cdot)$	minimal distance, p. 22
$\mathcal{B}(r, D)$	closed ball with radius r around D
Lim sup	limes superior for sets, p. 25
u	control or (internal) perturbation, p. 13
U	admissible values for u, p. 13
$\mathcal{U}$	space of functions u, p. 13
$\&_t$	concatenation of control functions, p. 13
$\mathcal{U}(B)$	set of functions $u \in \mathcal{U}$ depending on $x \in B$, p. 14
$\mathbf{U}$	sequences of functions in $\mathcal{U}(B)$, p. 131
w	(external) perturbation, p. 15
W	admissible values for w, p. 15
$\mathcal{W}$	space of perturbation functions, p. 15
p	(external) perturbation strategy, p. 15
$\mathcal{P}$	space of perturbation strategies, p. 15
$\mathcal{P}^\delta$	space of δ–nonanticipating strategies, p. 15
φ	trajectory of a continuous time system
Φ_h	trajectory of a discrete time system with time step h
Φ	trajectory of a continuous or discrete time system (depending on context)
φ^h	time-h map of a continuous time system

$\widetilde{\Phi}_h$ numerical one step approximation with time step h, p. 114

$\overline{\Phi}_h$ point–cell space discretization with time step h, p. 130

$\widehat{\Phi}_h$ cell–cell space discretization with time step h, p. 131

A attracting set

B attracted neighborhood

$\mathcal{D}(A)$ domain of attraction, p. 158

$\mathcal{R}(B)$ reachable set, p. 17 for perturbed, p. 190 for unperturbed systems

$P(\mathbb{R}^n)$ space of nonempty subsets of $\mathbb{R}^n$

D^-V viscosity subdifferential, p. 196

D^+V viscosity superdifferential, p. 196

$\mathcal{K}$ class of comparison functions, p. 201

$\mathcal{K}_\infty$ class of comparison functions, p. 201

$\mathcal{KL}$ class of comparison functions, p. 201

$\mathcal{KLD}$ class of comparison functions, p. 201

References

1. Abu Hassan, M., Storey, C. (1981): Numerical determination of domains of attraction for electrical power systems using the method of Zubov, Int. J. Control, **34**, 371–381
2. Albertini, F., Sontag, E.D. (1999): Continuous control–Lyapunov functions for asymptotically stable continuous time–varying systems, Int. J. Control, **72**, 1630–1641
3. Arnold, L. (1998): Random Dynamical Systems, Springer-Verlag, Heidelberg
4. Artstein, Z. (1983): Stabilization with relaxed controls, Nonlinear Anal., Theory Methods Appl., **7**, 1163–1173
5. Aubin, J.P. (1991): Viability Theory, Birkhäuser, Boston
6. Aubin, J.P., Frankowska, H. (1990): Set-Valued Analysis, Birkhäuser, Boston
7. Aulbach, B. (1983): Asymptotic stability regions via extensions of Zubov's method. I and II, Nonlinear Anal., Theory Methods Appl., **7**, 1431–1440 and 1441–1454
8. Bardi, M., Capuzzo Dolcetta, I. (1997): Optimal Control and Viscosity Solutions of Hamilton-Jacobi-Bellman equations, Birkhäuser, Boston
9. Beyn, W.J. (1987): The effect of discretization on homoclinic orbits, In: Bifurcation: Analysis, Algorithms, Applications, Küpper, T., Seydel, R., Troger, H. (eds), Prentice–Hall, 1–8
10. Beyn, W.J. (1987): On the numerical approximation of phase portraits near stationary points, SIAM J. Numer. Anal., **24**, 1095–1113
11. Beyn, W.J. (1991): Numerical methods for dynamical systems, In: Advances in numerical analysis. Volume I. Proceedings of the 4th summer school, held at Lancaster University, United Kingdom, 1990, Light, W.A. (ed), Oxford Science Publications, Clarendon Press, Oxford, 175–236
12. Beyn, W.J., Lorenz, J. (1987): Center manifolds of dynamical systems under discretization, Num. Func. Anal. and Opt., **9**, 381–414
13. Bhatia, N. (1967): On asymptotic stability in dynamical systems, Math. Syst. Theory, **1**, 113–127
14. Camilli, F. (1999): A note on convergence of level sets, Z. Anal. Anwendungen, **18**, 3–12
15. Camilli, F., Grüne, L., Wirth, F. (2000): A regularization of Zubov's equation for robust domains of attraction, In: Nonlinear Control in the Year 2000, Volume 1, Isidori, A., Lamnabhi-Lagarrigue, F., Respondek, W. (eds), Lecture Notes in Control and Information Sciences 258, NCN, Springer Verlag, London, 277–290
16. Camilli, F., Grüne, L., Wirth, F. (2000): Zubov's method for perturbed differential equations, In: Proceedings of the 14th International Symposium on

Mathematical Theory of Networks and Systems, Perpignan, France. CD-Rom, Article B100

17. Camilli, F., Grüne, L., Wirth, F. (2001): A generalization of Zubov's method to perturbed systems, SIAM J. Control Optim., **40**, 496–515

18. Capuzzo Dolcetta, I. (1983): On a discrete approximation of the Hamilton-Jacobi equation of dynamic programming, Appl. Math. Optim., **10**, 367–377

19. Capuzzo Dolcetta, I., Falcone, M. (1989): Discrete dynamic programming and viscosity solutions of the Bellman equation, Ann. Inst. Henri Poincaré, Anal. Non Linéaire, **6** (supplement), 161–184

20. Christofides, P.D., Teel, A.R. (1996): Singular perturbations and input-to-state stability, IEEE Trans. Autom. Control, **41**, 1645–1650

21. Coleman, C. (1965): Local trajectory equivalence of differential systems, Proc. Amer. Math. Soc., **16**, 890–892. *Addendum*, ibid., 17 (1966), 770

22. Colonius, F., Kliemann, W. (2000): The Dynamics of Control, Birkhäuser, Boston

23. Crandall, M.G., Evans, L.C., Lions, P.L. (1984): Some properties of viscosity solutions of Hamilton–Jacobi equations, Trans. Amer. Math. Soc., **282**, 487–502

24. Crandall, M.G., Lions, P.L. (1981): Conditions d'unicité pour les solutions generalises des equations d'Hamilton–Jacobi du premier ordre, C. R. Acad. Sci. Paris Sér. I Math., **292**, 487–502

25. Crandall, M.G., Lions, P.L. (1983): Viscosity solutions of Hamilton–Jacobi equations, Trans. Amer. Math. Soc., **277**, 1–42

26. Crauel, H., Flandoli, F. (1994): Attractors for random dynamical systems, Probab. Theory Relat. Fields, **100**, 365–393

27. Cyganowski, S., Grüne, L., Kloeden, P.E. (2001): MAPLE for stochastic differential equations, In: Theory and Numerics of Differential Equations, Blowey, J.F., Coleman, J.P., Craig, A.W. (eds), Springer–Verlag, 127–178

28. Dang, T., Maler, O. (1998): Reachability analysis via face lifting, In: Hybrid Systems: Computation and Control, Henzinger, T.A., Sastry, S. (eds), Lecture Notes in Computer Science 1386, Springer–Verlag, 96–109

29. Dellnitz, M., Hohmann, A. (1997): A subdivision algorithm for the computation of unstable manifolds and global attractors, Numer. Math., **75**, 293–317

30. Dellnitz, M., Junge, O. (1998): An adaptive subdivision technique for the approximation of attractors and invariant measures, Comput. Vis. Sci., **1**, 63–68

31. Deuflhard, P., Bornemann, F. (1994): Numerische Mathematik. II: Integration gewöhnlicher Differentialgleichungen, de Gruyter, Berlin

32. Falcone, M. (1987): A numerical approach to the infinite horizon problem of deterministic control theory, Appl. Math. Optim., **15**, 1–13. *Corrigenda*, ibid., 23 (1991), 213–214

33. Falcone, M., Ferretti, R. (1994): Discrete time high-order schemes for viscosity solutions of Hamilton-Jacobi-Bellman equations, Numer. Math., **67**, 315–344

34. Falcone, M., Giorgi, T. (1999): An approximation scheme for evolutive Hamilton-Jacobi equations, In: Stochastic analysis, control, optimization and applications, McEneaney, W.M., Yin, G., Zhang, Q. (eds), Birkhäuser, Boston, 288–303

35. Falcone, M., Grüne, L., Wirth, F. (2000): A maximum time approach to the computation of robust domains of attraction, In: EQUADIFF 99, Proceedings

of the International Congress held in Berlin, Germany, Fiedler, B., Gröger, K., Sprekels, J. (eds), World Scientific, Singapore, 844–849

36. Ferretti, R. (1997): High-order approximations of linear control systems via Runge-Kutta schemes, Computing, **58**, 351–364

37. Fiedler, B., Scheurle, J. (1996): Discretization of homoclinic orbits, rapid forcing and "invisible" chaos, Mem. Amer. Math. Soc. 119, no. 570

38. Garay, B.M. (1994): Discretization and Morse–Smale dynamical systems on planar discs, Acta Math. Univ. Comen., New Ser., **63**, 25–38

39. Garay, B.M. (1996): Hyperbolic structures in ODEs and their discretization with an appendix on differentiability properties of the inversion operator, In: Non linear analysis and boundary value problems for ordinary differential equations, Zanolin, F. (ed), CISM Courses and Lectures 371, Springer-Verlag, Wien, 149–173

40. Garay, B.M. (1996): On structural stability of ordinary differential equations with respect to discretization methods, Numer. Math., **72**, 449–479

41. Garay, B.M. (1996): Various closeness concepts in numerical ODE's, Comput. Math. Appl., **31**, 113–119

42. Garay, B.M., Kloeden, P.E. (1997): Discretization near compact invariant sets, Random Comput. Dyn., **5**, 93–123

43. Genesio, R., Tartaglia, M., Vicino, A. (1985): On the estimation of asymptotic stability regions: State of the art and new proposals, IEEE Trans. Autom. Control, **30**, 747–755

44. González, R.L.V., Tidball, M.M. (1991): On a discrete time approximation of the Hamilton-Jacobi equation of dynamic programming. INRIA Rapports de Recherche Nr. 1375

45. Gordon, R.A. (1994): The Integrals of Lebesgue, Denjoy, Perron, and Henstock, Graduate Studies in Mathematics, Vol. 4, American Mathematical Society, Providence, RI

46. Grüne, L. (1996): Numerische Berechnung des Lyapunov-Spektrums bilinearer Kontrollsysteme. Logos-Verlag, Berlin. Dissertation, Universität Augsburg

47. Grüne, L. (1997): An adaptive grid scheme for the discrete Hamilton-Jacobi-Bellman equation, Numer. Math., **75**, 319–337

48. Grüne, L. (1999): Input-to-state stability of exponentially stabilized semilinear control systems with inhomogenous perturbation, Syst. Control Lett., **38**, 27–35

49. Grüne, L. (1999): Stabilization by sampled and discrete feedback with positive sampling rate, In: Stability and Stabilization of Nonlinear Systems, Proceedings of the 1st NCN Workshop, Ayels, D., Lamnabhi-Lagarrigue, F., van der Schaft, A. (eds), Lecture Notes in Control and Information Sciences 246, Springer-Verlag, London, 165–182

50. Grüne, L. (2000): Attractors under perturbation and discretization, In: Proceedings of the 39th IEEE Conference on Decision and Control, Sydney, Australia, 2118–2122

51. Grüne, L. (2000): Convergence rates of perturbed attracting sets with vanishing perturbation, J. Math. Anal. Appl., **244**, 369–392

52. Grüne, L. (2000): Homogeneous state feedback stabilization of homogeneous systems, SIAM J. Control Optim., **38**, 1288–1314

53. Grüne, L. (2001): Input-to-state dynamical stability and its Lyapunov function characterization. Preprint, J.W. Goethe–Universität Frankfurt. submitted

54. Grüne, L. (2001): Persistence of attractors for one–step discretizations of ordinary differential equations, IMA J. Numer. Anal., **21**, 751–767

55. Grüne, L., Kloeden, P.E. (2001): Higher order numerical schemes for affinely controlled nonlinear systems, Numer. Math., **89**, 669–690

56. Grüne, L., Metscher, M., Ohlberger, M. (1999): On numerical algorithm and interactive visualization for optimal control problems, Comput. Vis. Sci., **1**, 221–229

57. Grüne, L., Sontag, E.D., Wirth, F.R. (1999): Asymptotic stability equals exponential stability, and ISS equals finite energy gain—if you twist your eyes, Syst. Control Lett., **38**, 127–134

58. Grüne, L., Wirth, F. (2000): Computing control Lyapunov functions via a Zubov type algorithm, In: Proceedings of the 39th IEEE Conference on Decision and Control, Sydney, Australia, 2129–2134

59. Häckl, G. (1992–1993): Numerical approximation of reachable sets and control sets, Random Comput. Dyn., **1**, 371–394

60. Häckl, G. (1995): Reachable Sets, Control Sets and Their Computation, Augsburger Mathematisch–Naturwissenschaftliche Schriften 7, Wißner Verlag, Augsburg. Dissertation, Universität Augsburg

61. Hahn, W. (1967): Stability of Motion, Springer-Verlag Berlin, Heidelberg

62. Hale, J.K. (1988): Asymptotic Behavior of Dissipative Systems, Mathematical Surveys and Monographs 25, American Mathematical Society, Providence, RI

63. Hirsch, M.W., Palis, J., Pugh, C.C., Shub, J. (1970): Neighborhoods of hyperbolic sets, Invent. Math., **9**, 121–134

64. Iserles, A. (1995): A First Course in the Numerical Analysis of Differential Equations, Cambridge Texts in Applied Mathematics, Cambridge University Press

65. Isidori, A. (1995): Nonlinear Control Systems. An Introduction, Springer-Verlag, Heidelberg

66. Isidori, A. (1996): Global almost disturbance decoupling with stability for non minimum-phase single-input single-output nonlinear systems, Syst. Control Lett., **28**, 115–122

67. Jiang, Z.P., Teel, A.R., Praly, L. (1994): Small-gain theorem for ISS systems and applications, Math. Control Signals Syst., **7**

68. Junge, O. (2000): Mengenorientierte Methoden zur numerischen Analyse dynamischer Systeme. Shaker Verlag, Aachen. Dissertation, Universität Paderborn

69. Junge, O. (2000): Rigorous discretization of subdivision techniques, In: EQUADIFF 99, Proceedings of the International Congress held in Berlin, Germany, Fiedler, B., Gröger, K., Sprekels, J. (eds), World Scientific, Singapore, 916–917

70. Kellett, C.M., Teel, A.R. (2000): Uniform asymptotic controllability to a set implies locally Lipschitz control–Lyapunov function, In: Proceedings of the 39th IEEE Conference on Decision and Control, Sydney, Australia, 3994–3999

71. Khalil, H.K. (1996): Nonlinear Systems, Prentice-Hall, 2nd ed.

72. Kirin, N.E., Nelepin, R.A., Bajdaev, V.N. (1982): Construction of the attraction region by Zubov's method, Differ. Equations, **17**, 871–880

73. Kloeden, P.E. (1975): Eventual stability in general control systems, J. Differ. Equations, **19**, 106–124

74. Kloeden, P.E. (1986): Asymptotically stable attracting sets in the Navier-Stokes equations, Bulletin Austral. Math. Soc., **34**, 37–52

75. Kloeden, P.E. (2000): A Lyapunov function for pullback attractors of nonautonomous differential equations, Electronic J. Differential Equ., Conference **05**, 91–102

76. Kloeden, P.E., Kozyakin, V.S. (2000): The inflation of attractors and their discretization: the autonomous case, Nonlinear Anal., Theory Methods Appl., **40**, 333–343

77. Kloeden, P.E., Lorenz, J. (1986): Stable attracting sets in dynamical systems and their one-step discretizations, SIAM J. Numer. Anal., **23**, 986–995

78. Kloeden, P.E., Lorenz, J. (1990): A note on multistep methods and attracting sets of dynamical systems, Numer. Math., **56**, 667–673

79. Kloeden, P.E., Platen, E. (1992): Numerical Solution of Stochastic Differential Equations, Springer–Verlag, Heidelberg. (3rd revised and updated printing, 1999)

80. Kloeden, P.E., Schmalfuß, B. (1996): Lyapunov functions and attractors under variable time-step discretization, Discrete Contin. Dynam. Systems, **2**, 163–172

81. Kloeden, P.E., Schmalfuß, B. (1997): Cocycle attractors of variable time-step discretizations of Lorenzian systems, J. Difference Equ. Appl., **3**, 125–145

82. Krasnosel'skii, M.A. (1968): The Operator of Translation along Trajectories of Differential Equations, Translations of Mathematical Monographs 19, American Mathematical Society, Providence, R.I.

83. Krstić, M., Deng, H. (1998): Stabilization of Nonlinear Uncertain Systems, Springer-Verlag, London

84. Lakshmikantham, V., Leela, S. (1969): Differential and Integral Inequalities Volume I, Mathematics in Science and Engineering Volume 55–I, Academic Press, New York and London

85. Lamba, H. (2000): Dynamical systems and adaptive timestepping in ODE solvers, BIT, **40**, 314–335

86. Lamba, H., Stuart, A.M. (1998): Convergence results for the MATLAB ode23 routine, BIT, **38**, 751–780

87. Lee, E.B., Markus, L. (1967): Foundations of Optimal Control, John Wiley & Sons, New York

88. Lin, Y., Sontag, E.D., Wang, Y. (1996): A smooth converse Lyapunov theorem for robust stability, SIAM J. Control Optim., **34**, 124–160

89. Lions, P.L. (1982): Generalized solutions of Hamilton-Jacobi equations, Pitman, London

90. Lorenz, E.N. (1963): Deterministic non–periodic flows, J. Atmospheric Sci., **20**, 130–141

91. Lorenz, J. (1994): Numerics of invariant manifolds and attractors, In: Chaotic Numerics, Kloeden, P.E., Palmer, K.J. (eds), Contemp. Math. 172, American Mathematical Society, Providence, R.I.

92. Lyapunov, A.M. (1892): The General Problem of the Stability of Motion, Comm. Soc. Math. Kharkow (in Russian). (reprinted in English, Taylor & Francis, London, 1992)

93. Nešić, D., Teel, A.R., Kokotović, P.V. (1999): Sufficient conditions for stabilization of sampled-data nonlinear systems via discrete-time approximations, Syst. Control Lett, **38**, 259–270

94. Ortega, R., Galaz-Larios, M., Bazanella, A.S., Stankovic, A. (2000): An energy–shaping approach to excitation control of synchronous generators. Preprint, LSS, CNRS–SUPELEC

95. Praly, L., Wang, Y. (1996): Stabilization in spite of matched unmodelled dynamics and an equivalent definition of input-to-state stability, Math. of Control, Signals, and Systems, **9**, 1–33

96. Rifford, L. (2000): Existence of Lipschitz and semiconcave control-Lyapunov functions, SIAM J. Control Optim., **39**, 1043–1064

97. Rosier, L., Sontag, E.D. (2000): Remarks regarding the gap between continuous, Lipschitz, and differentiable storage functions for dissipation inequalities, Syst. Control Lett., **41**, 237–249

98. Roxin, E.O. (1965): Stability in general control systems, J. Differ. Equations, **1**, 115–150

99. Saint-Pierre, P. (1994): Approximation of the viability kernel, Appl. Math. Optim., **29**, 187–209

100. Sepulchre, R., Jankovic, M., Kokotović, P. (1997): Constructive Nonlinear Control, Springer-Verlag, Berlin

101. Sontag, E.D. (1983): A Lyapunov-like characterization of asymptotic controllability, SIAM J. Control Optim., **21**, 462–471

102. Sontag, E.D. (1989): Smooth stabilization implies coprime factorization, IEEE Trans. Autom. Control, **34**, 435–443

103. Sontag, E.D. (1998): Comments on integral variants of ISS, Syst. Control Lett., **34**, 93–100

104. Sontag, E.D., Wang, Y. (1992): Generating series and nonlinear systems: analytic aspects, local realizability, and i/o, Forum Math., **4**, 299–322

105. Sontag, E.D., Wang, Y. (1995): On characterizations of the input-to-state stability property, Syst. Control Lett., **24**, 351–359

106. Sontag, E.D., Wang, Y. (1996): New characterizations of input-to-state stability, IEEE Trans. Autom. Control, **41**, 1283–1294

107. Soravia, P. (1995): Stability of dynamical systems with competitive controls: the degenerate case, J. Math. Anal. Appl., **191**, 428–449

108. Soravia, P. (1999): Optimality principles and representation formulas for viscosity solutions of Hamilton–Jacobi equations I. Equations of unbounded and degenerate control problems without uniqueness, Adv. Differ. Eq., **4**, 275–296

109. Soravia, P. (1999): Optimality principles and representation formulas for viscosity solutions of Hamilton–Jacobi equations II. Equations of control problems with state constraints, Differ. Integral Eq., **12**, 275–293

110. Stoer, J., Bulirsch, R. (1980): Introduction to Numerical Analysis, Springer Verlag, New York

111. Stuart, A.M. (1994): Numerical analysis of dynamical systems, Acta Numerica 1994, 467–572

112. Stuart, A.M. (1997): Probabilistic and deterministic convergence proofs for software for initial value problems, Numer. Algor., **14**, 227–260

113. Stuart, A.M., Humphries, A.R. (1996): Dynamical Systems and Numerical Analysis, Cambridge University Press

114. Szegö, G.P., Treccani, G. (1969): Semigruppi di trasformazioni multivoche, Lecture Notes in Mathematics 101, Springer–Verlag

115. Szolnoki, D. (1997): Berechnung von Viabilitätskernen. Diploma Thesis, Universität Augsburg

116. Szolnoki, D. (2000): Computation of control sets using subdivision and continuation techniques, In: Proceedings of the 39th IEEE Conference on Decision and Control, Sydney, Australia, 2135–2140

117. Szolnoki, D. (2000): Viability kernels and control sets, ESAIM Control Optim. Calc. Var., **5**, 175–185
118. Szolnoki, D. (2001): Algorithms for Reachability Problems. Shaker Verlag, Aachen. Dissertation, Universität Augsburg
119. Teel, A.R. (1996): A nonlinear small gain theorem for the analysis of control systems with saturation, IEEE Trans. Automat. Control, **41**, 1256–1270
120. Teel, A.R., Praly, L. (1999): Results on converse Lyapunov functions from class-$\mathcal{KL}$ estimates, In: Proceedings of the 38th IEEE Conference on Decision and Control, Phoenix, Arizona, USA, 2545–2550
121. Teel, A.R., Praly, L. (2000): A smooth Lyapunov function from a class-$\mathcal{KL}$ estimate involving two positive semidefinite functions, ESAIM Control Optim. Calc. Var., **5**, 313–367
122. Tsinias, J. (1997): Input to state stability properties of nonlinear systems and applications to bounded feedback stabilization using saturation, ESAIM Control Optim. Calc. Var., **2**, 57–85
123. Vannelli, A., Vidyasagar, M. (1985): Maximal Lyapunov functions and domains of attraction for autonomous nonlinear systems, Automatica, **21**, 69–80
124. Veliov, V. (1997): On the time discretization of control systems, SIAM J. Control Optim., **35**, 1470–1486
125. Wilson, F.W. (1967): The structure of the level surfaces of a Lyapunov function, J. Differ. Equations, **3**, 323–329
126. Wilson, F.W. (1969): Smoothing derivatives of functions and applications, Trans. Amer. Math. Soc., **139**, 413–428
127. Yoshizawa, T. (1966): Stability Theory by Lyapunov's Second Method, The Mathematical Society of Japan, Tokyo
128. Zou, Y.K., Beyn, W.J. (1997): Invariant manifolds for nonautonomous systems with application to One-Step Methods, J. Dyn. Differ. Equations, **10**, 379–407
129. Zubov, V.I. (1964): Methods of A.M. Lyapunov and their Application, P. Noordhoff, Groningen

Index

Recent Reprints and New Editions